PROGRESS IN CLINICAL AND BIOLOGICAL RESEARCH

RECENT TITLES

Vol 163: **Prevention of Physical and Mental Congenital Defects**, Maurice Marois, *Editor.* Published in 3 volumes: Part A: *The Scope of the Problem.* Part B: *Epidemiology, Early Detection and Therapy, and Environmental Factors.* Part C: *Basic and Medical Science, Education, and Future Strategies*

Vol 164: **Information and Energy Transduction in Biological Membranes,** C. Liana Bolis, Ernst J.M. Helmreich, Hermann Passow, *Editors*

Vol 165: **The Red Cell: Sixth Ann Arbor Conference**, George J. Brewer, *Editor*

Vol 166: **Human Alkaline Phosphatases,** Torgny Stigbrand, William H. Fishman, *Editors*

Vol 167: **Ethopharmacological Aggression Research**, Klaus A. Miczek, Menno R. Kruk, Berend Olivier, *Editors*

Vol 168: **Epithelial Calcium and Phosphate Transport: Molecular and Cellular Aspects,** Felix Bronner, Meinrad Peterlik, *Editors*

Vol 169: **Biological Perspectives on Aggression,** Kevin J. Flannelly, Robert J. Blanchard, D. Caroline Blanchard, *Editors*

Vol 170: **Porphyrin Localization and Treatment of Tumors,** Daniel R. Doiron, Charles J. Gomer, *Editors*

Vol 171: **Developmental Mechanisms: Normal and Abnormal,** James W. Lash, Lauri Saxén, *Editors*

Vol 172: **Molecular Basis of Cancer,** Robert Rein, *Editor*. Published in 2 volumes: Part A: *Macromolecular Structure, Carcinogens, and Oncogenes.* Part B: *Macromolecular Recognition, Chemotherapy, and Immunology*

Vol 173: **Pepsinogens in Man: Clinical and Genetic Advances,** Johanna Kreuning, I. Michael Samloff, Jerome I. Rotter, Aldur Eriksson, *Editors*

Vol 174: **Enzymology of Carbonyl Metabolism 2: Aldehyde Dehydrogenase, Aldo-Keto Reductase, and Alcohol Dehydrogenase,** T. Geoffrey Flynn, Henry Weiner, *Editors*

Vol 175: **Advances in Neuroblastoma Research,** Audrey E. Evans, Giulio J. D'Angio, Robert C. Seeger, *Editors*

Vol 176: **Contemporary Sensory Neurobiology,** Manning J. Correia, Adrian A. Perachio, *Editors*

Vol 177: **Medical Genetics: Past, Present, Future,** Kåre Berg, *Editor*

Vol 178: **Bluetongue and Related Orbiviruses,** T. Lynwood Barber, Michael M. Jochim, *Editors*

Vol 179: **Taurine: Biological Actions and Clinical Perspectives,** Simo S. Oja, Liisa Ahtee, Pirjo Kontro, Matti K. Paasonen, *Editors*

Vol 180: **Intracellular Protein Catabolism,** Edward A. Khairallah, Judith S. Bond, John W.C. Bird, *Editors*

Vol 181: **Germfree Research: Microflora Control and Its Application to the Biomedical Sciences,** Bernard S. Wostmann, *Editor*, Julian R. Pleasants, Morris Pollard, Bernard A. Teah, Morris Wagner, *Co-Editors*

Vol 182: **Infection, Immunity, and Blood Transfusion,** Roger Y. Dodd and Lewellys F. Barker, *Editors*

Vol 183: **Aldehyde Adducts in Alcoholism,** Michael A. Collins, *Editor*

Vol 184: **Hematopoietic Stem Cell Physiology,** Eugene P. Cronkite, Nicholas Dainiak, Ronald P. McCaffrey, Jiri Palek, Peter J. Quesenberry, *Editors*

Vol 185: **EORTC Genitourinary Group Monograph 2,** Fritz H. Schroeder, Brian Richards, *Editors.* Published in two volumes: Part A: *Therapeutic Principles in Metastatic Prostatic Cancer.* Part B: *Superficial Bladder Tumors*

Vol 186: **Carcinoma of the Large Bowel and Its Precursors,** John R.F. Ingall, Anthony J. Mastromarino, *Editors*

Vol 187: **Normal and Abnormal Bone Growth: Basic and Clinical Research,** Andrew D. Dixon, Bernard G. Sarnat, *Editors*

Vol 188: **Detection and Treatment of Lipid and Lipoprotein Disorders of Childhood,** Kurt Widhalm, Herbert K. Naito, *Editors*

Vol 189: **Bacterial Endotoxins: Structure, Biomedical Significance, and Detection With the Limulus Amebocyte Lysate Test,** Jan W. ten Cate, Harry R. Büller, Augueste Sturk, Jack Levin, *Editors*

Vol 190: **The Interphotoreceptor Matrix in Health and Disease,** C. David Bridges, *Editor*, Alice J. Adler, *Associate Editor*

Vol 191: **Experimental Approaches for the Study of Hemoglobin Switching,** George Stamatoyannopoulos, Arthur W. Nienhuis, *Editors*

Vol 192: **Endocoids,** Harbans Lal, Frank LaBella, John Lane, *Editors*

Vol 193: **Fetal Liver Transplantation,** Robert Peter Gale, Jean-Louis Touraine, Guido Lucarelli, *Editors*

Vol 194: **Diseases of Complex Etiology in Small Populations: Ethnic Differences and Research Approaches,** Ranajit Chakraborty, Emöke J.E. Szathmary, *Editors*

Vol 195: **Cellular and Molecular Aspects of Aging: The Red Cell as a Model,** John W. Eaton, Diane K. Konzen, James G. White, *Editors*

Vol 196: **Advances in Microscopy,** Ronald R. Cowden, Fredrick W. Harrison, *Editors*

Vol 197: **Cooperative Approaches to Research and Development of Orphan Drugs,** Melvin Van Woert, Eunyong Chung, *Editors*

Vol 198: **Biochemistry and Biology of DNA Methylation,** Giulio L. Cantoni, Aharon Razin, *Editors*

Vol 199: **Leukotrienes in Cardiovascular and Pulmonary Function,** Allan M. Lefer, Marlys H. Gee, *Editors*

Vol 200: **Endocrine Genetics and Genetics of Growth,** Costas J. Papadatos, Christos S. Bartsocas, *Editors*

Vol 201: **Primary Chemotherapy in Cancer Medicine,** D.J. Theo Wagener, Geert H. Blijham, Jan B.E. Smeets, Jacques A. Wils, *Editors*

Vol 202: **The 2-5A System: Molecular and Clinical Aspects of the Interferon-Regulated Pathway,** Bryan R.G. Williams, Robert H. Silverman, *Editors*

Vol 203: **Testicular Cancer,** Saad Khoury, René Küss, Gerald P. Murphy, Christian Chatelain, James P. Karr, *Editors*

Vol 204: **Tumor Markers and Their Significance in the Management of Breast Cancer,** Thomas Dao, Angela Brodie, Clement Ip, *Editors*

Vol 205: **Comparative Endocrinology: Developments and Directions,** Charles L. Ralph, *Editor*

Vol 206: **Genetic Toxicology of the Diet,** Ib Knudsen, *Editor*

Vol 207: **Monitoring of Occupational Genotoxicants,** Marja Sorsa, Hannu Norppa, *Editors*

Vol 208: **Risk and Reason: Risk Assessment in Relation to Environmental Mutagens and Carcinogens,** Per Oftedal, Anton Brøgger, *Editors*

Vol 209: **Genetic Toxicology of Environmental Chemicals,** Claes Ramel, Bo Lambert, Jan Magnusson, *Editors*. Published in two volumes: Part A: *Basic Principles and Mechanisms of Action*. Part B: *Genetic Effects and Applied Mutagenesis*

Vol 210: **Ionic Currents in Development,** Richard Nuccitelli, *Editor*

Vol 211: **Transfusion Medicine: Recent Technological Advances,** Kris Murawski, Frans Peetoom, *Editors*

Vol 212: **Cancer Metastasis: Experimental and Clinical Strategies,** D.R. Welch, B.K. Bhuyan, L.A. Liotta, *Editors*

Vol 213: **Plant Flavonoids in Biology and Medicine: Biochemical, Pharmacological, and Structure–Activity Relationships,** Vivian Cody, Elliott Middleton, Jr., Jeffrey B. Harborne, *Editors*

Vol 214: **Ethnic Differences in Reactions to Drugs and Xenobiotics,** Werner Kalow, H. Werner Goedde, Dharam P. Agarwal, *Editors*

Vol 215: **Megakaryocyte Development and Function,** Richard F. Levine, Neil Williams, Jack Levin, Bruce L. Evatt, *Editors*

Vol 216: **Advances in Cancer Control: Health Care Financing and Research,** Lee E. Mortenson, Paul F. Engstrom, Paul N. Anderson, *Editors*

Vol 217: **Progress in Developmental Biology,** Harold C. Slavkin, *Editor*. Published in two volumes.

Please contact the publisher for information about previous titles in this series.

Progress in Developmental Biology
Part A

PROGRESS IN DEVELOPMENTAL BIOLOGY

Part A

Proceedings of the Tenth International Congress of the International Society of Developmental Biologists, held in Los Angeles, CA, August 4–9, 1985

Editor

Harold C. Slavkin

Laboratory for Developmental Biology
School of Dentistry
University of Southern California
Los Angeles, California

ALAN R. LISS, INC. • NEW YORK

Address all Inquiries to the Publisher
Alan R. Liss, Inc., 41 East 11th Street, New York, NY 10003

Printed in the United States of America

Library of Congress Cataloging-in-Publication Data

International Society of Developmental Biologists.
International Congress (10th : 1985 : Los Angeles, Calif.)
Progress in developmental biology.

(Progress in clinical and biological research ; 217A)
Includes index.
1. Developmental biology—Congresses. I. Slavkin, Harold C. II. Title. III. Series.
QH491.I48 1985 574.3 86-7483
ISBN 0-8451-0192-7 (v. 1)
ISBN 0-8451-5067-7 (set)

Contents

INTERACTIONS IN THE DEVELOPING NERVOUS SYSTEM

SEX DETERMINATION

Contributors

Hansjuerg Alder, Department of Zoology, University of Basel, Basel, Switzerland **[27]**

Glen K. Andrews, Department of Biochemistry, University of Kansas Medical Center, Kansas City, KS **[371]**

Hermann J. Anton, Zoologisches Institut,Universität Köln, Köln, Federal Republic of Germany **[105]**

Phillip A. Arcuri, Department of Anatomical Sciences, University of Illinois, Urbana, IL **[75]**

J.C. Arents, Laboratory of Biochemistry, University of Amsterdam, Amsterdam, The Netherlands **[87]**

R.K. Ayer, Jr., Department of Anatomy and Cell Biology, Woodruff Health Science Center, Emory University, Atlanta, GA **[353]**

Adelbert Bacher, Lehrstuhl für Organische Chemie und Biochemie, Technical University of Munich, Garching, Federal Republic of Germany **[349]**

E. Barberá-Guillem, Department of Histology and Cell Biology, College of Medicine, University of the Basque Country, Vizcaya, Spain **[325]**

David Beebe, Department of Anatomy, USUHS, Bethesda, MD **[365]**

Jean-Claude Beetschen, Laboratoire de Biologie Générale, Université P. Sabatier, Toulouse, France **[403]**

B.N. Belintzev, Inst. Mol. Biol. Acad. Sci., Moscow, USSR **[101]**

L.V. Beloussov, Inst. Bioorg. Chem. Acad. Sci., Moscow, USSR **[101]**

Yehuda Ben-Shaul, Department of Microbiology, Tel Aviv University, Tel Aviv, Israel **[349]**

Stefan Berking, Zoologisches Institut, Universität, Heidelberg, Federal Republic of Germany **[79]**

Stephan Bilinški, Department of Zoology, Jagellonian University, Krakow, Poland **[345]**

John G. Bluemink, Hubrecht Laboratory, International Embryological Institute, Utrecht, The Netherlands **[345]**

A.M. Bolzern, Department of Biology, University of Milano, Milano, Italy **[331]**

James Bonner, Division of Biology, California Institute of Technology, Pasadena, CA **[3]**

The number in brackets is the opening page number of the contributor's article.

Nicole Bournias-Vardiabasis, Division of Cytogenetics, City of Hope Medical Center, Duarte, CA **[123]**

Peter J. Bryant, Developmental Biology Center, University of California, Irvine, CA **[297]**

S.V. Bryant, Developmental Biology Center, University of California, Irvine, CA **[61, 305]**

Pamela Burton, Center for Developmental Biology, Department of Zoology, University of Texas, Austin, TX **[301]**

M.A. Caligo, Istituto di Mutagenesi e Differenziamento–CNR, Pisa, Italy **[189]**

Antonio Caputo, Children's Medical Research Foundation, University of Sydney, Camperdown, Australia **[7]**

Nestor G. Carri, IMBICE - OC 403, La Plata, Argentina **[119]**

Andrew Carr, Department of Zoology, University of Nottingham, University Park, Nottingham, England **[253]**

T.C. Chambers, Department of Molecular and Cellular Biology, National Jewish Hospital and Research Center, Denver, CO **[93]**

M. Cigada-Leonardi, Department of Biology, University of Milano, Milano, Italy **[331]**

William C. Clark, Department of Genetics, University of Alberta, Edmonton, Alberta, Canada **[221]**

Esther Cohen, Department of Microbiology, Tel Aviv University, Tel Aviv, Israel **[349]**

John R. Coleman, Division of Biology and Medicine, Brown University, Providence, RI **[243]**

Marina Dan-Sohkawa, Department of Biology, Faculty of Science, Osaka City University, Osaka, Japan **[57]**

F. DeBernardi, Department of Biology, University of Milano, Milano, Italy **[331]**

R.L. DeHaan, Department of Anatomy and Cell Biology, Woodruff Health Science Center, Emory University, Atlanta, GA **[353]**

Arnold De Loof, Zoological Institute of the University, Leuven, Belgium **[359]**

David I. dePomerai, Department of Zoology, University of Nottingham, University Park, Nottingham, England **[253]**

Kourosh Derakhshanian, Developmental Biology Center, University of California, Irvine, CA **[293]**

Françoise Dieterlen-Lièvre, Institut d'Embryologie, Nogent-sur-Marne, France **[183]**

Marie T. Dimond, Department of Biology, Trinity College, Washington, D.C. **[159]**

Thomas Doetschman, Friedrich Miescher Laboratorium, Max Planck Gesellschaft, Tübingen, Federal Republic of Germany **[47]**

Grace Don, Laboratory for Developmental Biology, School of Dentistry, University of Southern California, Los Angeles, CA **[381]**

Sherry D. Doty, Division of Natural Sciences, New College, University of South Florida, Sarasota, FL **[97]**

Christine M. Duarte, Developmental Biology Center, University of California, Irvine, CA **[293]**

Don J. Durzan, Department of Pomology, University of California, Davis, CA **[199]**

Ted Ebendal, Department of Zoology, Uppsala University, Uppsala, Sweden **[119]**

David Ellis, Department of Zoology, University of Nottingham, University Park, Nottingham, England **[253]**

Laurence D. Etkin, Department of Genetics, University of Texas System Cancer Center, M.D. Anderson Hospital and Tumor Institute, Houston, TX **[11]**

David D. Eveleth, Jr., Developmental Biology Center, University of California, Irvine, CA **[221]**

Ortwin Faff, Lehrstuhl für Organische Chemie und Biochemie, Technical University of Munich, Garching, Federal Republic of Germany **[349]**

Scott Falkenthal, Department of Genetics and the Molecular, Cellular, and Developmental Biology Program, Ohio State University, Columbus, OH **[239]**

U. Fascio, Department of Biology, University of Milano, Milano, Italy **[331]**

Richard A. Firtel, Department of Biology, University of California–San Diego, La Jolla, CA **[17]**

Ira Fowler, Wenner Gren Lab and Department of Anatomy, University of Kentucky, Lexington, KY **[139]**

Vernon French, Department of Zoology, University of Edinburgh, Edinburgh, Scotland **[273]**

S. Fujii, Department of Anatomy and Cell Biology, Woodruff Health Science Center, Emory University, Atlanta, GA **[353]**

Jean Gautier, Laboratoire de Biologie Générale, Université P. Sabatier, Toulouse, France **[403]**

R. Daniel Gietz, Department of Genetics, University of Alberta, Edmonton, Alberta, Canada **[221]**

Olga N. Golubeva, Institute of General Genetics Ac. Sci. Moscow, USSR **[43]**

Richard H. Gomer, Department of Biology, University of California–San Diego, La Jolla, CA **[17]**

Achim Gossler, Friedrich Miescher Laboratorium, Max Planck Gesellschaft, Tübingen, Federal Republic of Germany **[47]**

Stephen Jay Gould, Museum of Comparative Zoology, Harvard University, Cambridge, MA **[205]**

Paul B. Green, Department of Biological Sciences, Stanford University, Stanford, CA **[195]**

Horst Grunz, Department of Zoophysiology, University GHS of Essen, Essen, Federal Republic of Germany **[135,143]**

Jin Guangqin, Guangxi Fisheries Research Institute, Nanning, China **[35]**

Willem J. Hage, Hubrecht Laboratory, International Embryological Institute, Utrecht, The Netherlands **[345]**

Khadija Hajji, Laboratoire de Zoologie, Faculté des Sciences, Agdal-Rabat, Morocco **[183]**

Anwar A. Hakim, Department of Histology, School of Dentistry, Loyola University Medical Center, Maywood, IL **[259]**

Lewis I. Held, Jr., Developmental Biology Center, University of California, Irvine, CA **[293]**

Ross B. Hodgetts, Department of Genetics, University of Alberta, Edmonton, Alberta, Canada **[221]**

Gregory Holler-Dinsmore, Developmental Biology Center, University of California, Irvine, CA **[61]**

Stephen D. Holmes, Department of Urology, Baylor College of Medicine, Houston, TX **[389]**

Heather Horton, Department of Anatomy, Medical College of Pennsylvania, Philadelphia, PA **[111]**

Yen-Ming Hsu, Department of Biochemistry, Michigan State University, East Lansing, MI **[393]**

Erwin Huebner, Department of Zoology, University of Manitoba, Winnipeg, Manitoba, Canada **[423]**

Katsumi Ishida, Department of Urology, Teikyo University School of Medicine, Tokyo, Japan **[407]**

Tina F. Jaskoll, Laboratory for Developmental Biology, School of Dentistry, University of Southern California, Los Angeles, CA **[381]**

Yan Jingzhi, Institute of Developmental Biology, Academia Sinica, Beijing, China **[35]**

Randall Johnson, Laboratory for Developmental Biology, School of Dentistry, University of Southern California, Los Angeles, CA **[381]**

Tetsuji Kakutani, Department of Botany, Faculty of Science, Kyoto University, Kyoto, Japan **[83]**

Hiroyuki Kaneko, Department of Biology, Faculty of Science, Osaka City University, Osaka, Japan **[57]**

Satoru Kaneko, Department of Obstetrics and Gynecology, Keio University School of Medicine, Tokyo, Japan **[179]**

K.B. Karim, Department of Zoology, Institute of Science, Nagpur, India **[427]**

Stella Keeble, National Institute for Medical Research, The Ridgeway, London, England **[309]**

Gregory M. Kelly, Department of Zoology, University of Manitoba, Winnipeg, Manitoba, Canada **[423]**

Rolf Kemler, Friedrich Miescher Laboratorium, Max Planck Gesellschaft, Tübingen, Federal Republic of Germany **[47]**

W.A. King, Université de Montréal, St. Hyacinthe, Quebéc, Canada **[149]**

Yoshinori Komiya, Department of Biology, Faculty of Science, Osaka City University, Osaka, Japan **[57]**

Eero Lehtonen, Department of Pathology, University of Helsinki, Helsinki, Finland **[411]**

B.H. Leichtling, Department of Molecular and Cellular Biology, National Jewish Hospital and Research Center, Denver, CO **[93]**

Georgia E. Lesh-Laurie, Department of Biology, Cleveland State University, Cleveland, OH **[53]**

Pat Levitt, Department of Anatomy, Medical College of Pennsylvania, Philadelphia, PA **[111]**

Paul J. Linser, C.V. Whitney Laboratory and Department of Anatomy, University of Florida, St. Augustine, FL **[127]**

C. Lynne Littlefield, Developmental Biology Center, University of California, Irvine, CA **[175]**

Cecilia W. Lo, Biology Department, University of Pennsylvania, Philadelphia, PA **[39]**

George V. Lopashov, Institute of General Genetics Ac. Sci., Moscow, USSR **[43]**

G. Luccarini, Istituto di Mutagenesi e Differenziamento–CNR, Pisa, Italy **[189]**

T.H. MacRae, Department of Biology, Dalhousie University, Halifax, Nova Scotia, Canada **[419]**

Malcolm Maden, National Institute for Medical Research, The Ridgeway, London, England **[309]**

Ken Marcu, Department of Biological Chemistry, SUNY at Stony Brook, Stony Brook, NY **[47]**

J. Lawrence Marsh, Developmental Biology Center, University of California, Irvine, CA **[221]**

C. Martin, Institut d'Embryologie, Nogent-sur-Marne, France **[183]**

G. Martini, Istituto di Mutagenesi e Differenziamento–CNR, Pisa, Italy **[189]**

Eric McCairns, Children's Medical Research Foundation, University of Sydney, Camperdown, Australia **[7]**

Stephen Meier, Center for Developmental Biology, Department of Zoology, University of Texas, Austin, TX **[301]**

Horacio Merchant-Larios, Instituto de Investigaciónes Biomedicas, Universidad Nacional Autonoma de Mexico, Mexico **[171]**

John R. Merriam, Department of Biology, University of California at Los Angeles, Los Angeles, CA **[289]**

Du Miao, Institute of Developmental Biology, Academia Sinica, Beijing, China **[35]**

Jay E. Mittenthal, Department of Anatomical Sciences, University of Illinois College of Medicine, Urbana, IL **[315]**

P. Mohanty-Hejmadi, Department of Zoology, Utkal University, Bhubaneswar, India **[159]**

Hideo Mohri, Department of Biology, University of Tokyo, Tokyo, Japan **[179]**

John B. Morrill, Division of Natural Sciences, New College, University of South Florida, Sarasota, FL **[97]**

Harald Mückter, Lehrstuhl für Organische Chemie und Biochemie, Technical University of Munich, Garching, Federal Republic of Germany **[349]**

Werner A. Müller, Zoologisches Institut, Universität, Heidelberg, Federal Republic of Germany **[79]**

Ken Muneoka, Developmental Biology Center, University of California, Irvine, CA **[61,305]**

James D. Murray, Centre for Math Biology, Math Institute, Oxford University, Oxford, England **[75]**

George E.O. Muscat, Children's Medical Research Foundation, University of Sydney, Camperdown, Australia **[7]**

B.N. Nagorcka, CSIRO, Division of Entomology, Canberra, Australia **[319]**

Wu Naihu, Institute of Developmental Biology, Academia Sinica, Beijing, China **[35]**

Tina P. Nair, Division of Neurosciences, City of Hope Beckman Research Institute, Duarte, CA **[123]**

Dean Nakamura, Center for Reproductive Biology, Spring Creek Ranch, Memphis, TN **[149]**

Wolfgang Nellen, Department of Biology, University of California–San Diego, La Jolla, CA **[17]**

Binhui Ni, Institute of Developmental Biology, Academia Sinica, Beijing, China **[397]**

V. Nuti Ronchi, Istituto di Mutagenesi e Differenziamento–CNR, Pisa, Italy **[189]**

David A. O'Brochta, Developmental Biology Center, University of California, Irvine, CA **[297]**

John D. O'Connor, Department of Biology, University of California, Los Angeles, CA **[235]**

Charles P. Ordahl, Department of Anatomy, University of California, San Francisco, CA **[243]**

Shigeru Oshio, Department of Biology, University of Tokyo, Tokyo, Japan **[179]**

A.P. Otte, Laboratory of Biochemistry, University of Amsterdam, Amsterdam, The Netherlands **[87]**

David S. Packard, Jr., Department of Anatomy and Cell Biology, State University of New York Upstate Medical Center, Syracuse, NY **[67]**

Anna-Maija Partanen, Institute of Dentistry, University of Helsinki, Helsinki, Finland **[375]**

Günter Plickert, Zoologisches Institut, Universität, Heidelberg, Federal Republic of Germany **[79]**

M.J.E. Plomp, Laboratory of Biochemistry, University of Amsterdam, Amsterdam, The Netherlands **[87]**

Carola Ponzetto, Department of Human Genetics and Development and Center for Reproductive Sciences, Columbia University College of P & S, New York, NY **[249]**

P. Rafiee, Department of Biology, Dalhousie University, Halifax, Nova Scotia, Canada **[419]**

S. Ranzi, Department of Biology, University of Milano, Milano, Italy **[331]**

B.B. Rawdon, Department of General Anatomy, School of Dentistry, University of the Witwatersrand, Johannesburg, South Africa **[385]**

Leena Rechardt, Department of Biomedical Sciences, University of Tampere, Tampere, Finland **[115]**

Christophe D. Reymond, Department of Biology, University of California–San Diego, La Jolla, CA **[17]**

H.V. Rickenberg, National Jewish Hospital and Research Center and University of Colorado Health Sciences Center, Denver, CO **[93]**

Susan R. Ross, Department of Biological Chemistry, University of Illinois at Chicago, Chicago, IL **[23]**

Peter B. Rowe, Children's Medical Research Foundation, University of Sydney, Camperdown, Australia **[7]**

Paul S. Salvaterra, Division of Neurosciences, City of Hope Beckman Research Institute, Duarte, CA **[123]**

D. Savic, Inst. Mol. Biol. Acad. Sci., Moscow, USSR **[101]**

Lauri Saxén, Department of Pathology, University of Helsinki, Helsinki, Finland **[411]**

Walter Schaffner, Institute of Molecular Biology II, University of Zürich, Zürich, Switzerland **[47]**

K.L. Schaller, Department of Molecular and Cellular Biology, National Jewish Hospital and Research Center and University of Colorado Health Sciences Center, Denver, CO **[93]**

Volker Schmid, Department of Zoology, University of Basel, Basel, Switzerland **[27]**

Stephen A. Schwartz, University of Chicago Department of Pathology, Columbus, OH **[341]**

Edgar Serfling, Institute of Molecular Biology II, University of Zürich, Zürich, Switzerland **[47]**

Marty Shankland, Department of Molecular Biology, University of California, Berkeley, CA **[71]**

Yan Shaoyi, Institute of Developmental Biology, Academia Sinica, Beijing, China **[35]**

Krishan K. Sharma, Department of Zoology, University of Rajasthan, Jaipur, India **[105]**

Yinghsien Shih, Institute of Developmental Biology, Academia Sinica, Beijing, China **[397]**

Charles F. Shuler, Ohio State University College of Dentistry, Columbus, OH **[341]**

Michael Silver, Department of Anatomy, USUHS, Bethesda, MD **[365]**

Betty F. Sisken, Wenner Gren Lab and Department of Anatomy, University of Kentucky, Lexington, KY **[139]**

Harold C. Slavkin, Laboratory for Developmental Biology, School of Dentistry, University of Southern California, Los Angeles, CA **[381]**

Roy G. Smith, Department of Urology, Baylor College of Medicine, Houston, TX **[389]**

Karla Snellings, Department of Anatomy, USUHS, Bethesda, MD **[365]**

Dexiu Song, Institute of Developmental Biology, Academia Sinica, Beijing, China **[397]**

C. Sotgia, Department of Biology, University of Milano, Milano, Italy **[331]**

Charlotte A. Spencer, Department of Genetics, University of Alberta, Edmonton, Alberta, Canada **[221]**

Gerald Spotts, Department of Urology, Baylor College of Medicine, Houston, TX **[389]**

Lawrence Stanton, Department of Biological Chemistry, SUNY at Stony Brook, Stony Brook, NY **[47]**

Teresa R. Strecker, Department of Biology, University of California at Los Angeles, Los Angeles, CA **[289]**

Paul E. Suchy, Department of Biology, Cleveland State University, Cleveland, OH **[53]**

Akira Suzuki, Department of Biology, School of Education, Waseda University, Tokyo, Japan **[407]**

Ruth E. Swiderski, Department of Biology, University of California, Los Angeles, CA; present address: Cell Biology Group, Worcester Foundation for Experimental Biology, Shrewsbury, MA **[235]**

Kristien Swinnen, Zoological Institute of the University, Leuven, Belgium **[359]**

Lothar Tacke, Department of Zoophysiology, University GHS of Essen, Essen, Federal Republic of Germany **[135,143]**

Noriko Takahashi, Department of Biochemistry, Nagoya City University Medical School, Nagoya, Japan **[131]**

Kenzo Takata, Radioisotope Center, Nagoya University, Nagoya, Japan **[131]**

Teruko Taketo, Center for Biomedical Research, Population Council, New York, NY **[171]**

Ikuo Takeuchi, Department of Botany, Faculty of Science, Kyoto University, Kyoto, Japan **[83]**

Irma Thesleff, Institute of Dentistry, University of Helsinki, Helsinki, Finland **[375]**

C. Tickle, Department of Anatomy and Biology as applied to Medicine, Middlesex Hospital Medical School, London, England **[335]**

K. Uriu, Department of Pomology, University of California, Davis, CA **[199]**

R. van Driel, Laboratory of Biochemistry, University of Amsterdam, Amsterdam, The Netherlands **[87]**

Herman Van Mellaert, Zoological Institute of the University, Leuven, Belgium **[359]**

Judson Van Wyk, Department of Pediatrics, University of North Carolina, Chapel Hill, NC **[365]**

Peter Verhaert, Zoological Institute of the University, Leuven, Belgium **[359]**

F. Vidal-Vanaclocha, Department of Histology and Cell Biology, College of Medicine, University of the Basque Country, Vizcaya, Spain **[325]**

Stephen Wachtel, Center for Reproductive Biology, Spring Creek Ranch, Memphis, TN; present address: University of Tennessee Center for the Health Sciences, Memphis, TN **[149]**

N. Wanek, Developmental Biology Center, University of California, Irvine, CA **[305]**

John L. Wang, Department of Biochemistry, Michigan State University, East Lansing, MI **[393]**

Jorma Wartiovaara, Departments of Electron Microscopy and Medical Biology, University of Helsinki, Helsinki, Finland **[115]**

S.E. Wedden, Department of Anatomy and Biology as applied to Medicine, Middlesex Hospital Medical School, London, England **[335]**

David A. Weisblat, Department of Zoology, University of California, Berkeley, CA **[71]**

Debra J. Wolgemuth, Department of Human Genetics and Development and Center for Reproductive Sciences, Columbia University College of P & S, New York, NY **[249]**

Zhang Xingzhong, Chang Jiang Fisheries Research Institute, China State Fisheries Institution, Shashi, China **[35]**

Hachiro Yamanaka, Kanebo Institute for Cancer Research, Osaka, Japan **[57]**

Kiyoko Yamazaki Yamamoto, Research Institute of Environmental Medicine, Nagoya, Japan **[131]**

Ikuo Yasumasu, Department of Biology, School of Education, Waseda University, Tokyo, Japan **[407]**

Qin Ying, Guangxi Fisheries Research Institute, Nanning, China **[35]**

Patrick Zaborski, CNRS, Centre de Cytologie Experimentale, Laboratorie de Biologie du Développement, University of Paris VI, Ivry-sur-Seine, France **[163]**

A.G. Zaraisky, Inst. Bioorg. Chem. Acad. Sci., Moscow, USSR **[101]**

Ketevan G. Zviadadze, Institute of Zoology Ac. Sci., Tbilissi, Georgia, SSR **[43]**

Contents of Part B

PATTERN FORMATION IN NERVOUS SYSTEM

EXTRACELLULAR MATRIX IN DEVELOPMENT

Foreword

I am so pleased to be a part of the Tenth Congress of the International Society of Developmental Biologists (ISDB), held in the unique city of Los Angeles, California, during the week of August 4–9th, 1985. Our Society, ISDB, is proud of sustaining the tradition to present the International Congress for these many years. The Congress, which is held every four years, is a highlight of the many activities sponsored by our Society. The Congress has long been appreciated as the major occasion for international communication between research scientists and teachers in the field of developmental biology. At every Congress we have not only enjoyed the true cosmopolitan atmosphere of each country, but have also learned and evaluated the "cutting edge" of the science of developmental biology.

Thanks to the tremendous effort of the Organizing Committee, chaired by the able hands of Professor Harold Slavkin, we met together in Los Angeles, perhaps the most cosmopolitan metropolis in the world; a creative environment in Southern California to complement our scientific experiences.

The theme of the Tenth Congress is "New Discoveries and Technologies," which clearly describes the present situation of rapid advances in the sciences of developmental biology. The scope and technologies in our field are dramatically expanding and changing in modern times. For instance, it is now very evident that Recombinant DNA Technology, immunochemistry and developmental biology studies are together unraveling the problem of gene structure and how genes may regulate major events in formation of an organism. Developmental Biology is expanding and impacting on many different areas of so-called biotechnology—the applications to improve the human condition.

At the same time, we also continue to advance scientific progress in the classical problems of developmental biology. I firmly believe that the scientific programme of the Tenth Congress, which consisted of 6 major symposia, 16 minisymposia, and 19 different topics for poster presentations, clearly was comprehensive to attract all colleagues with interests in developmental biology, ranging from molecular to classical approaches. On the other hand, the programme was exciting enough not only for expert researchers, but also for students, young scientists and educators. It is our tradition, indeed, to stimulate, nurture, and encourage young generations throughout the world to meet the challenges of developmental biology.

Tokindo S. Okada
President
International Society of Developmental Biologists

Preface

The Tenth Congress of the International Society of Developmental Biologists (ISDB) met in Los Angeles from August 4–9th, 1985 at The Westin Bonaventure Hotel. The Organizing Committee of the Tenth Congress, Board Members of ISDB and an international group of cellular, molecular and developmental biologists attempted to bring together the broad international community of scientists interested in problems of developmental biology. This publication, in part, transmits the content and texture of scientific contributions related to development in the 1980s. The theme of the Congress "New Discoveries and Technologies" suggests that developmental biology continues to represent hybridizations and rearrangements of many complementary scientific disciplines. The Congress, therefore, provided a forum to celebrate the achievements of contemporary developmental biology, to explore new and old questions of development, and to advance an outreach program to improve the teaching of developmental biologists to the future scientists of the 21st Century.

To facilitate optimal communication, the Organizing Committee for the Tenth Congress formed invaluable collaborations with private industry, public education and government. Through these collaborations and the vision of many people it was possible to provide Special Travel Fellowship Awards, to hold an International Science Film Competition, sponsor a High School Student Science Competition, sponsor a Youth Symposium for gifted high school students from the Los Angeles geographic area, sponsor a Breakfast Briefing for corporate and university executives, sponsor a Math/Science Teacher Fellowship Program, and to produce a film with National Public Broadcasting System called "A Lifetime of Change." ISDB collaborations with private and public interest groups clearly defined our deep commitment to nurture the scientists and citizenry of today and tomorrow.

Of course, the success of the Congress is the reflection of the untiring efforts of my colleagues who served on the Organizing Committee. The scientific contributions found in this publication, in part, represent the state-of-the-art of international investigators involved with developmental biology. It is our hope that these published proceedings will further foster renewed interests in the challenges of developmental biology.

As Editor, I am extremely grateful to those contributors who faithfully cooperated with the deadlines and preparatory requirements of this camera-ready publication. I am deeply indebted to Ms. Gwen Airkens who served as our very competent Congress Coordinator throughout the four years of pre-planning and now has served to assemble the manuscripts for publication. Finally, I wish to acknowledge the professional and thoughtful efforts of Alan R. Liss and his associates for their logistical support and rapid publication.

Harold C. Slavkin
Los Angeles

Genomic Structure and Function in Development

Progress in Developmental Biology, Part A, pages 3–4

INTRODUCTORY REMARKS CONCERNING DEVELOPMENTAL BIOLOGY

James Bonner

Division of Biology, California Institute of Technology
Pasadena, California 91125

Fellow Developmental Biologists:

Biology has spent a long time trying to find out how development works. We have accumulated a wealth of phenomenological information about development, starting with embryology, the dorsal lip of the blastopore, etc., and going on through hormonology, heat-shock-ology, phenocopies etc. However, these anecdotal facts, of which I have collected my fair share, haven't given us a great deal of understanding of the developmental process. They merely help to assure us that there is such a thing as development.

Developmental biology has been awaiting new methodology, new ways to find out how genes are turned off and on at the right time and in the right place. Even Gunther Stent, who bemoans his supposition that the golden age of biology is already over, admits that there is still something to discover in our field.

Well--we have now at least a part of the new methodology. During the past ten years we have acquired not only the technology for isolation of individual genes, but also the technology of causing them to be transcribed and expressed as mRNA as well as gene products--proteins. We even know a substantial amount about the gene regulatory sequences, the enhancer sequences which cause genes to be transcribed abundantly. We also know about the sequences which determine that a gene shall be expressed in response to a particular hormone, or heat shock, or light. More about these matters is being found out almost daily.

This new insight has been made possible to a considerable extent because we know how to insert specific genes into new hosts, Xenopus eggs, for example, a la John Gurdon, or even better to insert them into the genomes of various plants and animals. Gene isolation and subsequent transformation with them is therefore not only interesting, but useful, from the point of view of avarice, thus in a materialistic and commercial sense.

We have surmised for some time that genes are expressed in an orderly and appropriate manner because of control programs which see to this matter. Perhaps the homeobox is an example of a sub-routine of such a program.

My late colleague, Thomas Hunt Morgan, an embryologist as well as a geneticist, held many years ago that the road to the understanding of life would be through genetics. It has turned out that he was quite right. My colleague, Ed Lewis, from whom we will hear tomorrow, has held for many years that in order to understand development all we have to do is to make mutants of it and reduce it to genetics. What we will hear today, and all week, will show that this is true, and that molecular genetics is the key to much, perhaps not all--but much, of the understanding of development.

Transformation in Animals and Plants

Progress in Developmental Biology, Part A, pages 7–10

DIFFERENTIAL GENE EXPRESSION DURING LECTIN INDUCED TRANSFORMATION OF HUMAN LYMPHOCYTES

George E.O. Muscat, Antonio Caputo, Eric McCairns and Peter B. Rowe
Children's Medical Research Foundation,
The University of Sydney.
P.O. Box 61, Camperdown, 2050, Australia

INTRODUCTION

From the standpoint of the cell cycle the proliferative response of resting human peripheral blood lymphocytes (PBLs) to phytohemagglutinin (PHA) provides an ideal model for studying regulation of gene expression at the level of transcription and processing in normal resting and growing PBLs. The resting lymphocyte has all the characteristics of a mature differentiated T-cell, yet these cells which are in a dormant Go state, can readily re-enter the cell-cycle via a mitogenic or antigenic response. Many groups have described genes whose expression is cell cycle dependent. Genes which are preferentially expressed after mitogen induced lymphocyte proliferation include c-*myc* (Kelly et al., 1983), β- and γ- actin, α- and β-tubulin (McCairns et al., 1984), interleukin 2 (Efrat et al., 1982), γ-interferon (Efrat et al., 1982) and pNC5A and pNC8C from PHA and 12-0-tetradecanoyl phorbol 13-acetate (TPA) activation of PBLs (Chang et al., 1984). We had previously shown that 3-4% of the mRNA from proliferating lympocytes was unique to the stimulated state (McCairns et al., 1983), a result similar to those of groups who have used 3T3 cells (Williams and Penman, 1975). Our aim in this study was to construct a large cDNA clone bank from proliferating cells and by the use of an enriched probe and differential screening to isolate clones which were preferentially induced on activation of PBLs with PHA.

RESULTS

A cDNA library in λgt10 was constructed from the cytoplasmic poly(A)$^+$ RNA of human PBLs after 72 h of PHA stimulation. Flow cytofluorometric analysis (McCairns et al., 1984) indicated that this preparation of PBLs contained cells in the G_1 (50%), S (35%) and G_2 + M (15%) phases of the mitotic cell cycle. Hence these cells were an ideal source of mRNA for a cDNA library which contained sequences specific to and/or preferentially expressed in growing PBLs with respect to quiescent PBLs in G_o. An enriched probe was prepared by hydroxylapatite depletion (Williams and Penman, 1975) and used to differentially screen the library.

Three recombinants corresponding to discrete mRNAs whose expression was preferentially induced after lymphocyte activation were selected for further analysis. RNA blot analysis revealed that clones p9L2, p10L2F and p1L1 hybridised to single transcripts with sizes of 1700, 1000 and 700 bases respectively in the lectin stimulated PBL. The transcript sizes did not change on stimulation. Clones p1L1 and p9L2 also hybridised to mRNA transcripts of 700 and 1700 bases in (i) CCRF-CEM and MOLT-4 cells which are immature T cells derived from acute lymphoblastoid leukaemias and (ii) JP cells, a non-malignant normal B-lymphoblastoid cell line with evidence of Epstein-Barr virus infection. In contrast the mRNA transcript corresponding to clone p10L2F was not detectable in CCRF-CEM, MOLT-4 and JP cells and very rare in resting PBLs. Changes in the levels of the polyadenylated mRNAs represented by clones p10L2F, p9L2, p1L1 and β-actin (not shown) were determined by RNA dot blot analysis as the cells moved from the resting to the stimulated state (Fig.1). The mRNA for β-actin was increased 2-fold at 6 h and this was maintained up to 72 h. The mRNA for clone p1L1 essentially mimicked the response of β-actin, i.e. increased 2-fold at 6 h and maintained this level up to 72 h. The mRNA for clone p9L2 had increased 2- to 3-fold by 6 h; by 24 h (coincident with the onset of DNA replication) the level had increased to approximately 6- to 7-times the level in the resting cell and this level is maintained up to 72 h. The mRNA for clone p10L2F was up 25- to 30-fold by 6 h (prior to overall transcriptional increases); after 24 h the level had decreased slightly to approximately 19- to 24-fold the level in the quiescent cell. By 72 h however (when a large proportion of the cells were in the S and G_2 + M phases) the level had increased 50- to 60-fold greater than the level in the resting cell. Growing T-lymphoblasts (CEM cells) have a slightly lower level of p1L1 mRNA but a higher level of p9L2 mRNA than the fully stimulated lymphocytes (72 h).

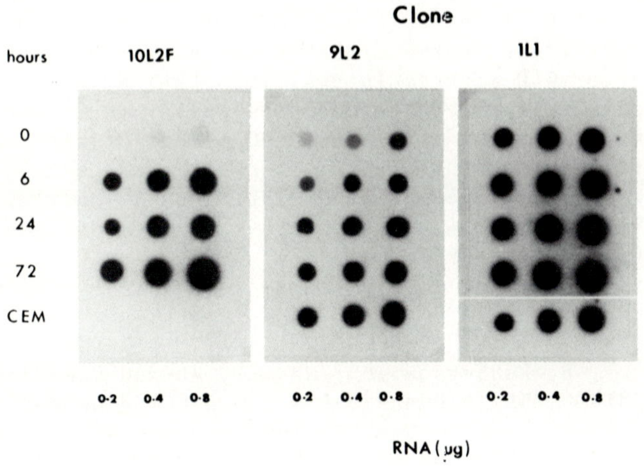

Fig. 1. Dot blot analysis of lymphocyte mRNA. Increasing amounts of mRNA isolated from PBLs after 0, 6, 24 and 72 h of PHA stimulation were dotted onto filters as shown. Samples of mRNA from growing leukaemic T-lymphoblasts (CEM cells) are also included. Filters were hybridised with nick-translated p10L2F, p9L2 and p1L1 probes at 2 x 10^6 cpm/ml.

All of the above increases are expressed relative to a constant amount of polyadenylated RNA. There is a net 6-fold increase in polyadenylated RNA per cell after 72 h of lectin stimulation (McCairns et al., 1983). The results (taking into account insert sizes) implied that the mRNAs for clones p1L1, p9L2 and p10L2F were in a similar abundance class to β-actin in the transformed cell after 72 h of stimulation. However, in the resting cell the mRNAs coded by p9L2 and p10L2F were present at approximately 10% and 2% of the level of β-actin respectively.

A second class of clones was isolated which hybridised to a smear on RNA blots in the size range 4-30s as determined by agarose/formaldehyde gel electrophoresis. A representative of these clones, p6L2 was hybridsed to poly(A)$^+$ RNA from PBLs 0, 6, 24 and 72 h after lectin activation to examine the kinetics of induction. The mRNA smear which hybridised to p6L2 was transiently induced approximately 10-fold 3-6 h after stimulation co-incident with the onset of the G_1 phase. The level of the poly(A)$^+$ RNA hybridising to these clones rapidly decayed to resting levels after 6 h of stimulation. These recombinant clones contained Alu-like repeat sequences which hybridised under stringent conditions to a sequenced member of the Alu repeat family known as BLUR 8. These clones containing Alu-like repeat sequences also hybridised to poly(A)$^+$ RNA from CCRF-CEM, MOLT-4 and JP cells in agreement with Calabretta et al., 1981.

REFERENCES

Calabretta, B, Robberson, DL, Maizel, AL and Saunders, GF (1981) mRNA in human cells contains sequences complementary to the Alu family of repeated DNA. Proc Natl Acad Sci USA 78: 6003-6007

Chang, NT, Tam, SH, Kung, PC and Chang, TW (1984) A cDNA clone encoding a product of activated human T-lymphocytes. Mol Biol Med 2: 151-165

Efrat, S, Pilo, S and Kaempfer, R (1982) Kinetics of induction and molecular size of mRNAs encoding human interleukin-2 and γ-interferon. Nature 297: 236-239

Kelly, K, Cochran, BH, Stiles, CD and Leder, P (1983) Cell specific regulation of the c-myc gene by lymphocyte mitogens and platelet-derived growth factor. Cell 35: 603-610

McCairns, E, Fahey, D, Muscat, GEO and Rowe, PB (1983) The effects of lectin transformation on cytoplasmic polyadenylated RNA from human lymphocytes. Mol Cell Biochem 56: 165-175

McCairns, E, Fahey, D, Muscat, GEO, Murray, M and Rowe, PB (1984) Changes in the levels of actin and tubulin mRNAs upon the lectin activation of lymphocytes. Mol Cell Biol 4: 1754-1760

Williams, JG and Penman, S (1975) The messenger RNA sequences in growing and resting mouse fibroblasts. Cell 6: 197-206.

Progress in Developmental Biology, Part A, pages 11–16

GENE EXPRESSION IN TRANSGENIC *XENOPUS LAEVIS*

Laurence D. Etkin

Department of Genetics, The University of Texas System Cancer Center M. D. Anderson Hospital & Tumor Institute, 6723 Bertner Ave., Houston, TX 77030

INTRODUCTION

Xenopus laevis has been used extensively in the analysis of molecular and embryological events during development and cellular differentiation. The organism is amenable to the use of techniques such as microsurgery, microinjection, and biochemical analysis. The microinjection of purified genes into *Xenopus* oocytes has yielded a plethora of results revealing the various factors involved in the transcriptional regulation of gene expression (for review see Gurdon and Melton, 1981; Etkin, 1982, Etkin and DiBerardino, 1983). Recently, cloned genes have been injected into the fertilized *Xenopus* egg in hopes of producing transgenic frogs (Bendig, 1981; Rusconi and Schaffner, 1981; Etkin and Roberts, 1983; Etkin et al, 1984). We will summarize the results obtained dealing with the fate and expression of genes injected into fertilized *Xenopus* eggs.

FATE OF INJECTED EXOGENOUS DNA SEQUENCES

Xenopus laevis embryos have been transformed by microinjection of exogenous DNA into the cytoplasm of fertilized eggs. The injected DNA is replicated, expressed, and possibly integrated in the recipient embryos (Bendig, 1981; Rusconi and Schaffner, 1981; Etkin and Roberts, 1983; Bendig and Williams, 1983; Etkin et al, 1984). Injected circular DNA molecules remain extrachromosomal during the early developmental stages in the initial transformants. They persist until the gastrula stage then gradually diminish, probably as a result of degradation.

We have followed the fate of exogenous DNA embryos by injecting radiolabelled plasmids. An M13 clone containing the Drosophila melanogaster alcohol dehydrogenase gene was labelled by the synthesis of the complementary strand using the Klenow fragment of E. coli polymerase I and tritiated TTP and ATP. Approximately 1 ng of the labelled M13 DNA (2 x 10^6 cpm/ug) was injected into the cytoplasm of artificially inseminated eggs of Xenopus laevis. The embryos at stages 1 and 6.5 were fixed, embedded in paraffin, sectioned and the location of the injected DNA was determined by autoradiography. In the stage 1 embryos (approximately 15 minutes after injection) silver grains were present in the cytoplasm around the site of injection over an area 100-200 μm in diameter. By stage 6.5 essentially all the grains appeared over the nuclei of the cells with only background labelling over the cytoplasm. The number of grains over the nuclei ranged from zero to greater than 50. Most of the nuclei (55%) had 25 or more grains over them, while 14% of the nuclei were scored as being unlabelled (0-5 grains). In some cases grains were seen localized over metaphase chromosomes. We feel that the majority of molecules are not integrated but may be sticking to the chromatin. The presence of unlabelled nuclei indicates that labelling was not a consequence of degradation and the re-incorporation of the labelled nucleotides into newly synthesized DNA. The injected labelled DNA remained as discrete circular forms in cleavage-stage embryos as determined by electrophoresis and autoradiography.

INTEGRATION AND GERM-LINE TRANSMISSION

By the late-blastula stage, a significant amount of injected exogenous DNA comigrates with the high-molecular-weight cellular DNA. While much of this DNA appears to consist of tandem repeats or concatemers, we believe that some of the exogenous DNA sequences are integrated into the frog genome.

To determine whether injected DNA was integrated into the frog genome, we analyzed DNA from injected larvae and young frogs by digestion with various restriction enzymes. The evidence for integration consists of: (1) generation of putative junction fragments differing in size from the input DNA fragment when DNA from injected embryos is digested with an enzyme which cuts once in a plasmid; (2) formation of bands which migrate differently from input DNA upon digestion of DNA from larvae and embryos with an enzyme that does not cut the plasmid; (3) migration of input DNA sequences with high-molecular-weight frog DNA in injected embryos in which the DNA is untreated, and (4) the reduction in size of this DNA upon digestion with enzymes that do not cut the plasmid DNA.

Injected genes persist in tissues of adult frogs but exhibit a mosaic pattern of distribution with regard to the presence or absence of the gene and to copy number between different tissues. When linear pSV2CAT was injected, (contains the SV40 early gene promoter and the gene coding for chloramphenicol acetyltransferase) sequences were detected in 60% of the young frogs. These sequences comigrated with the high-molecular-weight DNA and did not exhibit gross structural rearrangements as determined by restriction digestion analysis. Data from restriction enzyme digestion of the DNA favors the organization of the injected DNA into tandem head-to-tail multimers. The copy number variation was from 2-30 copies/cell in various tissues of different frogs.

The state of the injected DNA may also be an important factor in the efficiency of integration. Linearized plasmids persist in a much higher percentage of injected embryos than do injected circular plasmids and presumably are integrated at a higher rate.

The presence of exogenous DNA sequences in the offspring of transformed animals demonstrate transmission of the exogenous DNA sequences through the germ-line. Plasmid DNA containing the gene for Drosophila alcohol dehydrogenase (ADH, sAC-1) was injected into Xenopus laevis eggs. The embryos were maintained in the laboratory until they reached sexual maturity. Sperm from a transformed male was used to inseminate eggs from a "wild type" female. The F1 generation embryos were allowed to develop for 1-3 months. The DNA was extracted from isolated tissues, digested with either PstI or BamHI restriction endonucleases, separated by agarose gel electrophoresis, blotted onto nitrocellulose, and probed with ^{32}P-labelled sAC-1 DNA.

Six offspring were examined for the presence of sAC-1 sequences. Four of the six animals were positive for the exogenous DNA. DNA from two F1 frogs was analyzed by restriction enzyme digestion and Southern blotting. BamHI which cleaves the injected sAC-1 plasmid at two sites, produced two fragments of 2.8 and 6.2 kb in frog F1-c similar to the fragments produced in control digests with this enzyme. Digestion of the DNA with PstI, which cuts plasmid sAC-1 at a single site, produced the predicted 9.0 kb DNA fragment and a putative junction of approximately 2.8 kb in each of the F1-c tissues. This fragment presumably consists of sAC-1 sequences joined to frog DNA. The presence of the predicted restriction fragments and junction fragments demonstrated that there were no major rearrangements of the injected DNA and that the DNA was integrated as tandem head-to-tail concatemers. The presence of a similar restriction pattern from different tissues suggested that the sAC-1 DNA was integrated into a single site in F1-c. In each tissue the 9.0 kb fragment was present in a 2:1 relationship with the junction fragment. Assuming that the junction fragment is present as a single copy per genome then the two-fold higher intensity of the 9.0 kb fragment indicates that two complete copies plus a partial copy (as junction fragments) exist per genome equivalent of frog F1-c. Samples from a second frog showed a different pattern of junction fragments, indicating that the site of integration was different between these two frogs even though they were from the same mating. This suggests that the male which produced the sperm, must have had multiple sites of integration within its germ cells.

EXPRESSION OF EXOGENOUS GENES IN TRANSFORMED EMBRYOS

The Early Embryo

In the amphibian embryo, transcription is not detectable until the mid-blastula stage of development. At this time, there is an increase in transcription from the embryonic genome. Exogenous DNA injected into the fertilized egg also follows this same pattern of expression in that there is no detectable transcription until the blastula stage of development. Injected genes coding for rabbit β-globin (Rusconi and Schaffner, 1981), sea urchin histones (Bendig, 1981), Drosophila alcohol dehydrogenase (Etkin et al, 1984), yeast tRNA (Newport and Kirschner, 1981a, b), ribosomal RNA (Busby and Reeder, 1983), adult Xenopus α- and β-globin (Bendig and Williams, 1983, 1984) and chloramphenical acetyl transferase (CAT) (Etkin and Balcells, 1985) are expressed at the mid-late blastula stage of development. The early embryo therefore, may be viewed as a transient expression system in which detectable transcription on injected templates peaks at the blastula-gastrula stage of development and then declines in parallel with the loss of the extrachromosomal plasmid molecules.

Expression of Genes in Tissues of Young Frogs

Following the blastula stage of development, injected DNA sequences which are extrachromosomal are degraded. A residual fraction of injected DNA, however, persists and is believed to be integrated into the frog genome (Etkin et al, 1984). Many of the initial transformants are found to be mosaic organisms which contain the gene integrated into different positions and/or have different copy number in various tissues. We have analyzed expression of a plasmid, pSV2CAT, (contains the CAT gene with SV40 early promoter), which in principal, should not contain any frog regulatory sequences on the DNA since it is a hybrid containing both viral and prokaryotic DNA sequences. This gene, however, does exhibit a variety of patterns of expression in different transformed frogs. Expression is not dependent on copy number since tissues containing two-five copies of the gene may express relatively large amounts of the CAT gene product while other tissues containing ten or more copies do not express the gene. Restriction enzyme digestion analysis of the DNA from those tissues not expressing the gene indicate that large rearrangements have not occurred, suggesting that the lack of expression of the gene may be due to the effects of the site of integration.

REFERENCES

Bendig MM (1981). Persistence and expression of histone genes injected into *Xenopus laevis* eggs in early development. Nature (London) 292:65.

Bendig MM, Williams JG (1983). Replication and expression of *Xenopus laevis* globin genes injected into fertilized *Xenopus* eggs. Proc Natl Acad Sci USA 80:6197.

Bendig MM, Williams JG (1984). Differential expression of the *Xenopus laevis* tadpole and adult β-globin genes when injected into fertilized *Xenopus laevis* eggs. Mol and Cell Biol 4:5670.

Busby SJ, Reeder RH (1983). Spacer sequences regulate transcription of ribosomal gene plasmids injected into *Xenopus laevis* embryos. Cell 34:989.

Etkin LD (1982). Analysis of the mechanisms involved in gene regulation and cell differentiation by microinjection of purified genes and somatic cell nuclei into amphibian oocytes and eggs. Differentiation 21:149.

Etkin LD, DiBerardino MA (1983). Nuclear transplantation and gene injection systems. In MacLean N, Gregory SP,

Flavell RA (eds): " Eukaryotic genes: their structure, activity, and regulation," UK, Butterworths.

Etkin LD, Roberts M (1983). Transmission of integrated sea urchin histone genes by nuclear transplantation in *Xenopus laevis*. Science 221:67.

Etkin LD, Pearman B, Roberts M, Bektesh S (1984). Replication, integration, and expression of exogenous DNA injected into fertilized eggs of *Xenopus laevis*. Differentiation 26:194.

Etkin LD, Balcells S, (1985). Transformed *Xenopus* embryos as a transient expression system to analyze gene expression at the mid-blastula transition. Develop Biol 108:173.

Gurdon JB, Melton DA (1981). Gene transfer in amphibian eggs and oocytes. Ann Rev Genet 15:189.

Newport J, Kirschner M (1982a). A major developmental transition in early *Xenopus*: I. Characterization and timing of cellular changes at the mid-blastula stage. Cell 30:675.

Newport J, Kirschner M (1982b). A major developmental transition in early *Xenopus* embryos: II. Control of the onset of transcription. Cell 30:687.

Rusconi S, Schaffner W (1981). Transformation of frog embryos with a rabbit β globin gene. Proc Natl Acad Sci USA 78:5051.

Progress in Developmental Biology, Part A, pages 17–21

REGULATION OF THE DICTYOSTELIUM RAS GENE DURING DEVELOPMENT AND IN TRANSFORMANTS

Christophe D. Reymond, Wolfgang Nellen,
Richard H. Gomer and Richard A. Firtel

Department of Biology, B-022
University of California, San Diego
La Jolla, CA 92093

We have analyzed a Dictyostelium gene encoding a protein homologous to the mammalian ras proteins. The Dictyostelium, yeast, and mammalian ras proteins have highly homologous N-terminal sequences and less conserved C-terminal regions (Reymond et al., 1984). At positions where mutations have been associated with a transformed phenotype (Fasano et al., 1984) and tumorigenicity (Zarbl et al., 1985) in mammalian cells, the Dictyostelium ras is identical to the mammalian cellular proteins. In particular a glycine is present at position 12.

We have used DNA probes and a polyclonal antibody to examine the differential accumulation of ras RNA and protein through development. The gene encodes two mRNAs (0.9 and 1.2 kb) which are differentially expressed. The 1.2 kb RNA is found in vegetative cells and disappears rapidly upon initiation of development. Later, both RNAs accumulate preferentially in prestalk cells.

Using the λgt11 vector system, we have produced rabbit polyclonal antibodies against the carboxy-terminal two-thirds of the Dictyostelium ras gene. By immunostained Western blots, we have shown that the ras protein is present in vegetative cells. Upon starvation, the level decreases slightly until 15 hours then more rapidly. At the fruiting body stage the ras protein is no longer detectable (Reymond et al., 1984). Two ras proteins can be seen on Western blots: a major 23 kd and a minor 24 kd variant. Throughout development, the 23 kd variant is found predominantly in the cytosol fraction of fractionated cells. The 24 kd variant is found mainly in the microsomal fraction and can only be detected in vegetative cells and during the first few hours of development (Gomer, Reymond and Firtel, submitted for publication), indicating a change in the subcellular location of ras protein during Dictyostelium development. In migrating pseudoplasmodia, ras protein can only be detected in prestalk cells.

Both ras RNAs appear during the formation of pseudoplasmodia (~10 hours) concomittantly with new synthesis of the protein, as shown by others (Pawson et al., 1985). A single ras protein is present at that time during development as seen on 2-D gels (Gomer, Reymond and Firtel, submitted for publication), suggesting either that only one of the RNAs encodes the protein or that both RNAs share the same protein coding region. S1 experiments have shown that the two RNAs are not a result of differential splicing. Both RNAs are observed when using parts of the coding region, thus suggesting that the difference between the two transcripts must lie within the 5' or 3' ends.

A single cell culture system has been developed in which prestalk specific mRNAs accumulate only upon addition of external cAMP (Mehdy et al., 1983). Like other prestalk genes, Dd-ras can be induced with cAMP in the absence of cell contact under such conditions (Reymond et al., 1984). Further indication of the role of cAMP during late development was obtained by disaggregation experiments. The level of prestalk specific transcripts decreases when aggregated cells are dissociated but increases again when external cAMP is added (Mehdy et al., 1983). Under such conditions both ras mRNAs decrease. Upon addition of cAMP, the 1.2 kb mRNA reaccumulates at a higher level than that in normal developing cells (Reymond et al., 1984).

We used the Dictyostelium transformation system to analyze which ras sequences are responsible for this cAMP dependent regulation (Nellen et al., 1984). Constructs were made in which ~500 nts of the 5' flanking region of the Dictyostelium ras gene and variable amounts of the coding region were linked to a ras cDNA in a transformation vector. These constructs were then transformed into Dictyostelium cells and the regulation of the ras constructs examined. In transformants, transcripts from the ras gene constructs were found at high levels in cells in single cell cultures containing cAMP, whereas transcripts were either not detectable or present at very low levels in cultures lacking exogenous cAMP. When normal developing aggregates were dissociated, RNA from the ras constructs decreased rapidly but then re-accumulated in the presence of cAMP. These results show that the sequences necessary for the response to external cAMP are present within ~650 bp upstream from the ATG start codon and/or within portions of the coding region. Moreover, the proper regulation of ras gene expression in transformants containing a high copy number of the construct DNA suggests that trans-acting factors which may control transcription are not limiting (Reymond et al., 1985).

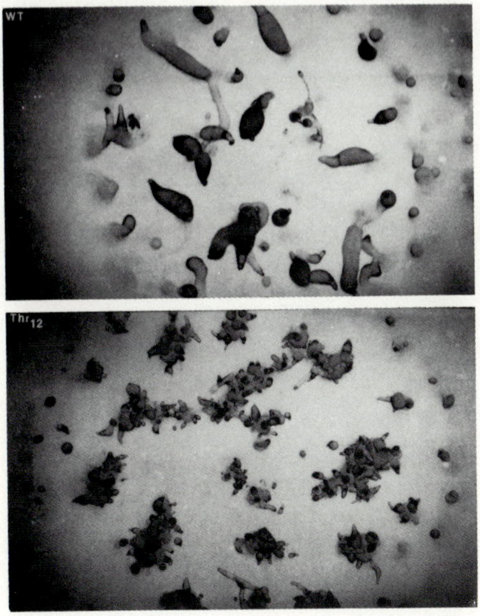

Figure 1. Ax-3 Dictyostelium cells (wt) and cells transformed with a $Gly_{12} \rightarrow Thr_{12}$ substitution (Thr) were plated with Klebsiella aerogenes as a food source on SM agar. After exhaustion of the food source, the cells were allowed to develop. 70% of the single colonies from the transformants developed multiple tips. Analysis of the copy number of the vector DNA in transformants showed a direct correlation between the extent of the phenotype and copy number of the mutated gene. All colonies from Ax-3 cells formed single fruiting bodies under the same conditions.

Vector constructs were also examined in which the cDNA was present in the opposite orientation compared to the gene fragment (anti-sense orientation) (Reymond et al., 1985). When these were transfected into cells, no transformants were obtained suggesting that, as in yeast (Kataoka et al., 1984), ras gene expression is essential for vegetative growth in Dictyostelium.

In order to obtain insights on the function of the ras protein in Dictyostelium, the glycine codon at position 12 of the protein was changed by in vitro mutagenesis to encode a threonine instead. Such a mutation has been shown to be sufficient to produce a transformed phenotype in 3T3 cells (Seeburg et al., 1984). DNA fragments encoding the whole ras protein with a glycine or threonine at position 12 were inserted in a transformation vector and reintroduced into Dictyostelium. Western blots show an elevated level of the ras protein in transformants containing the reintroduced DNA. Transformants containing the threonine codon have a tendency to form large aggregates developing multiple tips (see Figure 1). This phenotype is dependent upon the copy number of the DNA. Tips are thought to correspond to organizing centers in the formation of the slugs during Dictyostelium development. cAMP is thought to play a role during this process (for review see Loomis, 1982). Work in yeast has shown that ras might modulate adenylate cyclase activity. A link between this phenotype and adenylate cyclase activity in Dictyostelium still remains to be analyzed.

REFERENCES

Fasano O, Aldrich T, Tamanoi F, Taparowski E, Furth M, Wigler M (1984). Analysis of the transforming potential of the human H-ras gene by random mutagenesis. Proc. Natl. Acad. Sci. USA 81: 4008-4012.

Kataoka T, Powers S, McGill C, Fasano O, Strathern J, Broach J, Wigler M (1984). Genetic Analysis of Yeast RAS1 and RAS2 Genes. Cell 37: 437-445.

Loomis WF (1982). In "The Development of Dictyostelium discoideum". New York: Academic Press.

Mehdy MC, Ratner D, and Firtel R (1983). Induction and modulation of cell-type-specific gene expression in Dictyostelium. Cell 32: 763-771.

Nellen W, Silan C, Firtel RA (1984). DNA-Mediated Transformation in Dictyostelium discoideum: Regulated Expression of an Actin Gene Fusion. Mol. Cell. Biol. 4: 2890-2898.

Pawson T, Amiel T, Hinze, E, Auersperg N, Neave N, Sobolewski A, Weeks G (1985). Regulation of a ras-Related Protein During Development of Dictyostelium discoideum. Mol. Cell. Biol. 5: 33-39.

Reymond CD, Gomer RH, Mehdy MC, Firtel RA (1984). Developmental Regulation of a Dictyostelium Gene Encoding a Protein Homologous to Mammalian ras Protein. Cell 39: 141-148.

Reymond CD, Nellen W, Firtel RA (1985). Regulated Expression of ras gene constructs in Dictyostelium transformants. Proc. Natl. Acad. Sci. USA, in press.

Seeburg PH, Colby WW, Capon DJ, Goeddel DV, Levinson AD (1984). Biological properties of human c-Ha-ras1 genes mutated at codon 12. Nature 312: 71-75.

Zarbl H, Sukumar S, Arthur AV, Martin-Zanca D, Barbacid M (1985). Direct mutagenesis of Ha-ras-1 oncogenes by N-nitrosa-N-methylurea during initiation of mammary carcinogenesis in rats. Nature 315: 382-385.

Progress in Developmental Biology, Part A, pages 23-26

EXPRESSION OF MTV SEQUENCES INTRODUCED INTO THE MOUSE GERMLINE

Susan R. Ross

University of Illinois at Chicago
Department of Biological Chemistry
1853 W. Polk St.
Chicago, IL 60612

Mouse mammary tumor virus (MTV), an endogenous murine retrovirus, is expressed in the mouse primarily in lactating mammary gland and mammary tumors (Varmus et al., 1973; Marcus et al., 1981; Hu et al., 1984). Many hormones are involved in the development of this tissue; as of yet, however, the only hormones documented to be directly involved in the regulation of MTV transcription are the glucocorticoids, which appear to exert their effects by the binding of hormone-bound receptor complexes to specific sequences in the viral genome (Payvar et al., 1983; Scheidereit et al., 1983). We wish to determine why the expression of MTV is restricted to the lactating mammary gland in the mouse, given that functional glucocorticoid receptors are found in most tissues, and whether the virus encodes tissue-specific enhancers in addition to the glucocorticoid regulatory elements (GRE).

We introduced a plasmid containing the long terminal repeat (LTR) of MTV, previously shown to encode several GRE's (Payvar et al., 1983), linked to the HSV I thymidine kinase (tk) gene (Colbere-Garapin et al., 1979), into the mouse germline by microinjection into one cell mouse embryos, to create transgenic mice bearing these plasmid sequences. Six lines, derived from independent injections, were obtained (Table I). Two lines resulted from the microinjection of supercoiled plasmid; the other four lines were derived by microinjection of plasmid linearized in the pBR322 sequences. The lines contain from about five to fifty copies of the injected DNA.

TABLE I

Mouse Plasmid	# Strain	# Injected	# Implanted	# Born	# Containing Injected DNA*
pLTR2tk	ICR	450	311(0.69)	28(0.09)	2(0.07)
HIII-pLTR2tk	C57Bl/6	139	79(0.57)	4(0.05)	1(0.25)
SalI/HIII-pLTR2tk	C57Bl/6	425	310(0.73)	19(0.06)	3(0.16)

*Dot blots of tail DNA were performed and hybridized with nick-translated tk probe (Ross and Solter, 1985).

These mice were then mated with normal mates in order to produce sublines and to test whether the injected DNA had integrated into the germline. One of the mice, Ltk 12, a female, died before puberty; two other male mice, Ltk 15 and HS Ltk 2, despite extensive attempts to mate them failed to produce any offspring. Another male, HS Ltk 2, and one female (HS Ltk 3) are genetic mosaics, transmitting the microinjected gene to approximately 11% of their offspring. Both Ltk 3 and H/S Ltk 3 appear to have single insertions of the injected sequences, since approximately 50% of the G1 and G2, respectively, inherit the injected DNA; Southern blot analysis using several restriction enzymes of DNA isolated from these mice indicates that the integration sites contain tandemly integrated copies of the plasmid (Ross and Solter, 1985).

We have examined the tissues of several of these mice and/or their offspring for the presence of RNA hybridizing to the microinjected plasmid DNA. RNA was isolated from at least 12 tissues from lactating, virgin and dexamethasone-treated virgin (100 μg injected intraperitoneally 4-5 hours prior to sacrifice) female offsprint of Ltk 3 and analyzed by dot blots or Northern gels. A representative dot blot of the tissue RNA hybridized to tk probe from a lactating female is shown in Figure 1a. Only the ovary/oviduct and the mammary gland in this mouse appear to synthesize significant amounts of tk-homologous RNA. Little transcription of these sequences is found in the same tissues from untreated or dexamethasone-injected virgins less than one year of age. Injection of lactating mice for three consecutive days with 100 μg of 2-bromo-α-ergocryptene, to reduce serum prolactin levels, appeared to have no affect on the level of tk transcription in mammary glands (not shown).

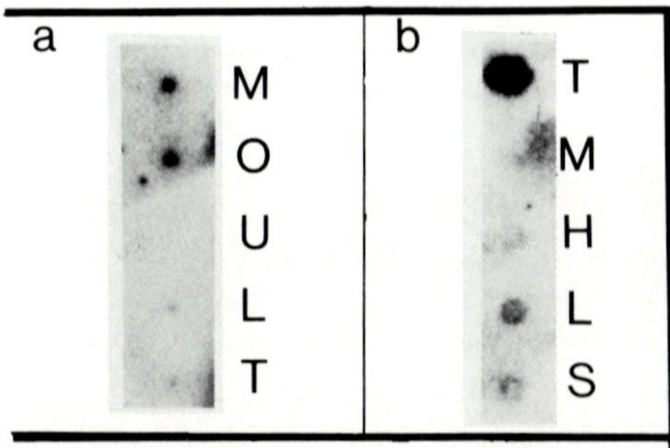

FIG. 1 Dot blot analysis of organ RNA. Five µg of total RNA was spotted onto nitrocellulose and hybridized with tk probe. Control filters were hybridized with MTV envelope gene and tubulin probes (not shown). a. thymus (T), liver (L), uterus (U), ovary and oviduct (O), lactating mammary gland (M). b. spleen (S), liver (L), heart (H), skeletal muscle (M), testes (T).

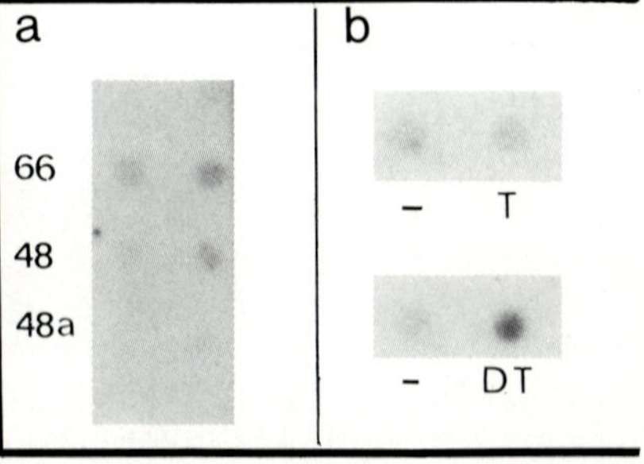

Fig. 2 Dot blot analysis of testes RNA. a. testes RNA from male 66 and a littermate, 48 before (B) and after (A) adrenalectomy. b. testes RNA from male before (-) and after injection with dihydrotestosterone (T) and a male littermate before (-) and after injection with dexamethasone and dihydrotestosterone (DT).

We have also looked for the presence of tk RNA in males. In males, the major site of expression is in the testis (Fig. 1b), although no transcription of the endogenous MTV sequences are seen in this tissue (Hu et al., 1984; Ross, unpublished observation). This expression is induced by glucocorticoids; twenty to thirty times more tk homologous RNA is produced in the testis of an offspring of Ltk 3 before than after adrenalectomy (Figure 2a). Treatment with dexamethasone also generally increases these levels; however, the basal level of expression varies from mouse to mouse, probably due to the levels of circulating glucocorticoids. Adrenalectomy reduces these basal levels drastically. Dihydrotestosterone has little affect on the level of expression. The level of tk transcription remained unchanged in the testis of a mouse analyzed before and five hours after injection with 100 µg of this androgen, whereas the testis of a littermate analyzed before and after injection with both dexamethasone and testosterone shows about a thirty-fold increase in tk RNA levels (Figure 2b).

Our results indicate that MTV LTR encodes sequences responsible for its expression in lactating mammary glands in the mouse. In addition, since the newly introduced sequences are also found in the testes of at least four strains generated from independent injections, we believe that there are sequences encoded by the LTR which specify transcription in this tissue as well. Perhaps the endogenous sequences are not accessible to the necessary transcription factors, such as the glucocorticoid receptor, in the testes; by reintroducing MTV into a new genomic location and at a high copy number, such factors may now interact with the requisite DNA sequences in the MTV genome.

(Supported by grant HD 19444 from the National Institute of Health).

REFERENCES

Colbere-Garapin, F, Chousterman, S, Horodniceau, F, Kourilsky, P, Garapin, A-C (1979). Cloning of the active thymidine kinase gene of herpes simplex virus type 1 in Escherichia coli K-12. Proc Natl Acad Sci USA 76: 3755-3759.

Hu, W-S, Fanning, TG, Cardiff, RD (1984). Mouse mammary tumor virus: specific methylation patterns of proviral DNA in normal mouse tissues. J Virol 49: 66-71.

Marcus, SL, Smith, SW, Sarkar, NH (1981). Quantitation of murine mammary tumor virus-related RNA in mammary tissues of low- and high-mammary tumor-incidence mouse strains. J Virol 40: 87-95.

Payvar, F, DeFranco, D, Firestone, GL, Edgar, B Wrange, O, Okret, S, Gustafsson, J-A, Yamamoto, KR (1983). Sequence-specific binding of glucocorticoid receptor to MTV DNA at sites within and upstream of the transcribed region. Cell 35: 381-392.

Ross, SR, Solter, D (1985). Glucocorticoid regulation of mouse mammary tumor virus sequences in transgenic mice. Proc Natl Acad Sci USA 82, in press.

Schedereit, C, Geisse, S, Westphal, HM, Beato, M (1983). The glucocorticoid receptor binds to defined nucleotide sequences near the promoter of mouse mammary tumor virus. Nature 304: 249-252.

Varmus, HE, Quintrell, N, Medeiros, E, Bishop, JM, Nowinski, RC, Sarkar, NH (1973). Transcription of mouse mammary tumor virus genes in tissues from high and low tumor incidence mouse strains. J Mol Biol 79: 663-679.

Progress in Developmental Biology, Part A, pages 27–34

THE ROLE OF THE CELL CYCLE IN TRANSDIFFERENTIATION OF MONONUCLEATED CROSS STRIATED MUSCLE CELLS INTO SMOOTH MUSCLE AND SENSORY CELLS IN VITRO

Hansjuerg Alder and Volker Schmid

Department of Zoology, University of Basel,
Rheinsprung 9, CH-4051 Basel, Switzerland

INTRODUCTION

In nuclear transplantation experiments (DiBerardino, 1984) and in some transdifferentiation systems in vivo (Yamada, 1977) and in vitro (Okada, 1983) it has been shown that stably repressed genes can be reactivated. This also occurs in a novel in vitro transdifferentiation and regeneration system that is based entirely on isolated, mononucleated, striated muscle of a jellyfish (Schmid and Alder, 1984). The isolated and cultivated muscle can undergo transdifferentiation and even form a complex regenerate consisting of 6-8 new cell types.

The process of cellular commitment in ontogeny and the regulatory events governing transdifferentiation are presumably to some extent the same. In both cases number and quality of cell cycles seem to play a major role (Holtzer et al., 1979; Satoh and Ikegami, 1981). The determined state of a gene is possibly changed during the S phase of a cell cycle (Brown, 1984). Although our results in general confirm the necessity of cell cycles for change in cellular commitment, we provide evidence of exceptions to this rule.

RESULTS AND DISCUSSION

If isolated, mononucleated, striated muscle tissue from the anthomedusa, Podocoryne carnea, is treated with bacterial collagenase and then cultered, it is able not only to undergo complete transdifferentiation into several new cell types but also to regenerate the feeding and sexual organ of the animal, the manubrium (Figure 1; Schmid and Alder, 1984; Schmid and Alder, 1985) or tentacles (unpublished). The evidence for this pluripotent transdifferentiation is based on the facts that the isolated muscle tissue is reliably uncontaminated with other cell types and that the newly formed cell types occasionally show remnant bundles of cross striated myofibrils (Figure 5). These latter are thought to be remnant structures since they are not contained in vacuoles (and are therefore not phagocytosed) and are seen in various stages of degradation. Since debris of cross striated myofibers are never observed normally in the other cell types (besides striated muscle cells) we concluded that the observed new cell types originate from striated muscle cells (Schmid and Alder, 1984).

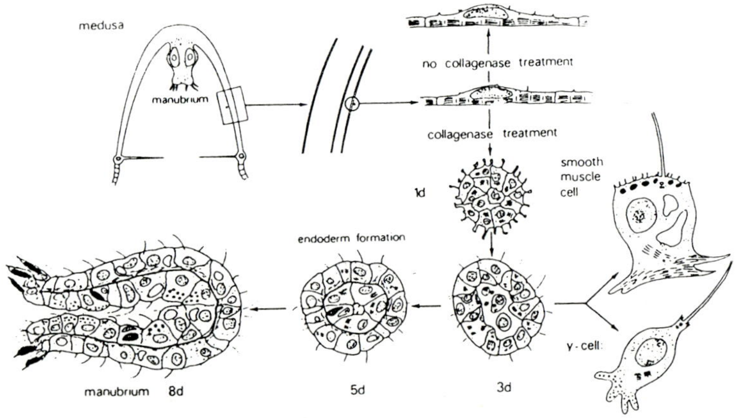

Figure 1. Schematic drawing summarizing the transdifferentiation potential of isolated striated muscle of the anthomedusa Podocoryne carnea.

DNA-synthesis is initiated at about 56 hours after isolation of the striated muscle (Figure 6). At about 62 hours 80% of the cells are labeled. We estimate that the remaining 20% of the cells are dying or are unable to

Figure 2. Cross section through a muscle fragment 10 days after isolation. Smooth muscle cells and sensory cells are well differentiated. 6'100x.
Figure 3. Sensory cell with its basal body of the flagellum and small vesicles 10 days after isolation. 8'500x.
Figure 4. Basal portion of smooth muscle cell and sensory cell 10 days after isolation. 18'700x.
Figure 5. Muscle fragment 6 days after isolation. The fragment has formed an outer (ectodermal) layer and an inner (endodermal) layer separated by a basal lamina; the first nematoblasts differentiate, the thread is still outside the capsule which is adjacent to remnants of striated myofibrils. 6'800x.
Abbrevations to figures 2-5: by, basal body of flagellum of sensory cell; ec, outer (ectodermal) layer; en, inner (endodermal) layer; c, capsule of nematoblast; f, flagellum; m, mesoglea; n, nucleus; ny, nucleus of sensory cell; pv, peripheral vacuoles typical of smooth muscle; s, smooth muscle-like arrangement of myofibrils; sn, nucleus of smooth muscle-like cell type; sv, small vesicles of sensory cells; t, thread of nematoblast.

initiate DNA-synthesis due to damage in the isolation procedure (Schmid and Alder, 1984) or collagenase treatment (collagenase concentration, temperature and duration of digestion). At 66 hours the fragments start swimming due to newly formed flagellar action. This is before completion of DNA-synthesis, which occurs at about 70 hours. All the cells demonstrate the typical features of smooth muscle cells including flagella, peripheral vacuoles, and smooth muscle-like arrangement of contractile filaments at the cell base (Figure 2,4; Schmid and Alder, 1984). A second replication phase occurs at 90 to 101 hours with a labeling rate of about 40% (Figure 6). In addition to smooth muscle cells, a new cell type, previously called y-cell (Figure 2,3; Schmid and Alder, 1984) appears during the S phase of this second cell cycle. Based on ultrastructure and positive staining with antisera to FMRFamide (unpublished, a gift of G. O. Mackie) we concluded this second cell type to be sensory cells (Figure 2,3).

The influence of specific cell cycles on the formation of smooth muscle and sensory cells was investigated with the drug aphidicolin. Aphidicolin is a highly specific and

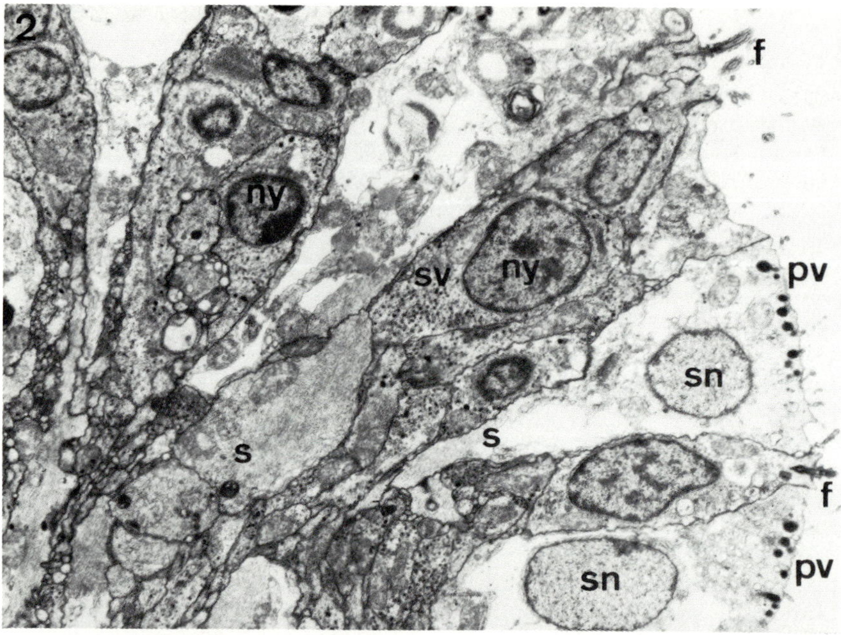
2
f
ny
ny
sv
pv
sn
s
s
f
sn
pv

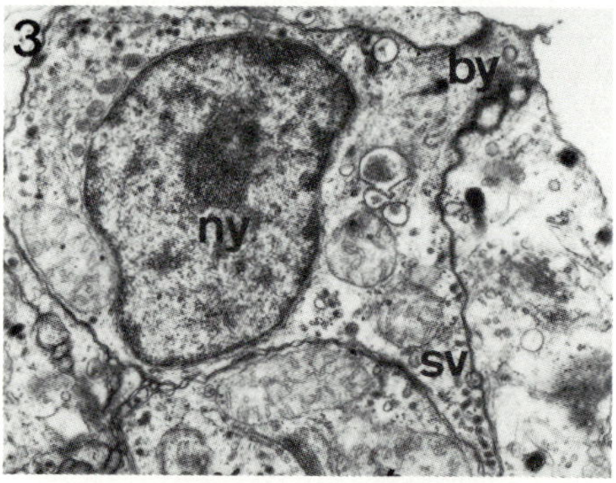
3
by
ny
sv

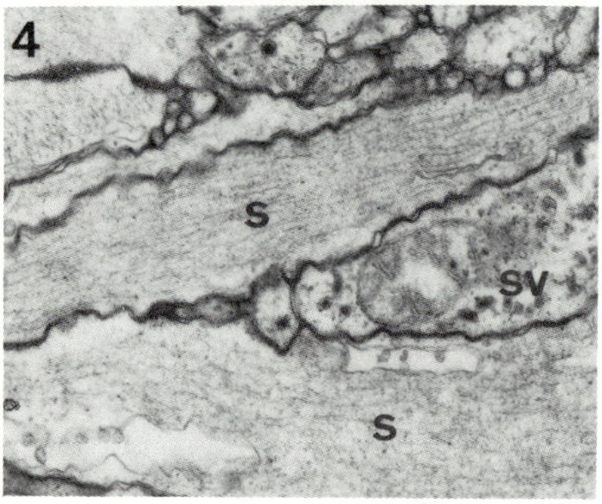
4
s
sv
s

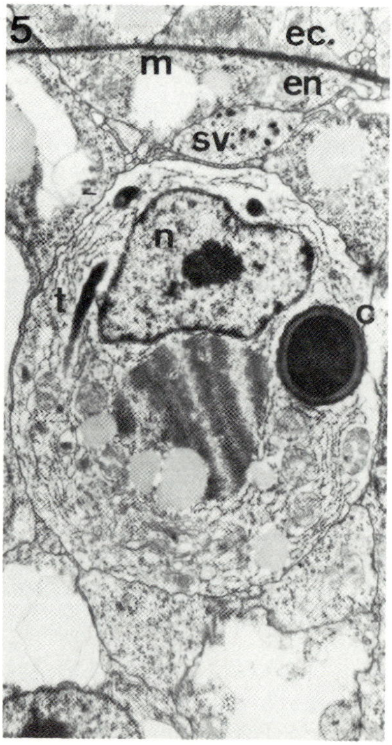
5
ec.
m
en
sv.
n
t
c

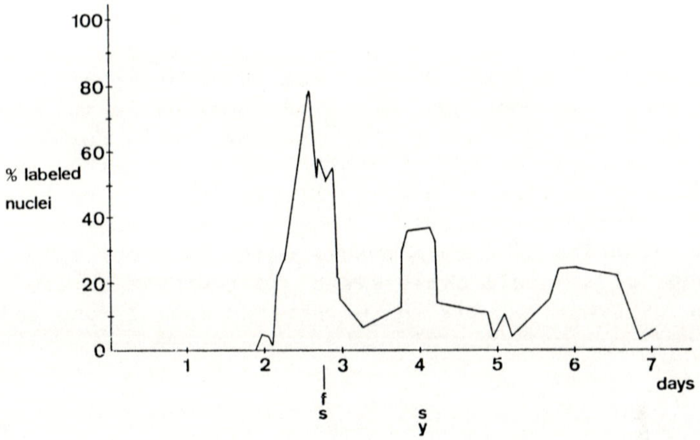

Figure 6. DNA-synthesis in transdifferentiating striated muscle cells. Striated muscle was incubated in 3 μCi/ml ^{3}H-thymidine for 2 hours and treated with 0.1 m citric acid to get rid of the cytoplasm. The liberated nuclei were mounted on slides coated with gelatine. After autoradiography the nuclei were stained with DAPI (Schmid and Alder, 1984) and checked for label. Points represent the average of 4 isolates. Bottom: First appearance of cell types (Schmid and Alder, 1984). f, flagellum of smooth muscle cell; s, smooth muscle cell; y, sensory cell.

reversible inhibitor of DNA polymerase α (Pedrali-Noy, 1979; Ogura, 1980) which blocks DNA-synthesis, arresting the cells at the border G1 -/S phase (Ikegami et al., 1979; Spadari et al., 1982). Striated muscle which has been continously treated after isolation with aphidicolin (controls show no ^{3}H-thymidine incorporation) transdifferentiate normally into smooth muscle cells, however no sensory cells are found. Since transdifferentiation into smooth muscle cells requires both transcription and translation (unpublished data), these results show that neither of these processes is influenced by the drug treatment.

Aphidicolin was used to investigate the effect of a

single cell cycle on the transdifferentiation events. The drug was applied and then removed from the isolates at the expected initiation time of the first S phase (Figure 7a). The isolates thereby entered the cell cycle at the same time as the controls. When aphidicolin was added again after the S phase, the cells completed the first cell cycle and sensory cells appeared at the end of the 3 rd day, as is normally the case. The ratio of smooth muscle cells to sensory cells and the number of labeled nuclei (Figure 7a) seem to be the same as in untreated controls. These results demonstrate that the formation of smooth muscle cells is completely independent of any cell cycle event, in contrast to the formation of sensory cells which requires exactly one cell cycle.

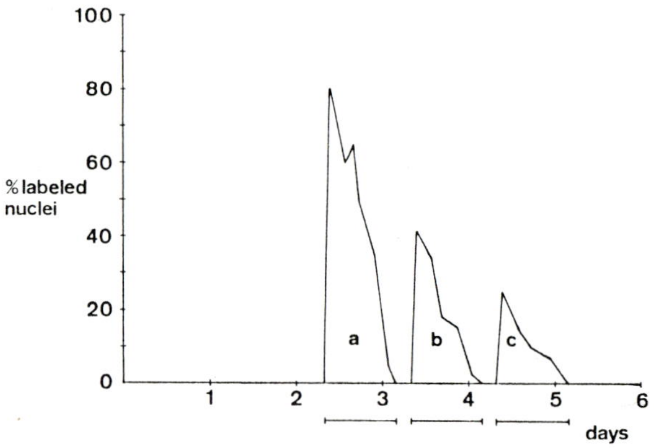

Figure 7. Delay of the first cell cycle by aphidicolin. After isolation of the tissue and collagenase treatment aphidicolin was added (10 μg/ml). In the first experiment (7a) the cells were washed free of the drug at 56 hours after isolation to let the cells initiate and complete the first cell cycle. At 76 hours aphidicolin was again added. In the second (7b) and third (7c) experiment the drug treatment was prolonged for 1 and 2 days respectively and the initiation and completion time of the first cell cycle was delayed for 1 and 2 days. The bars on the bottom indicate the time in each experiment where the cells were free of aphidicolin.

Following the same drug treatment protocol the initiation of the first cell cycle can be delayed one day, i.e. between the 3 rd and 4 th day after isolation (Figure 7b). Nuclear labeling decreases to a maximum of about 40% and the number of sensory cells is strongly reduced. The labeling maximum further drops to 25% and the number of sensory cells decreases to less than 1% when the cell cycle is delayed for 2 days (Figure 7c). These observations suggest that for optimal transdifferentiation of striated muscle cells to sensory cells DNA has to be replicated once within the second and third day.

REFERENCES

Brown DD (1984). The role of stable complexes that repress and activate eucaryotic genes. Cell 37: 359-365.

DiBerardino MA, Hoffner NJ, Etkin LD (1984). Activation of dormant genes in specialized cells. Science 224: 946-952.

Holtzer H, Rubinstein N, Fellini S, Yeoh G, Chi J, Birnbaum J, Okayama M (1979). Lineages, quantal cell cycles and the generation of cell diversity. Quart Rev Biophys 8: 523-557.

Ikegami S, Amemiya S, Oguro M, Nagano H, Mano Y (1979). Inhibition by aphidicolin of cell cycle progression and DNA replication in sea urchin embryos. J Cell Physiol 100: 439-444.

Oguro M, Shioda M, Nagano H, Mano Y (1980). The mode of action of aphidicolin on DNA synthesis in isolated nuclei. Biochem and Biophys Res Comm 92(1): 13-19.

Okada TS (1983). Recent progress in studies of the transdifferentiation of eye tissue in vitro. Cell Diff 13: 177-183.

Pedrali-Noy G, Spadari S (1979). Effect of aphidicolin on viral human DNA polymerases. Biochem and Biophys Res Comm 88(4): 1194-1202.

Satoh N, Ikegami S (1981). On the clock mechanism determining the time-specific enzyme development during ascidian embryogenesis. J Embryol Exp Morph 64: 61-71.

Schmid V, Alder H (1984). Isolated, mononucleated, striated muscle can undergo pluripotent transdifferentiation and form a complex regenerate. Cell 38: 801-809.

Schmid V, Alder H (1985). The potential for transdifferentiation of differentiated medusa tissues in vitro. In Moscona A (ed): "Current Topics in Developmental Biology", Volume 20, Okada TS (ed): "Commitment and Instability in

Cell Differentiation", New York: Academic Press, in press.
Spadari S, Sala F, Pedrali-Noy G (1982). Aphidicolin: a specific inhibitor of nuclear DNA replication in eucaryotes. TIBS: 29-32.
Yamada T (1977). Control mechanisms in cell-type conversion in newt lens regeneration. In Awalsky (ed): "Monographs in Developmental Biology", Basel: Karger, p 13.

Progress in Developmental Biology, Part A, pages 35–38

IDENTIFICATION OF INTERGENERIC NUCLEO-CYTOPLASMIC HYBRID FISH OBTAINED FROM THE COMBINATION OF CARP NUCLEUS AND CRUCIAN CYTOPLASM

Yan Shaoyi, Du Miao, Wu Naihu, Yan Jingzhi
(Institute of Developmental Biology,Academia Sinica,Beijing,China)
Jin Guangqin, Qin Ying
(Guangxi Fisheries Research Institute,Nanning, China)
Zhang Xingzhong
(Chang Jiang Fisheries Research Institute,The China State Fisheries Institution,Shashi,China)

INTRODUCTION

In a previous paper,we reported that hybrid fish are obtained from the combination of nucleus and cytoplasm from different genera of fresh-water teleosts using the technique of nuclear transplantation,i.e. by combinating nucleus of the carp(Cyprinus carpio) and cytoplasm of the crucian(Carassius auratus).The survival percentage of those adult hybrid fish is about 3.2%. Morphological characteristics of those hybrid fish showed that some features are inherited from the nucleus doner fish such as barbs and pharyngeal teeth,some seem to come from the cytoplasm host fish such as the number of vertebrae,and some are intermediates such as the number of scales along the literial line. Those results indicate that both nucleus and cytoplasm can influence the expression of genetic information on the hybrid fish. Those hybrid fish are sexually mature and can produce normal offspring by artificial insemination. (Research Group of Cytogenetics,Institute of Zoology,Academia Sinica,1980).

In another previous paper,we also reported that hybrid fish are obtained from the reversed combination of nucleus and cytoplasm from two genera of fresh-water teleosts using the technique of nuclear transplantation,i.e.the combination of nucleus of the crucian(Carassius auratus)and cytoplasm of the carp(Cyprinus carpio). The survival percentage of those adult hybrid fish is 0.9%. Morphological characteristics of those hybrid fish that have been examined so far indicate that some features such as barbs,pharyngeal teeth, the number of scales along the lateral line,and the number of vertebrae are similar to those of crucian. Some of the hybrid fish grow to normal adult fish(Yan Shaoyi et al,1984).

The above results obtained so far indicate that there is a inconsistency of some morphological characteristics of the hybrid fish(CyCa) obtained from the combination of nucleus of the carp(Cy) and cytoplasm of the crucian(Ca) and those of the hybrid fish(CaCy) obtained from the combination of nucleus of the crucian and cytoplasm of the carp.

Therefore, it seems very interesting to repeat those experiments and make further investigations on both CyCa and CaCy hybrid fish.

RESULTS AND DISCUSSIONS

In 1984,we repeated the nuclear transplantation experiments for obtaining the hybrid fish from the combination of nucleus of the carp and cytoplasm of the crucian,i.e.the CyCa hybrid fish. The experiments were carried out at the Guangxi Fisheries Research Institute in Nanning,Guangxi Zhuang Autonomous Region during the spawing season of both carp and crucian(March to April). The local red carp and crucian were used as materials. The technique for nuclear transplantation has been described before(Tung,T.C. et al, 1963 and 1980).

Thirteen adult CyCa hybrid fish are obtained and the survival percentage of these adult hybrid fish is 2.0%(see Table 1).

TABLE 1
The Survival Percentages Of Embryonic Stages And Adult Fish From The Nuclear Transplantation Experiments Between Carp Nucleus And Crucian Cytoplasm(CyCa Hybrid Fish).

The No.of Nuclear Trans-planted Eggs and their De-veloping stages	No. of Transplanted Eggs	Arrested at Blastula Stage	Arrested at Gastrula Stage	Arrested at Larval Fish Stage	No. of Adult Hybrid Fish
	705	409	155	28	14
Survival Percentages	100%	58%	22%	3.9%	2.0%

According to the taxonomic criteria,some main morphological characteristics of carp,crucian and the thirteen adult CyCa hybrid fish are shown in Table 2.

TABLE 2
Comparision Of Some Morphological Characteristics
Of Carp,Crucian And CyCa Hybrid Fish

Type of Fish	Morphological Characteristics					
	Pair of Barbs	Pharyngeal Teeth	No. of Scale on Lateral Line	No. of Vertebrae	Body Colour	Shape of Air Bladder
Carp (Cy)	2	Molar Shaped,5 teeth in 3 rows, 3-1-1	34-36 Majority 35	4+30 to 4+31	Red	Divides into Anterior(A) & Posterior(P) parts. A > P The long axis of $\frac{A}{P} = \frac{1}{0.8}$
Crucian (Ca)	None	Wedge Shaped,4 teeth in one row	27-31 Majority 28	4+24 to 4+29	Black	Divides into Anterior(A) & Posterior(P) parts, A < P The long axis of $\frac{A}{P} = \frac{1}{1.25}$
CyCa Hybrid Fish	2	Molar Shaped,5 teeth in 3 rows, 3-1-1	29-34 Majority 33	4+22 to 4+31	Red (Some fish with black pigment spots	Divides into Anterior(A) & Posterior(P) parts A > P The long axis of $\frac{A}{P} = \frac{1}{0.68}$

It can be seen in Table 2:(1) Some features of the CyCa hybrid fish are inherited from the nucleus doner fish (carp) such as barbs,pharyngeal teeth and body colour(some hybrid fish with pigment sports which was identified as carp origin because in the nucleus doner red carp,similar pigment sports consisted by similar pigment cells are observed),(2)some seem to come from the influence of cytoplasm host fish(crucian) such as the number of vertebrae,(3)some are intermediates such as the number of scales along the literal line and(4)some are different either from the nucleus doner and cytoplasm host fish such as the shapes of air bladder.

The results of analysis of some characteristics of carp, crucian and CyCa hybrid fish at biochemical level were also obtained as follows:

1.The starch gel electrophoretic patterns of CyCa hybrid fish is similar to that of carp(nucleus doner fish).

2.The starch gel electrophoretic pattern of LDH isozyme of one CyCa hybrid fish is different from that of carp or crucian either in banding and their migration rates.

3.The Starch gel electrophoretic patterns of MDH isozyme of three CyCa hybrid fish are different from that of carp or crucian either in banding and their migration rates.

4.The polyacrylamide gel electrophoretic patterns of serum proteins of CyCa hybrid fish is basically similar to that of carp but with some obvious modifications in their banding and migration rates.

The growth rates of red carp and CyCa hybrid fish at early stages of breeding are measured after bred them for 107 days in the similar conditions and the results are shown in Table 3A and Table 3B.

TABLE 3A

Growth rate of Red Carp bred for 107 days

No. of Fish	1	2	3	4	5	6	7	8	9	10	Average
Total length (cm)	27.0	27.8	26.5	25.4	28.1	27.5	25.7	26.0	27.8	24.6	26.6
Body length (cm)	22.0	22.7	21.9	19.8	22.5	22.3	20.6	20.7	22.2	19.6	21.4
Body weight (g.)	250	250	200	200	250	250	200	250	250	200	230

TABLE 3B

Growth rate of CyCa Hybrid Fish bred for 107 days

No. of Fish	1	2	3	4	5	6	7	8	9	10	Average
Total length (cm)	30.8	31.1	29.3	26.4	30.0	24.8	22.6	25.7	23.3	23.9	26.8
Body length (cm)	25.8	25.8	24.0	21.0	25.2	19.5	17.6	19.9	18.0	19.0	21.6
Body weight (g.)	250	300	300	250	250	250	200	250	250	150	245

Obviously,the average body weight of CyCa hybrid fish is 6.5% higher than that of red carp.

Some of the CyCa hybrid fish have been grown to maturity in the spring of 1985(male).

The results obtained from these experiments are basically consitent with those we reported in 1980 and it seems that the technique of nuclear transplantation may become one of the methods of cultivating new breeds of fish.

REFERENCES

Tung,T.C. et al(1980). Scientia Sinica,23(4):518-523.
Yan Shaoyi et al(1984).Scientia Sinica,27(10):1029-1033.
Tung,T.C. et al(1963). Scientia Sinica,14(11):1244-1245.

Progress in Developmental Biology, Part A, pages 39–42

THE STUDY OF GENE DOSAGE AND POSITION EFFECTS WITH EXOGENOUS GLOBIN GENES IN DNA TRANSFORMED MICE

Cecilia W. Lo

Biology Department, University of Pennsylvania
Philadephia, Pa. 19104

INTRODUCTION

DNA microinjection have made it possible to insert exogenous DNA into the genome of eukaryotic cells. Using this approach, DNA transformation of mouse embryos has been obtained at a relatively high frequency of 3-30% (for reveiw see Gordon and Ruddle, 1985). The exogenous DNA in such transgenic mice have been observed to be stablely transmited through the germ line.

To carry out such microinjections, two methods have been used, including pressure injection and iontophoretic injection. With the former method, the DNA is introduced into cells by the net displacement of fluid from the injection pipette. With the latter injection method, there is no net fluid displacement but rather the DNA is introduced by microelectrophoresis. Using the latter method, we have obtained tissue culture cell transformants as well as transgenic mice (Lo, 1983). Our previous results indicated that iontophoretic DNA injection may provide a different pattern of DNA integration from that obtained by pressure injection. Thus unlike the prevalence of tandem arrays observed with pressure injection (for reveiw see Gordon and Ruddle, 1985), we found that transformants generated by iontophoresis exhibited multiple independent insertions of exogenous DNA with no detectable tandem arrays.

In this study, we further characterized the integration of mouse β-globin genes introduced into mouse embryos by iontophoretic microinjection. Given that the germ line of these transgenic mice demonstrated varying degrees of mosaicism, we also used Southern restriction analysis and in situ hybridization to characterize the extent of tissue mosaicism throughout the body of these transgenic mice. In addition, we examined whether the presence of additional copies of the mouse β-globin genes can bring about trans effects on the expression of the endogenous β-globin genes. Previously, a mutant mouse strain containing a duplication of the β-globin locus was observed to undergo suppression of β-globin gene expression (Popp, Stratton, Hawley and Effron, 1979).

RESULTS

Using iontophoretic injection, we obtained three transgenic animal which we refer to as 59, 43 and 83. These three parent transformants gave rise to four separate strains of mice containing 1, 5, 150, and over 1,000 copies of the mouse β-major globin gene. We refer to these four transgenic mouse strains as 59a, 59b, 43 and 83, respectively. As the recipient mouse embryos contain the β-single allele, we were able to distinguish the endogenous from the exogenous globin genes in Southern blot hybridizations (due to differences in their restriction pattern). Using Southern analysis, we established that multiple independent insertions must have mediated the integration of the injected globin plasmids in animal 59b, 43 and 83. In the parental transformant 59, two chromosomes were involved in the original integration events. This was revealed by breeding experiments which demonstrated that the exogenous globin restriction fragments of animal 59 assorted in two independent linkage groups, 59a and 59b. In situ hybridization of metaphase chromosome spreads revealed that in 59a, the exogenous DNA sequences are integrated close to the centromere in chromosome 5 and in 59b, in the distal portion of chromosome 3. In 59b, the exogenous globin inserts were integrated via a minimum of six integration events. Note that among these multiple globin inserts, there was no evidence of any tandem arrays. However, similar analysis of transformant 43 and 83 revealed that the integration of exogenous DNA was accompanied by the insertion of large tandem arrays. In animal 43, the tandem array consisted of approximately 150 copies of the injected globin plasmid and in situ hybridization analysis revealed that the integration occurred near the centromere of chromosome 17. In animal 83, the tandem arrays consisted of over 1000 copies of the globin inserts and they were found to be localized in the telomere of chromosome 3. Interestingly, it appears that in the high copy tandem inserts, the integration event mediating the insertion of the globin sequences occurred very late. This is indicated by the high level of mosaicism detected in the tissues of animal 43 and 83 and also the observation that the placental tissues of animal 83 did not contain any of the exogenous globin inserts. Thus perhaps the formation of tandem arrays in these animals is facilitated by the prolonged presence of the injected DNA in the unintegrated state. Note that our previous finding of no tandem arrays in transgenic mice (Lo, 1983) were obtained from the analysis of transformed embryos of 14-16 days of gestation and our results indicated little or no mosaicism in these animals as compared with animal 43 and 83.

To further examine the extent of tissue mosaicism in these animals, we first used Southern analysis to quantitate the percentage of cells in any tissue that is derived from the original transformed cell in the embryo. Our analysis revealed a large amount of mosaicism in all tissues. This result suggests that the transformed cell in each case was capable of participating in the development of all tissues and organs. To characterize the exact distribution of these transformed cell derivatives in the various tissues and organs, we used in situ hybridization of paraffin tissue sections to precisely localize the progeny of the transformed cell. This analysis was carried out for the transformant 83. The results of this analysis agree with the Southern blotting data, demonstrating a large amount of mosaicism. Moreover, we observed that often the transformed cell derivatives appear to be sprinkled throughout the tissue in an apparently random fashion. This suggests that mouse embryonic development involves a large amount of cell mixing, unlike the coherent pattern of development observed in insects (Garcia-Bellido, Rippoll, and Morata, 1973).

We have examined various hematological parameters in these animals to determine if the presence of the extra copies of mouse β-globin genes can elicit globin suppression as observed previously in the mouse mutant containing the duplicated β-globin locus (Popp, Stratton, Hawkins, and Effron, 1979). Thus we determined the red blood cell count, the hematocrit, the hemoglobin content, and the mean red cell volume of offspring of animals 59a, 59b, 43 and 83. Our analysis revealed little or no difference between control samples and blood samples from these transgenic mice.

REFERENCES

Garcia-Bellido, A, Rippoll, P, and Morata, G (1973). Developmental compartmentalization of wing disk of Drosophila. Nature 245:251-253.

Gordon, J, Ruddle, F (1985). DNA-mediated genetic transformation of mouse embryos and bone marrow-a review. Gene 33:121-136.

Lo, C (1983). Transformation by iontophoretic microinjection of DNA: multiple integrations without tandem insertions. Mol. Cell. Biol. 3:1803-1814.

Popp, R, Stratton, L., Hawley, D and Effron, K (1979). Hemoglobin of mice with radiation-induced mutations at the hemoglobin loci. J. Mol. Biol. 127:141-148.

Progress in Developmental Biology, Part A, pages 43–46

COMMON DETERMINANTS OF EMBRYONIC INDUCTION AND TRANSDIFFERENTIATION

George V.Lopashov, Olga N.Golubeva and Ketevan G.Zviadadze

Institute of General Genetics Ac.Sci.USSR, Moscow 117809 (G.V.L., O.N.G.) and Institute of Zoology Ac.Sci.Georg.SSR, Tbilissi 380079 (K.G.Z.)

Transdifferentiation proceeding spontaneously in eye cells as a consequence of their isolation (Moscona and Degenstein 1981, Okada et. al.1983) results in the appearance of a set of cell types, i.e. pigmented epithelium (PE), retina (R) (containing neurons and glial cells), lenses or lentoids (L) and melanophores (M). The problem we want to discuss is how the competence to convert into these different cell types appears in cells undergoing the transformation. At the late gastrula-early neurula stage, the whole anterior head endomesoderm is capable to induce the formation of a set of similar cell types including L, although it is located outside of the neural plate normally appearing over this area (Lopashov 1983b). This leads to the assumption that the competence to transformations appears in the course of induction and can depend on determinants or factors similar to these which are responsible for the induction of eye cells.

This assumption was subjected to direct verification. Early gastrula ectoderm (EGE) is capable of yielding all (almost 200) cell types of the vertebrate organism. If eye cells of the adult animals induce similar cell types in the EGE, this can be taken as the most serious argument that they still contain determinants which appeared during the induction and that their presence is critical for their capability to transdifferentiate.

Sheets of isolated PE cells or whole PE layers, pieces of R and parts of the forebrain of the adult <u>Xenopus laevis</u> were placed onto a Millipore TH filter (thickness 25 μ, pore size 0.8 μ). Pieces of R and of the brain were oriented to the filter always by their inner side; the sheet of hyaloid vessels was removed from R. A piece of EGE was placed onto the other side of the filter, the EGE inner side towards the filter. To avoid EGE wrapping on itself, a piece of a thicker filter (HA) was placed onto it and lightly pressed by placing some weight. The main filter was placed onto the inducer 3-4 h prior to placing the EGE; the inducing exposure was for 4-6 h; thereafter EGE was released and incubated in the medium for 5-6 days. The experiment was performed in standing droplets, about 0.1 ml in volume, in small sealed chambers in the standard Niu-Twitty solution. The explants were fixed using a Bouin solution and processed histologically by routine techniques or examined by indirect immunofluorescence.

To evaluate the results, we did not classify them as usually according to types of differentiation, but instead determined the combination frequency of various cell types which appeared in each explant. This approach demonstrates not only the frequency of occurrence of various cell types, but in addition provides data about their combinations with each other (Table 1). It can be seen from this Table that more complex differentiations such as R, PE, L and brain fragments (N) are found less often than M and ectomesenchyme (EM), and as a rule alongside with the latter ones.

TABLE 1. Cell types complexes induced by eye cells of adult Xenopus laevis in their EGE

Sets of cell types in explants	Inducer	
	PE	R
AE + E + EM	35	11
E + EM + M	10	8
E + EM + M + N	13	6
E + EM + M + R	5	5
E + EM + M + R + PE	4	4
E + EM + M + N + R + PE + L	5	4
Total number of explants	74	35

A typical (AE) or typical (E) epidermis was found in all explants. Nasal placodes generally found upon the induction of forebrain and eyes were absent. In control experiments with the forebrain, the eye tissues were not found, only N, M and EM were present. Fig.1 shows a typical picture of induction of EM and M, whereas Fig.2 shows EM, M, R, PE and L. L exhibited a positive reaction with antiserum against chicken crystallins. R and PE of white mutants (a^p/a^p) induced M similar to that induced by R and PE of +/+ animals.

Figure 1. EM and M induced in EGE

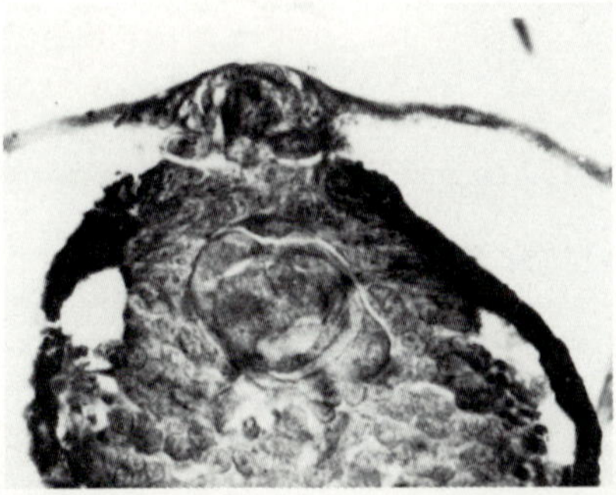

Figure 2. Eye-like structure containing R, PE and L induced in EGE

Our experiments have thus confirmed the initial point: sets of cell types appearing during the transdifferentiation of PE and R as well as the set of cell types induced by the head endomesoderm coincide with the spectrum of cell types induced in the EGE by eye tissues of adult animals. This leads to the conclusion that the competence of eye cells to transdifferentiate indeed eppears during induction. This competence then survives and in the case of induced transdifferentiation can manifest for the whole lifetime, as follows from the experiments with PE of adult frogs (Lopashov 1983 a,b) and rats (Lopashov, Golubeva and Zviadadze, unpubl.).

It can thus be concluded that those determinants of PE and R cells, which appear during the induction of gastrula-neurula stages then continue to be synthesized in eye cells throughout lifetime. The number of such determinants cannot be estimated at the moment, since PE and R induce a similar spectrum of cell types. This spectrum is similar to the set of cell types induced by neuralizing inducing factor (Tiedemann 1984). Therefore the mechanism underlying the stability of cell types in cell generations may involve the continuing production of inducing determinants in the cells throughout the whole life of the animal. Perhaps protein inducers could support the activity of regulator genes coding the synthesis of the inducers themselves and in addition the activity of genes responsible for terminal products of cell differentiation. Such a stability, however, is not absolute and depends also on normal cell interactions in the body. Duality of the operating mechanisms can explain the balance of the stability of cell types and their reversibility observed under conditions of transdifferentiation.

REFERENCES

Lopashov G V (1983a). Transdifferentiation of pigmented epithelium induced by the influence of lens epithelium in frogs. Differentiation 24: 27-32.

Lopashov G V (1983 b). Induced and spontaneous transdifferentiation and their relation to induction in development. In.: Developmental Biology: An Afro-Asian Perspective. Poona, Ind Soc Devel Biol: 87-96.

Moscona A A and Degenstein L (1981). Lentoids in aggregates of embryonic retina cells. Cell Differ 10: 39-46.

Okada T S, Nomura K and Yasuda K (1983). Commitment to transdifferentiation into lens occurs in neural retina cells after brief spreading culture of dissociated cells. Cell Differ 12: 85-92.

Tiedemann H (1984). Neural embryonic induction. In.: The Role of Cell Interactions in Early Neurogenesis. New York and London, Plenum Publ Co: 89-105.

Progress in Developmental Biology, Part A, pages 47–50

INTRODUCTION OF GENES INTO MOUSE EMBRYONIC STEM CELLS

Thomas Doetschman, Achim Gossler, Edgar Serfling, Walter Schaffner, Ken Marcu, Lawrence Stanton and Rolf Kemler

F. Miescher Lab., Max Planck Soc., Tübingen, GFR. (TD,AG,RK); Mol. Biol. II, Univ. Zürich (ES,WS); Dept. of Biochem., Stony Brook, NY (KM,LS).

INTRODUCTION

In order to investigate the proper expression of genes introduced into embryonic mammalian cells, an experimental developmental system must be available (preferably with in vitro capabilities) in which the corresponding endogene is properly expressed, and vectors to which the cells are permissive must be found to deliver the foreign genes.

We have characterized the developmental potential of mouse blastocyst-derived embryonic stem (ES) cells (Doetschman et al, 1985) and have shown that after 8–10 days of culture under differentiation conditions (removal from a mouse embryonic fibroblast feeder layer) they can develop into cystic embryoid bodies (CEB) which have similarities to visceral yolk sac in that they contain alpha-fetoprotein, transferrin and several lipoproteins. Often myocard and occasionally blood islands are seen. When the newly formed CEB are incubated in the presence of Iscove's medium supplemented with 20% human cord serum, about 30% of them produce blood islands. As in the 11-day embryo the nucleated erythrocytes only express mouse embryonic globins. If kept in culture for longer periods of time, smooth and skeletal muscle occasionally appear. It is, therefore, clear that ES cells can develop in vitro into highly organized embryonic-like structures and thereby provide a developmental system which can be used to investigate the expression of embryonic genes in embryonic mammalian cells.

This paper assesses the amenability of ES cells to

transfection by infection with retroviral-based vectors and by calcium phosphate precipitation. The efficiency of transfection and the in vitro developmental potentiality of the transfected cells will be given.

RESULTS AND DISCUSSION

An advantage of retroviral-based vectors is that they usually permit the integration of only one gene copy per cell. A disadvantage is that embryonic stem cells are not permissive to infection by retroviruses. In order to determine the feasibility of infecting ES cells with such vectors, the efficiency of integration of the neo resistance gene by two vectors was measured and compared to that of embryonal carcinoma F9 (EC-F9) cells and to differentiated cells, embryonic fibroblasts. One of the vectors (pLSras) was based on the pZIP-Neo vector produced by Cepko et al. (1983), and in addition to the neo resistance gene, also contains v-ras^{Ha} cDNA inserted into the Bam H1 site. The second vector is pMMuLV-SVtkneo (Rubenstein et al, 1984, here called pTKneo) which contains only the neo resistance gene. Both vectors have MuLV LTRs at each end and infect the cells as virus particles. The two genes in pLSras are under LTR promotion alone whereas the neo resistance gene in pTKneo is also under SV40 TK promotion.

Various cell lines were infected with the retroviral vectors basically as described (Cepko et al, 1984), and were screened for resistance to the antibiotic G-418 which is inactivated by the neo resistance gene product. Table 1 shows that the two vectors were equally capable of delivering G-418 resistance to embryonic fibroblasts (differentiated cells) at very high efficiency. Consistant with the non-permissivity of early embryonic stem cells to retroviruses, the efficiency of infection of EC-F9 and ES-D3 cells was lower than that for the differentiated cells by 5-6 orders of magnitude. The pTKneo vector was more efficient in infecting both types of stem cells than the vector containing only LTRs. This is consistent with the results of Rubenstein et al. (1984) and suggests that the strength of promotion can overcome to some extent the block to early embryonic infection. That the EC-F9 cells were infected at a higher rate than the ES-D3 cells suggests that the latter are more "embryonic" in nature. These results indicate that the use of retroviral vectors for infecting ES cells will be limited unless the viral sequences can be changed in such a way as to make the viruses more infectious on ES cells.

TABLE 1. Retroviral-based Vectors: Efficiency of Infection.

Cell lines	G-418-Resistant Colonies	
	pLSras	pTKneo
Emb. Fibs.	$1\text{-}5\times10^{-1}$ (1)	$1\text{-}5\times10^{-1}$
EC-F9	4×10^{-6} (2)(3)	6.5×10^{-4}
ES-D3	0	4×10^{-6}

1) Ras expression confirmed by indirect immunofluorescence.
2) Neo and ras gene integration confirmed by Southern blots.
3) One of 4 resistant lines had deleted the ras gene.

In another set of experiments ES-D3 cells were co-transfected (16 hr CaPhos precipitation followed by a 2 min 25% DMSO shock) with pSVtkneoß (Nicolas and Berg, 1983) to provide G-418 resistance, plus one of 3 plasmid vectors containing the human ε, γ^A or ß-globin genes or a cosmid vector containing the human ε-globin gene plus about 30kB of upstream and 5 kB of downstream sequences.

Two kinds of feeder cells were used during the transfection and selection procedures: primary and secondary embryonic fibroblasts, both of which had been transfected with G-418 resistance. As can be seen in Table 2, no G-418 resistant colonies were found in 8 cultures with secondary fibroblastic feeder cells, whereas 5 cultures with primary feeder cells contained a total of 68 colonies. These results indicate that ES cells are much easier to maintain under severe selection conditions when grown on primary embryonic fibroblasts. The efficiency of transfection was about 3x better than with the best retroviral vector (compare to Table 1). Of the 31 cell lines established at least 30 contained the neo resistance gene and 13 contained human globin genes. Most had from 1-3 copies per cell. Nine of the 11 transfected ES cell lines have been karyotyped and were found to have a normal complement of 40 chromosomes and 7 of the 11 lines still had the capacity to produce CEBs in culture, indicating that the selection procedure did not significantly reduce the *in vitro* developmental potential of the ES cells. Tests are underway to determine whether the globin genes are properly expressed *in vitro* and in chimeric mice which have been made from some of the cells.

TABLE 2. Transfection of ES cells with human globin genes.

	Feeder Cells	
	Pri. Emb. Fibs.	Sec. Emb. Fibs.
# Cult. transfected	5	8
# Cult. w/resist. colonies	5	0
# Resist. colonies	68	
Efficiency of transfection	1.3×10^{-5}	
# Cell lines established	31	
# G-418 resistant lines	30-31	
# w/integ. globin genes	13 (5-ε; 2-cos-ε; 1-$^{A}\gamma$; 5-ß)	
# producing CEB	7 of 11 tested	
# w/40 chromosomes	9 of 9 tested	

In summary, we have demonstrated that foreign genes can be integrated into the genome of ES cells both by infection with retroviral-based vectors and by CaPhos-mediated transfection and that after these procedures most of the cells continue to display a high degree of *in vitro* developmental potential. Therefore, ES cells provide a good experimental system for investigating the regulation of embryonic genes in embryonic mammalian cells.

REFERENCES

Cepko CL, Roberts BE, Mulligan R; (1984). Construction and applications of a highly transmissible murine retrovirus shuttle vector. Cell 37:1053.

Doetschman TC, Eistetter H, Katz M, Schmidt W, Kemler R (1985). The *in vitro* development of blastocyst-derived embryonic stem cell lines: formation of visceral yolk sac, blood islands and myocardium. J Emb exp Morph. 87:

Nicolas J-F, Berg P (1983). Regulation of expression of genes transduced into embryonal carcinoma cells. In Silver,Martin, Strickland (eds): "Teratocarcinoma Stem Cells," Cold Spring Harbor Laboratory, p 469.

Rubenstein JLR, Nicolas J-F, Jacob F (1984). Construction of aretrovirus capable of transducing and expressing genes in multipotential embryonic cells. Proc Natl Acad Sci USA. 81:7137.

Pattern Formation in Lower Organisms

Progress in Developmental Biology, Part A, pages 53–56

REGULATION IN METAGENIC CNIDARIAN PLANULAE

Georgia E. Lesh-Laurie and Paul E. Suchy

Department of Biology, Cleveland State University, Cleveland, Ohio 44115

Organized from two germ layers, many cnidarians exhibit a complex, metagenic life cycle. Investigators have long sought to understand the factors affecting metagenic transformations, as well as the developmental potential expressed by tissue so stimulated. A significant metagenic transformation occurs when free-living planula larvae become sessile, polypoid individuals. Both hydrozoan and scyphozoan planulae have a limited number of morphologically well characterized cell types. Therefore, if one could predictably alter the developmental potential expressed during the planula → polyp transformation, it should be possible to compare, quantitatively, the number and relative proportion of each cell type under both normal and perturbed conditions.

This report defines developmental parameters that will permit comparisons of cell population dynamics associated with regulation during the planula → polyp transformation in *Eudendrium carneum*, a hydrozoan, and *Cassiopea xamachana*, a scyphozoan.

METHODS AND MATERIALS

Mature male and female fronds of *Eudendrium* were collected at Flatts Inlet, Bermuda and maintained in aerated, filtered sea water (glass fiber; 0.22 μ Millepore). Planulae were released discontinuously, with 55-70% of the planulae liberated in the 12-hour period before harvesting. *Cassiopea* medusae were collected from a marine pond on Coney Island, Bermuda. Medusae were maintained and planulae were collected and cultured as described by Neumann (1979) and Martin and Chia (1982).

Data from the experimental manipulations were compared in an RxC test of independence using the G statistic. To determine probability, the G value was compared to the X^2 distribution with (R-1)(C-1) degrees of freedom.

Planulae were sectioned transversely into one-half and one-fifth fractions. Cuts were made so that each fraction contained similar amounts of tissue as measured by DNA content. To clarify any relationship between DNA synthesis and the developmental potential of planulae, larvae were incubated in hydroxyurea (HU) for 24 hours. During the period of exposure DNA synthesis was measured using procedures described and verified by Lesh-Laurie (1976). The efficacy of HU as a temporary antagonist of DNA synthesis in cnidarians has been well documented. *Eudendrium* planulae were exposed to 2.5 mg/ml HU, whereas *Cassiopea* planulae were administered 1 mg/ml HU.

RESULTS AND DISCUSSION

Following a brief free-living period planula larvae settle, attach and transform into a polypoid form. In this transformation the anterior regions of planulae form the stolon or basal portions of the polyps and the posterior planular regions develop the head/oral regions of the polyps. Generally, planulae exhibit considerable regulative capacity (e.g., Freeman, 1981).

Eudendrium (Table 1) and *Cassiopea* (Table 2) larvae likewise display predictable regulation when surgically reduced in size. One-half planulae of both species may develop into complete polyps, with all structures present and appropriately proportioned. Small numbers of anterior-half *Eudendrium* planulae form only stolons, whereas small numbers of posterior half-planulae give rise to unattached hydranths. Reducing the *Eudendrium* cell number further reveals that one-fifth larvae remain capable of forming normally-proportioned polyps, so long as the isolated fraction is taken from the anterior or middle fifths of the animal. Posterior one-fifth larvae, although surviving and swimming 3 weeks following amputation, never transform into any portion of the polyp.

Inhibition of DNA synthesis for a 24 hour period dramatically alters polypoid transformation. Following removal from HU *Eudendrium* planulae display exaggerated stolon growth. The normally compact initial stolon mat is replaced with extreme horizontal outgrowths that extend over 3 mm in the 5 day transformation period. Exaggerated stolon growth occurs in a large proportion (55.7%) of the animals exposed to HU (Table 1). *Cassiopea* polyp formation is also modified following HU treatment. However, the morphogenesis expressed is the elaboration of unattached calyxes in 25.6% of the transforming individuals (Table 2).

TABLE 1. Developmental Potential of *Eudendrium* Planulae

Experimental Condition	n	Developmental Potential St/Up/Hyd	St/Up	St	Hyd
Normal, untreated	120	103	17	-	-
Anterior one-half					
0 day	84	65	16	3	-
1 day	64	46	14	4	-
Posterior one-half					
0 day	53	44	7	-	2
1 day	42	36	5	-	1
Anterior one-fifth	22	3	11	8	-
Middle one-fifth	6	2	3	1	-
Posterior one-fifth	(51)	-	-	-	-
				Exaggerated Stolon Growth >2mm	>1mm
Hydroxyurea-treated	165	156	9	59	33

Table 2. Developmental Potential of *Cassiopea* Planulae

Experimental Condition	n	Developmental Potential 8T Polyp	Scyphula	Calyx
Normal, untreated	133	107	26	-
One-half planulae	33	23	-	10
Hydroxyurea-treated	172	128	-	44

Data analysis confirms the veracity of the experimental treatments to produce significant modifications in the developmental outcome of planulae. For Eudendrium G = 224.84. Since the X^2_{10} = 29.59, the G value is significant at $P \ll 0.001$. For Cassiopea G = 103.41. Again as the X^2_4 = 18.47, the G value is significant at $P \ll 0.001$.

Cell maceration was performed by a modification of the techniques developed for marine cnidarians (e.g., Schmid et al., 1981). Stages to be macerated were placed in a solution of glacial acetic acid: glycerol: distilled water (1:1:8) for 30 minutes at 33°C; agitated; allowed to incubate 30 additional minutes at 33°C; agitated again and fixed with 20% formaldehyde (added to a volume 10% of the macerating fluid).

These data, therefore, describe developmental parameters of both hydrozoan and scyphozoan planula → polyp transformations that are amenable to quantitative cell population analyses. Perturbation resulting in regulation to normal developmental outcomes may be compared to perturbations yielding predictable modifications in morphogenetic expression.

This paper represents Contribution No. 1058 from the Bermuda Biological Station for Research. The authors acknowledge with appreciation a Wright Fellowship during the field portions of this investigation.

REFERENCES

Freeman G (1981). The role of polarity in the development of the hydrozoan planula larva. Roux's Arch 190:168-184.

Lesh-Laurie GE (1976). Stolon vs hydranth determination in Pennaria tiarella planulae: A role for DNA synthesis. In Mackie, GO (ed): "Coelenterate Ecology and Behavior," New York: Plenum, pp 365-375.

Martin VJ, Chia F-S (1982). Fine structure of a scyphozoan planula, Cassiopea xamachana. Biol Bull 163:320-328.

Neumann, R (1979). Bacterial induction of settlement and metamorphosis in the planula larvae of Cassiopea andromeda. Mar Ecol Prog Ser 1:21-28.

Schmid V, Stidwill R, Bally A, Marcum B, Tardent P (1981). Heat dissociation and maceration of marine cnidaria. Roux's Arch 190:143-149.

Progress in Developmental Biology, Part A, pages 57–60

MORPHOGENETIC CAPABILITY OF EPITHELIAL AND MESENCHYMAL CELLS DISSOCIATED FROM SWIMMING STARFISH EMBRYOS

Marina Dan-Sohkawa, Yoshinori Komiya, Hiroyuki Kaneko and Hachiro Yamanaka.
Department of Biology, Faculty of Science, Osaka City University, Osaka 558 (M.D-S., Y.K., H.K.) and Kanebo Institute for Cancer Research, Osaka 534 (H.Y), Japan.

It was found recently that cells dissociated from swimming embryos (blastula to late gastrula) of the starfish, Asterina pectinifera, reaggregate and develop into the larval form, the bipinnaria, through specific morphogenetic steps (Fig. 1) (Dan-Sohkawa, Yamanaka and Watanabe, 1985; Yamanaka, Tanaka-Ohmura and Dan-Sohkawa, 1985). This experimental system provides us with a model system which allows us to search more deeply into the morphogenetic capabilities of cells. Moreover, it has an advantage over the sea urchin (Giudice, 1962; Giudice and Mutolo, 1970) by being simpler in form and, therefore, more capable of successful reconstruction.

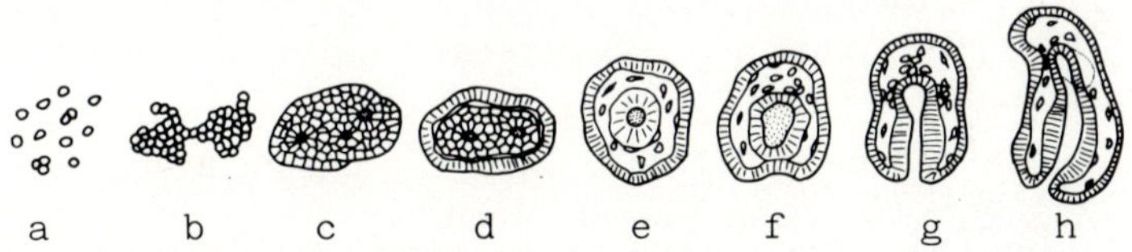

Fig. 1 Process of reconstruction
a: dissociated cells b: aggregation c: compaction
d: formation of external epithelium e: "blastula"
f: "gastrulation" g: mesenchymal gastrula
h: formation of mouth (from Dan-Sohkawa et al., 1985)

In the present study, cells were dissociated from late gastrulae by treating 0.2 ml of packed embryos briefly (5 min) with the dissociation medium (1.2 M glycine supplemented with 1% sea water and 6% newborn bovine serum: NBS) in a total volume of 4 ml and to subsequent pipetting with a Pasteur pipette. When dissociated cells are suspended in reaggregation medium (2 ml sea water containing 4% NBS) and left standing in a test tube for 15 min, they segregate sharply into two fractions, namely the quickly aggregating (q.a.) fraction, which sediments to the bottom, and the slowly aggregating (s.a.) fraction, which stays suspended.

Although either of these fractions are capable of producing bipinnariae, they differ from one another in several respects. That is, 1) q.a. cells aggregate faster and detach from the substratum within a shorter period of time (2.5 to 3 hrs at 15 C) than s.a. cells (5 to 6 hrs) when plated onto a Petri dish (Fig. 2). 2) While q.a. cells require 4% NBS in

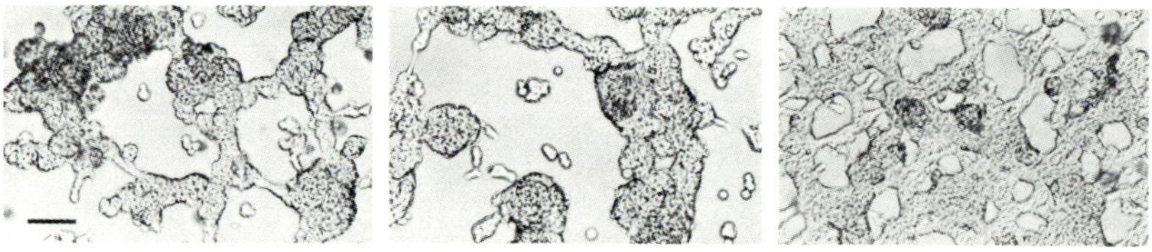

a b c

Fig. 2 Aggregation on glass surface (1.5 hr after plating) a: non-fractionated cell population. Bar indicates 50 um. b: q.a. fraction c: s.a. fraction.

the medium during the initial reaggregation process, i.e., until the external epithelium is completed (24 hrs), s.a. cells require an additional 24 hrs in sea water containing 2% NBS. This means that s.a. cells take twice as long to form the epithelium (Fig.3).

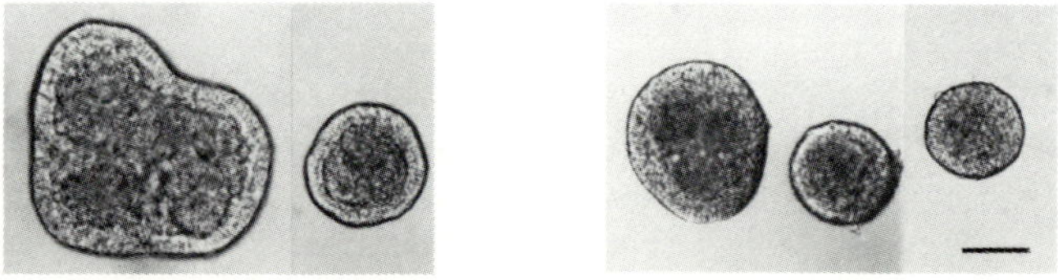

a b

Fig. 3 35-hr-reaggregate While the external epithelium is prominent in reaggregates of the q.a. fraction (a) it is not clear in those of s.a. fraction (b). Bar: 50 um.

The poorer capacity of s.a. cells to aggregate, their lesser affinity towards the substratum, the prolonged serum requirement and the less well-developed epithelium all point to the mesenchymal nature of s.a. cells, while the readiness of q.a. cells to aggregate and the ease by which they form the epithelium indicate the epithelial nature of these cells. We do not know, however, whether s.a. and q.a. fractions are composed purely of mesenchymal and epithelial cells, respectively. As a matter of fact, we think that the s.a. fraction is one that is relatively rich in mesenchymal cells, whereas the q.a. fraction is rich in epithelial cells. This speculation is partly based on the fact that 20 % of the dissociated cells will go into the s.a. fraction, a proportion which appears to be too large for what is expected of the mesenchymal cells at the late gastrula stage (probably not exceeding 5%).

Further differences between reaggregates of the two cell fractions is the timing of "gastrulation". Probably owing to the difference in timing required to complete the external epithelium, reaggregates of the s.a. fraction "gastrulate" about 24 hrs later than those of the q.a. fraction (Fig. 4).

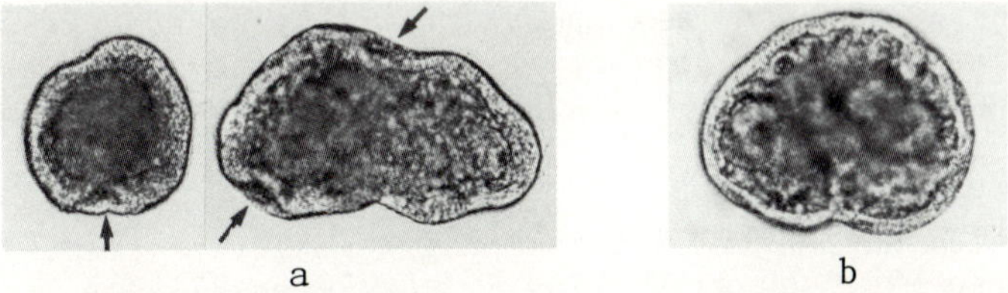

Fig. 4 68-hr-reaggregate "Gastrulation" has taken place only in reaggregates of the q.a. fraction (a) and not in those of the s.a. fraction (b). Arrows indicate the sites of invagination. Magnification same as Fig. 3.

As to the morphology of the resultant bipinnariae, those deriving from the non-fractionated cell population are almost normal (Fig. 5 b). In comparison, those reconstructed from the q.a. fraction are 1) more spherical and 2) relatively small in size owing to 3) a poorly developed blastocoel containing fewer mesenchymal cells. What characterizes them best, however, is the prominent, and often oversized intestine (Fig. 5 a, arrows). Bipinnariae reconstructed from the s.a. fraction, on the other hand, are characterized by 1) their opacity as a result of 2) blastocoels overpopulated by mesenchymal cells, and 3) the lack of an intestine (Fig. 5 c). Digestive tracts of bipinnariae from q.a. and s.a. fractions are proportionately larger than normal.

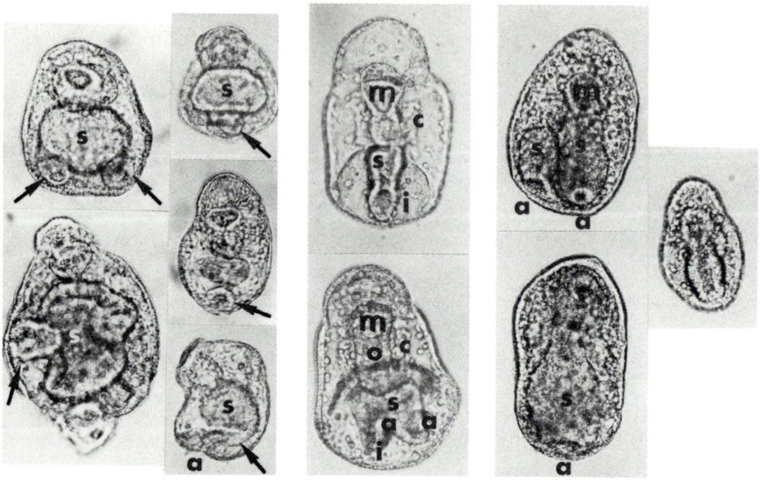

a b c

Fig. 5 Bipinnariae reconstructed from q.a. fraction (a), non-fractionated cell population (b) and s.a. fraction (c). a: anus c: coelomic pouch i: intestine m: mouth o: oesophagus s: stomach Magnification same as Fig. 3.

Among the three cell populations, the q.a. fraction is the most stable to go through the process, while s.a. fraction requires the most delicate control of conditions. As far as the degree of normal morphogenesis is concerned, however, the q.a. fraction produces more irregularly shaped bipinnariae (Fig. 5 a) than s.a. fraction (Fig. 5 c). The non-fractionated population often produces bipinnariae indistinguishable from the normal ones (Fig. 5 b).

These facts indicate that a cell population rich in epithelial cells is capable of forming a stable embryo, which, however, is irregular in shape. On the other hand, a cell population rich in mesenchymal cells achieves more normal morphogenesis, but the resultant bipinnariae often lack the intestine and are generally less stable.

Efforts are being made to obtain pure populations of both mesenchymal and epithelial cells in order to find out more about the morphogenetic capabilities of the two types of cells, about the optimal proportions of the two and the reason for the unusual behavior of cells which constitute the intestine

Progress in Developmental Biology, Part A, pages 61–65

REGENERATION FROM DISCONTINUOUS CIRCUMFERENCES IN AXOLOTL LIMBS

K. Muneoka, G. Holler-Dinsmore and S. Bryant

Developmental Biology Center, University of California, Irvine, California 92717

INTRODUCTION

The most characteristic feature of regulating systems is that pattern continuity is restored during the process of regulation, regardless of how abnormal the final morphology of the structure may appear. This feature of regulating system is at the heart of most models which attempt to account for pattern regulation. It is of some interest, therefore, that recent reports have uncovered situations in the regenerating urodele limb in which the principle of continuity appears not to be followed (Maden & Mustafa, 1982; Holder & Weekes, 1984). Here we report on an analysis of pattern and cellular contribution to regenerates from surgically constructed limbs containing dorsal/ventral (D/V) discontinuities between digits 2 and 3. The results of this study demonstrate that the original D/V discontinuities are of two types, defined by the polarity of the tissues, and that one type is resolved during regeneration, whereas the second type is not.

MATERIALS & METHODS

All experiments were carried out on *Ambystoma mexicanum*. Surgical procedures and maintenance were as described previously (Muneoka *et al.*, 1985). Grafting operations to create stumps containing D/V discontinuities were carried out between sibling pairs of axolotls (1 diploid, 1 triploid) of similar size. Amputation of the lower arms was performed immediately after grafting. Diagrams of the limb stump at amputation are shown in Figure 1. When regeneration was complete, limbs were analyzed to determine the anterior-posterior (A-P) pattern of the skeleton, the pattern of D and V muscles, and the cellular contribution from each half of the stump to the dermis of the regenerate.

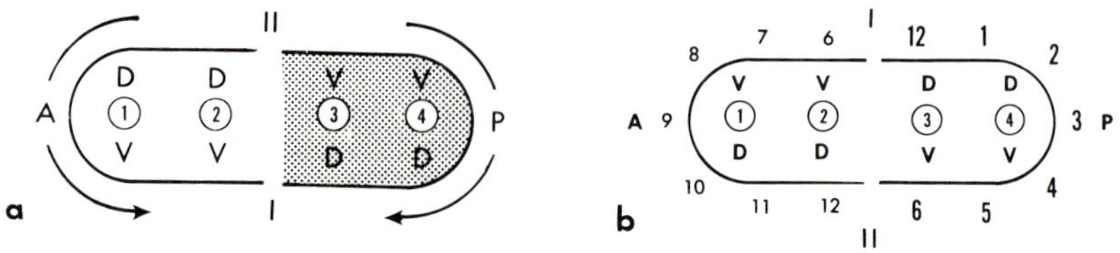

Figure 1. Diagrams of experimental limb stumps. Circled numbers indicate digital elements. Type I & II discontinuities are indicated. D, dorsal and V, ventral muscles; A, anterior; P, posterior. A: triploid grafted tissue is stippled. Arrows indicate directionality of intercalation based on previous experiments. B: circumferential positional values indicated by numbers outside the outline.

RESULTS AND DISCUSSION

The results of this study are presented in detail elsewhere and are summarized here. In the descriptions which follow, the term discontinuity refers to the abnormal positioning of D muscles next to V muscles. The original discontinuities created by grafting are termed central discontinuities. In each chimeric limb stump there is one AV/PD central discontinuity (Type I) and one AD/PV central discontinuity (Type II; Fig. 1). Discontinuities in the regenerate which appear between digits 1 and 2 or 3 and 4 are termed peripheral discontinuities.

Of 16 Type I discontinuities analyzed, only one was maintained as a central discontinuity in the regenerate. The remaining 95% of Type I discontinuities became resolved in various ways during regeneration leading to pattern continuity (Table 1).

Table 1. Patterns After Regeneration

	Type I	Type II
Supernumerary Digits	3	1
Peripheral Discontinuity	6	3
No Discontinuity	6	1
Central Discontinuity	1	11
Total	16	16

At 3 Type I discontinuities, supernumerary outgrowths developed which can be seen from Figure 2 to restore pattern continuity at that junction.

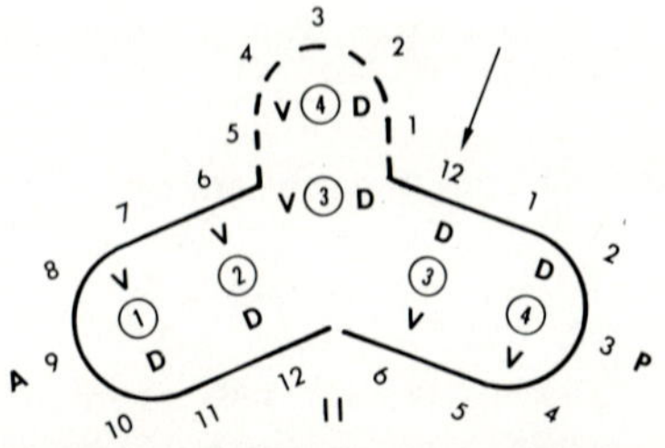

Figure 2. Diagram of limb with supernumerary digits at site of original Type I discontinuity. Intercalated positional values (dotted line) restore pattern continuity. Symbols as in Fig. 1.

Six Type I discontinuities were lost, as illustrated in Figure 3a. Pattern continuity in these cases could have been brought about by incomplete participation of the stump in blastema formation, followed by intercalation to form a continuous sequence of positional values. Alternatively, intercalation at the original Type I discontinuity followed by midline loss of positional values along the line of symmetry created by the intercalation (see Bryant et al., 1981) could also generate this result. The remaining 6 Type I discontinuities regenerated as peripheral discontinuities. As Figure 3b illustrates, the sequence of positional values at peripheral discontinuities can be interpreted as being continuous, despite the "apparent" discontinuity between dorsal and ventral muscle masses.

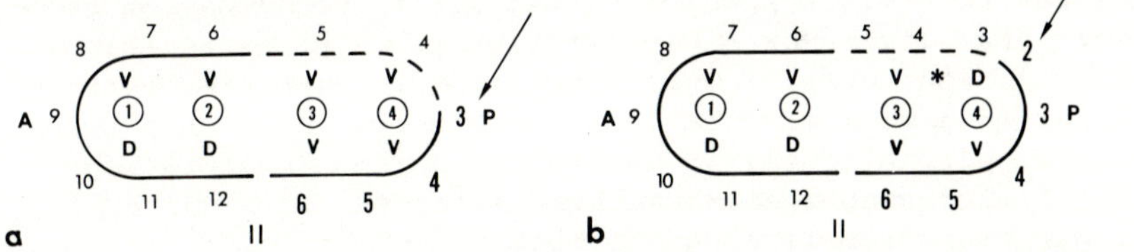

Figure 3. A: Diagram of limb which lost Type I discontinuity during regeneration. Intercalated positional values combined with loss on line of symmetry (arrow) or incomplete participation of stump plus intercalation could lead to resolution of the Type I discontinuity. B: Diagram of limb in which a Type I discontinuity became peripheral during regeneration. Pattern continuity could have been restored by mechanisms described in A. Symbols as in Figure 1.

As in the previous cases where the Type I discontinuity is lost altogether, the formation of a peripheral "discontinuity" could be brought about either by incomplete participation of the stump plus intercalation, or intercalation at the Type I discontinuity and midline loss during outgrowth. Additional support for the conclusion that interactions leading to pattern regulation occur at Type I discontinuities comes from an analysis of cellular distribution after regeneration (Table 2).

Table 2. Cellular Displacement During Regeneration

	Original Limb Half	Other Limb Half
Normal Limbs	76%	24%
Central Discontinuity	84%	16%
Peripheral Discontinuity	77%	23%
No Discontinuity	52%	48%

In normal limbs, without any discontinuity, 24% of the cells which originate in one limb half can be found in the other limb half after regeneration (Muneoka et al., 1985). When peripheral discontinuities or no discontinuities are formed, as is typical of Type I discontinuities, the same or a greater amount of cellular displacement away from the original limb half is observed. We interpret this displacement to be a consequence of cellular interactions and intercalation at the Type I discontinuity.

As can be seen from Table 1, 69% of all Type II discontinuities were maintained unchanged during regeneration suggesting that little or no interaction takes place at this type of discontinuity. This conclusion is supported by the cellular displacement data in Table 2 which shows that when a central discontinuity is maintained during regeneration, less cellular displacement occurs across the original graft/host junction than is found in normal limbs without discontinuities.

In conclusion, the original central discontinuities created in this experiment are clearly of two types, based on their different behaviors during regeneration. Although at both types of discontinuity D tissue is apposed to V tissue, the polarity of the tissue at the discontinuities is different. At Type I discontinuites AV is opposed to PD, and at Type II, AD is opposed to PV. In a previous study (Muneoka & Bryant, 1984) evidence was presented to suggest that intercalation in the limb circumference has a directional bias, with A tending to contribute to V tissue and

P tending to contribute to D tissue. When this directionality is indicated by arrows on the mixed-handed limb stumps of this study (Fig. 1), with arrow heads pointing in the direction in which contribution tends to occur, it can be seen at that Type I discontinuities intercalation is expected and our results support this idea. However, at Type II discontinuities, intercalation is not expected and our results show that it seldom occurs. Further analysis of this apparent tissue polarity is necessary in order to understand its cellular basis.

ACKNOWLEDGEMENTS

Research supported by PHS Grant HD 06082.

REFERENCES

Bryant SV, French V, Bryant PJ (1981). Distal regeneration and symmetry. Science 212:993.

Holder N, Weekes C (1984). Regeneration of surgically created mixed-handed axolotl forelimbs: pattern formation in the dorsal-ventral axis. J Embryol exp Morph 82:217.

Maden M, Mustafa K (1982). The structure of 180° supernumerary limbs and a hypothesis of their formation. Develop Biol 98:257.

Muneoka K, Bryant SV (1984). Cellular contribution to supernumerary limbs in the axolotl, *Ambystoma mexicanum*. Develop Biol 105:166.

Muneoka K, Holler-Dinsmore GV, Bryant S (1985). A quantitative analysis of regeneration from chimeric limb stumps in the axolotl. J Embryol exp Morph (in press).

Progress in Developmental Biology, Part A, pages 67–70

CHANGE IN SOMITE NUMBER FOLLOWING INTERSPECIFIC GRAFTS OF AVIAN SEGMENTAL PLATE MESODERM

David S. Packard, Jr.

Department of Anatomy and Cell Biology, State University of New York Upstate Medical Center, Syracuse, New York 13210

INTRODUCTION

Recent studies have shown that the segmental plate mesoderm of chicken and Japanese quail embryos is committed to forming a definite number of somites in specific locations (Packard, 1978; Packard, 1980a), and is morphologically segmented into a series of tandem cellular domains, termed somitomeres (Meier, 1979; 1980; Packard and Meier, 1983). In view of the importance of the segmental pattern to the future anatomy of the embryo, it is reasonable to suggest that some mechanism exists to maintain the segmental pattern from the time of its specification through the completion of somite formation. If such a mechanism exists, embryos should be able to regulate disturbances of the somitomere pattern in the segmental plate. An attempt was made to learn whether such regulation exists by asking whether chick embryos could change the number of somites that a grafted quail segmental plate would make. It was found that chick embryos could significantly alter the number of somites formed by grafted quail segmental plates.

MATERIALS AND METHODS

Fertile White Leghorn chicken and Japanese quail eggs were incubated at 38° C. for about 48 hours, removed from the yolks, and washed briefly in calcium- and magnesium-free Tyrode's solution. The embryos were placed in a second dish with 5 ml of calcium- and magnesium-free

Tyrode's solution containing 1% trypsin (1-250, hog pancreas) and 0.1% EDTA at room temperature. After 5 to 15 minutes, each embryo was transferred to a 35 mm plastic culture plate containing a nutritive agar substratum and about 1 ml of fresh, whole-egg supernatant (Packard and Jacobson, 1976). The excess supernatant was drawn off and microsurgery was performed with tungsten wire needles. The ectoderm over the right segmental plate was cut into a flap and folded laterally. The segmental plate was cut away from the neural tube and surrounding mesoderm and then peeled from the endoderm. The node region of the embryo, including all 3 germ layers, was removed to prevent further segmental plate formation (Packard and Jacobson, 1976). Grafting was accomplished by transferring the segmental plate to the host embryo's dish with fine forceps. The graft, in its original orientation, was then inserted into the host's removal site and the ectodermal flap was closed over it. If, once the graft was positioned in the host, it was found to be shorter than the host's segmental plate, the host was completely transected at the caudal extremity of the quail segmental plate. Thus the two segmental plates were always the same length. The embryos were cultured as previously described (Packard and Jacobson, 1976) for 16 to 20 hours, and were fixed, embedded in paraffin, sectioned, and stained according to the Feulgen-Rossenbeck technique. Probability was calculated with the paired T test.

RESULTS

A series of control experiments was performed by treating chick or quail embryos having 8 to 21 pairs of somites with trypsin. The right segmental plate of each embryo was removed and then replaced immediately with its original orientation preserved. The embryos were cultured for a period of time and then the number of somites formed by the left and right segmental plates of each embryo were counted. As expected, the segmental plates of each embryo almost always formed the same number of somites despite the microsurgical manipulation ($r=0.899$, $N=34$, $P<0.01$).

In order to determine whether host embryos can alter the number of somites formed by a grafted segmental plate, right segmental plates were removed from trypsinized donor quail embryos having 5 to 23 pairs of somites. These segmental plates were then grafted into trypsinized host chicken embryos containing 9 to 21 pairs of somites in place of the host's right segmental plate. Donor and host embryos were cultured, fixed, and sectioned. Since Packard (1980) and Packard and Meier (1983) have shown that segmental plates removed simultaneously from the same embryo will form the same number of somites, the number of somites that the grafted segmental plate was originally committed to form could be determined by counting the number of somites ultimately formed by the donor's left, unoperated, segmental plate. The number of somites actually formed by the graft was determined by counting the number of quail somites in the chicken host. This experiment differed from the previous control experiment only in that the excised segmental plate was grafted into another embryo rather than being returned to the donor. When the number of somites formed by the donor's left segmental plate was compared with the number formed by the graft, the correlation that was found ($r=0.383$, $N=27$) was significantly lower than that found in the control experiments. When the number of somites formed by the host's unoperated, left segmental plate was compared with the number of somites formed by the graft, a significant correlation was found ($r=0.592$, $N=27$, $P=<0.01$). It was concluded that the host embryos demonstrated a significant ability to alter the number of somites formed by the grafted segmental plates.

DISCUSSION

Menkes and Miclea (1962) and Menkes and Sandor (1969) disrupted the segmental plates and associated ectoderm and endoderm of chicken embryos with glass needles. Within 3 hours, the segmental plates had reformed and somite formation continued with only minor abnormalities. Sandor (1972) excised areas of future intersomitic furrows from the segmental plates of explanted chicken embryos and found that the somite pattern still formed normally. The present study has extended our knowledge of the ability of avian embryos to regulate disturbances of the segmental plate mesoderm by showing that host embryos can alter the number of somites formed by a grafted segmental plate to a number that is similar, but rarely equal to, the number of somites formed by the unoperated segmental plate of the host embryo. It may be that whatever mechanisms originally specify the pattern of segmentation in the segmental plate continue to function after segmental plate formation to insure that any disturbances of the pattern are at least partially corrected before somite formation is complete.

ACKNOWLEDGEMENT

This study was supported by grants HD 13396 and HD 17419 from the U.S. National Institutes of Health. I am grateful to Marisa Martini, Sandra McGillis and Steven Falen for their help.

REFERENCES

Meier S (1979). Development of the chick embryo mesoblast. Devel Biol 73:5-45.

Meier S (1980). Development of the chick embryo mesoblast: pronephros, lateral plate, and early vasculature. J Embryol Exp Morph 55: 291-306.

Menkes B, Miclea C (1962). Cercetari asupra formarii organelor exiale III. Posibilatile de vindecare si reorganizare a mezodermului axial presomitic in urma disocierii mecanice. Formarea de somite supranumerare. St Cerc St Med Baza Acad R.P.R. 9:203-228.

Menkes B, Sandor S (1969). Researches on the development of axial organs V. Rev Roum d'Embryol Cytol, Serie d'Embryol 6:65-71.

Packard, DS (1978). Chick somite determination: the role of factors in young somites and the segmental plate. J Exp Zool 203:295-306.

Packard DS (1980) Somitogenesis in cultured embryos of the Japanese quail, <u>Coturnix coturnix japonica</u>. Am J Anat 158:83-91.

Packard DS, Jacobson AG (1976). The influence of axial structures on chick somite formation. Devel Biol 53:36-48.

Packard DS, Meier S (1983). An experimental study of the somitomeric organization of the avian segmental plate. Devel Biol 97:191-202.

Sandor S (1972). Researches on the formation of axial organs in the chick embryo VIII. Some new aspects of regulation potencies during somitogenesis. Rev Roum d'Embryol 8:113-121.

Progress in Developmental Biology, Part A, pages 71–74

POSITIONAL SPECIFICATION OF BLAST CELLS IN THE LEECH EMBRYO INVOLVES A STEPWISE COMMITMENT

Marty Shankland and David A. Weisblat

Departments of Molecular Biology and Zoology, University of California, Berkeley, California 94720.

Once fertilized the egg of the glossiphoniid leech develops into its mature form via a stereotyped sequence of embryonic cell divisions (Stent et al., 1982). Most of these divisions exhibit a fixed orientation and degree of daughter cell asymmetry, and as a result the individual blastomeres can be uniquely identified either by (i) size and position or (ii) their lines of descent from the fertilized egg. The developmental fate of the individual blastomeres has been characterized through the injection of histologically identifiable lineage tracers, and the results show that each lineally defined blastomere gives rise to a very specific complement of descendant cells in the mature leech.

The fate of any given blastomere could be constrained either by its cell lineage or by its position within egg, and a number of experimental studies have attempted to assess the relative contributions of these two factors. Some blastomeres appear to be born with a lineage-specific identity which influences their subsequent development independent of cell interactions. However, there are also a few instances in which cells are born with equivalent developmental potential and subsequently follow divergent developmental pathways as a result of their relative positions within the egg (Weisblat and Blair, 1984; Shankland, 1984).

Our recent studies have focused on one of these instances of positional specification in the leech species *Helobdella triserialis*. The O and P cell lines arise from a pair of sister blastomeres known collectively as the O/P teloblasts, and each O/P teloblast undergoes a series of highly asymmetrical cell divisions to generate a column or bandlet of much smaller o/p blast cells. These two o/p bandlets merge with the bandlets arising from three other, lineally distinct teloblasts to form a germinal band, and it is this band which gives rise to the ectodermal and mesodermal tissues for either the right or left half of the mature leech body. Lineage tracer injection has revealed that sister O/P teloblasts give rise to distinct O and P patterns of descendant tissue in the mature leech, and that these patterns correlate with the relative positions of their bandlets within the germinal band (Weisblat and Blair, 1984). Once they are fated to take on particular positions within the germinal band, the two o/p bandlets are designated as o and p bandlets accordingly.

Previous studies have shown that both the o and p blast cells are capable of following either an O or a P developmental pathway, and that they normally diverge as the result of an interaction between the two bandlets (Weisblat and Blair, 1984; Zackson, 1984; Shankland, 1986). If this interaction is prevented by elimination of the p bandlet, blast cells in the surviving o bandlet will change fate and follow the P pathway instead. If this elimination is performed early enough in o blast cell development--for instance, by ablation of the p bandlet's parental teloblast--then the o blast cell will produce a descendant clone which contains only those pattern elements normally derived from a p blast cell.

A different method can be used to eliminate the p bandlet within the germinal band. By injecting the parental teloblast with a fluorescein-conjugated dextran, the p blast cells can be selectively labeled and sensitized to light energy at the fluorescein excitation wavelength (Shankland, 1984). If the p bandlet is so ablated, the developmental fate of the neighboring o blast cells depends upon their age at the time of the ablation (Shankland and Weisblat, 1984). Blast cells which are less than 20 hr old respond by giving rise exclusively to P pattern elements, but subsequent to this age there is a series of three sequential restrictions on the o blast cell's ability to respond. Each of these restrictions leads to a reduction in the number of P pattern elements produced, and at least two of the three events also lead to a concomitant formation of certain O pattern elements. Thus, each event brings about a partial commitment of the o blast cell's descendant clone to the O developmental pathway, and only when an o clone has undergone all three partial commitments is its fate independent of any subsequent p bandlet ablation.

The o blast cell begins a stereotyped sequence of cell divisions soon after it enters the germinal band (Zackson, 1984). The observed commitment events occur around the time of the first few cell divisions and are separated by intervals comparable to those which separate successive mitoses (Shankland and Weisblat, 1984). These considerations sugggest that the observed commitment events may in fact be associated with particular blast cell divisions, and if this were the case one might expect that the partial nature of the commitment derives from the selective restriction on the fate of only one of the two sublineages arising from daughter cells of that division.

The developmental fate of the different sublineages has been determined by injecting various o blast cell progeny with lineage tracer and characterizing the cellular composition of their descendant clones (Shankland, 1986). Those subsets of O pattern elements whose fates are determined in unison at a particular commitment event do in fact arise during normal development from different o blast cell sublineages. For instance, the first event brings about a commitment to produce the pattern elements of one particular sublineage, but does not commit the o blast cell clone to produce those pattern elements which normally arise from the sister sublineage. The pattern elements associated with the latter sublineage are determined at the two succeeding commitment events, and the segregation into committed and uncommitted subunits at the second event parallels the segregation of sister sublineages at the next cell division (Shankland, 1986).

These findings support the idea that the three observed commitment events are associated with particular o blast cell divisions, and that at least the first two events bring about a selective commitment of only one of the two daughters of a division. If the p bandlet is present during such a division, it apparently interacts with the premitotic cell in the o bandlet in such a way as to insure that one of the two daughters of the division will be incapable of taking on a P fate--and hence be committed to the O pathway--while the other daughter will retain the ability to take on either an O fate or a P fate (Fig. 1). On the other hand, if the p

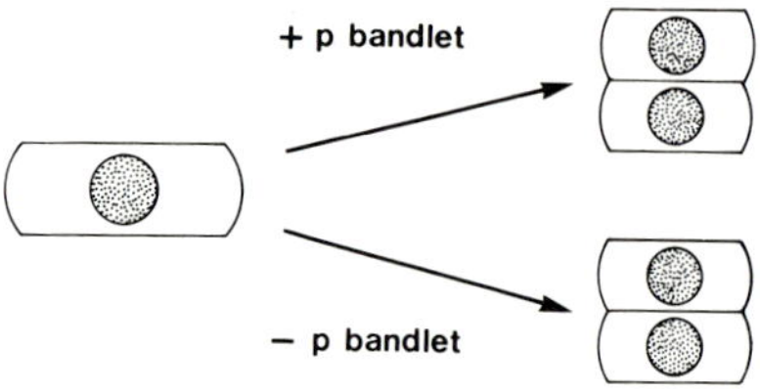

Figure 1. Influence of the p bandlet on daughter cell commitment at certain o blast cell divisions.

bandlet is eliminated prior to this interaction, both daughters of the division will retain the ability to take on P fates. Unfortunately, it is not known whether daughter cells produced in the latter circumstance retain the ability to take on O fates.

REFERENCES

Shankland M (1984). Positional determination of supernumerary blast cell death in the leech embryo. Nature 307:541.

Shankland M (1986). Position-dependent cell interactions and commitments in the formation of the leech nervous system. In Hunt RK (ed): "Current Topics in Developmental Biology, Neural Development Part IV," New York: Academic, in press.

Shankland M, Weisblat DA (1984). Stepwise commitment of blast cell fates during the positional specification of the O and P cell lines in the leech embryo. Dev Biol 106:326.

Stent GS, Weisblat DA, Blair SS, Zackson SL (1982). Cell lineage in the development of the leech nervous system. In Spitzer NC (ed): "Neuronal Development," New York: Plenum, p 1.

Weisblat DA, Blair (1984). Developmental indeterminacy in embryos of the leech *Helobdella triserialis*. Dev Biol 101:326.

Zackson SL (1984). Cell lineage, cell-cell interaction and segment formation in the ectoderm of a glossiphoniid leech embryo. Dev Biol 104:143.

Progress in Developmental Biology, Part A, pages 75–78

PATTERN SENSITIVITY TO BOUNDARY AND INITIAL CONDITIONS IN REACTION DIFFUSION MODELS OF PATTERN FORMATION.

Phillip A. Arcuri and James D. Murray

Department of Anatomical Sciences, University of Illinois, 506 S. Mathews Ave., Urbana, Illinois, 61801 (PAA), Centre for Math Biology, Math Institute, 24-29 St. Giles, Oxford OX1 3LB (JDM)

INTRODUCTION

Reaction diffusion mechanisms (for example, Turing, 1952; Meinhardt and Gierer, 1974; Thomas, 1976; etc.) with zero flux boundary conditions have been criticized because of the sensitivity of the generated spatial patterns to changes in the initial conditions, model parameters, and domain geometry and size (see, for example, Bard and Lauder, 1974). In contrast, our results show that other equally plausible boundary conditions may give rise to spatial patterns which are, in fact, quite robust (Arcuri and Murray, 1986; Arcuri, 1984) implying that simple Turing-type mechanisms based on two reacting and diffusing morphogens are capable of precise pattern formation in randomly perturbed morphogenetic systems. Of course, the question regarding the nature and degree of modulation of such mechanisms by other factors in particular cases remains unknown. Nevertheless, reaction diffusion mechanisms are so physically simple, being only one degree more complex than passive diffusion mechanisms, that inevitably they must play some role in pattern formation and morphogenesis.

METHODS AND RESULTS

Turing-type reaction diffusion models take the form

$$\frac{\partial u}{\partial t} = \frac{\partial^2 u}{\partial x^2} + \gamma f(u,v) \qquad \frac{\partial v}{\partial t} = \beta \frac{\partial^2 v}{\partial x^2} + \gamma g(u,v)$$

plus initial and boundary conditions, where $u=u(x,t)$, $v=v(x,t)$ are morphogen concentrations in, for example, a one-dimensional domain $0 \leqslant x \leqslant 1$ (results generalize to higher dimensions); and where $\partial u/\partial t$, $\partial v/\partial t$ represent the rate of change with time of the morphogen concentrations; $\partial^2 u/\partial x^2$, $\partial^2 v/\partial x^2$ represent (Fickian) diffusion; and $f(u,v)$, $g(u,v)$ model the chemical reaction between u and v. We require f and g to admit a unique kinetic steady state (u^*,v^*) so that $f(u^*,v^*)=g(u^*,v^*)=0$.

The two free parameters β and γ are common to all reaction diffusion models. β represents the ratio of the diffusion rates: $\beta = D_v/D_u$; and γ is a scale factor relating the size of the domain, the absolute size of the diffusivities and the overall rate of the reactions.

The key factor which determines pattern sensitivity in such systems is the boundary conditions, which in general specify a rate of flux across the domain boundary. For simplicity we consider the flux rate to be proportional to the difference between internal and external concentrations at the boundary. For example, for the u-species, the boundary flux rate of u $= -k_u(u_{int} - u_{ext})$ so that $u_{ext} > u_{int}$ implies a net influx $(k_u > 0)$, and $u_{ext} < u_{int}$ implies a net efflux $(k_u > 0)$. The special cases of zero flux conditions (modelling an insulated or isolated domain) and fixed boundary conditions are given by $k_u=0$ and $k_u=\infty$ respectively.

The results of many computer simulations of several model mechanisms show that robust patterns are generated for boundary conditions which bias the polarity of the patterns. This occurs when (u_{ext},v_{ext}) are different from (u^*,v^*). Such "non-homogeneous" conditions generate patterns which have their polarity determined by the boundary conditions. Moreover, the patterns are insensitive to small random perturbations in the initial conditions, and to changes in the domain size, shape, and other model parameters. In contrast, "homogeneous" boundary conditions, including zero flux conditions, or flux conditions where $u_{ext}=u^*$ and $v_{ext}=v^*$, give rise to patterns which are very sensitive to these quantities.

These results are clearly demonstrated in Figures 1 and 2 which summarize computer simulations of the Thomas substrate-inhibition mechanism (Thomas, 1976). Figure 1 depicts the distribution of patterns obtained under zero flux conditions for various values of β and γ (axes in each graph) and for two different sets of small random perturbations about the steady state (Figures 1a and 1b). Each dot in the graphs represents a computer simulation.

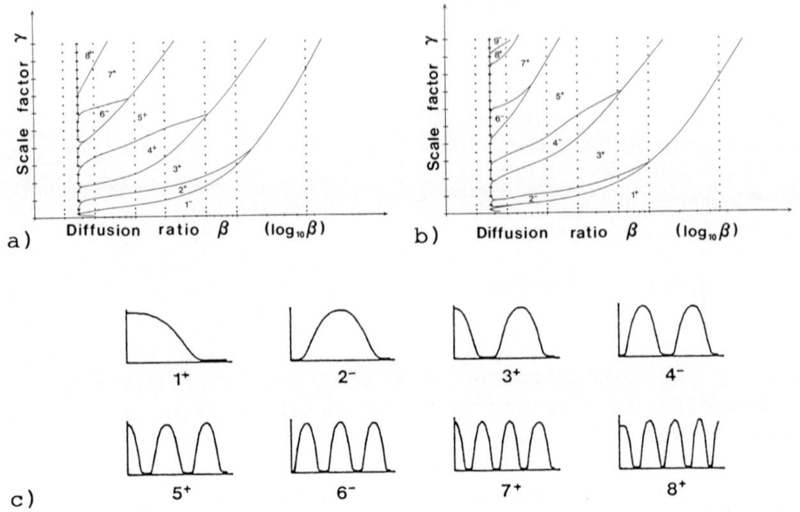

Figure 1: a,b: Distribution of patterns for zero flux boundary conditions and different random perturbations. c: Examples of patterns generated. Reproduced with permission.

Note that only certain patterns are generated, and that each pattern occupies a particular contiguous region in the β-γ plane. A comparison between Figures 1a and 1b shows the sensitivity of the pattern polarity (denoted by + or - superscripts on pattern numbers) and of the patterns themselves to small variations in the initial state of the system.

In contrast, Figure 2 shows the distribution of patterns obtained under the (non-homogeneous) fixed boundary conditions $u_{ext}=1.1u^*$, $v_{ext}=1.1v^*$. These conditions determine the pattern's polarity and increase the size of the regions in the β-γ plane where a given pattern will develop. In addition, the distribution of patterns is virtually unaffected by noise in the initial data.

a)

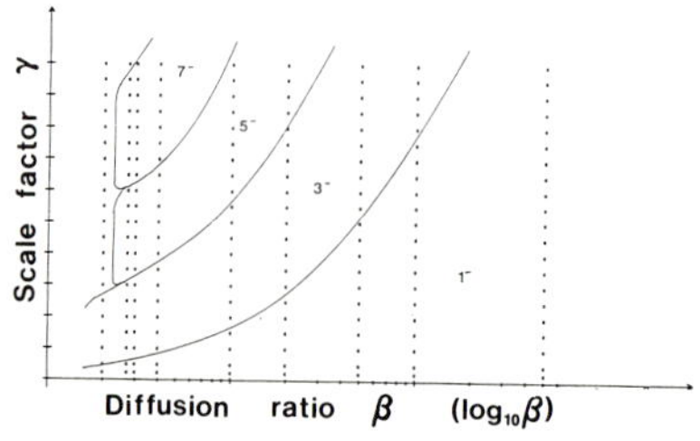

b)

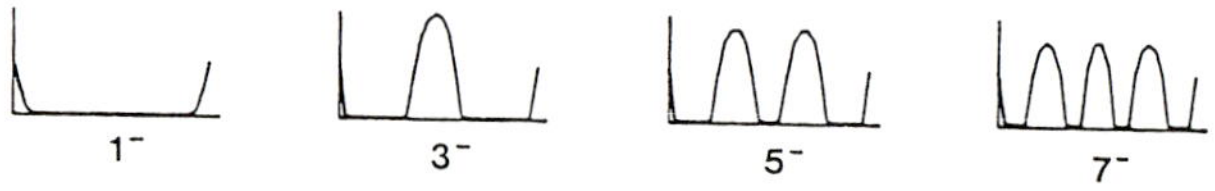

Figure 2. a: Distribution of patterns for non-homogeneous fixed boundary conditions. b: Examples of patterns generated. Reproduced with permission.

REFERENCES

Arcuri PA (1984) "Reaction Diffusion Mechanisms for Biological Pattern Formation." D. Phil. thesis, Oxford University.

Arcuri PA, Murray JD (1986) Pattern sensitivity to boundary and initial conditions in reaction diffusion models. To appear in J Math Biol.

Bard JBL, Lauder I (1974) How well does Turing's theory of morphogenesis work? J Theor Biol 45: 501

Meinhardt H, Gierer A (1974) Application of a theory of biological pattern formation based on lateral inhibition. J Cell Sci 15: 321

Thomas D (1975) Artificial enzyme membranes: transport, memory and oscillatory phenomena. In: Thomas D, Kernevez JP (eds): "Analysis and Control of Immobilized Enzyme Systems." p. 115, Springer Verlag, New York.

Turing AM (1952) The chemical basis of morphogenesis. Phil Trans Roy Soc London B 237: 37

Progress in Developmental Biology, Part A, pages 79–82

ENDOGENOUS LOW MOLECULAR WEIGHT FACTORS, VITAMIN A AND PHORBOL ESTERS DRAMATICALLY AFFECT PATTERN FORMATION AND POSITIONAL VALUES IN MARINE HYDROIDS

Stefan Berking, Günter Plickert and
Werner A. Müller
Zoologisches Institut, Universität, Im Neuenheimer Feld 230, D-6900 Heidelberg, Germany

The architecture of a hydrozoan colony is designed and controlled by several pattern forming systems: In the planula undergoing metamorphosis into a founder polyp, and likewise, in each new member of the growing colony, an apico-basal system provides positional information to define the length of the body and to subdivide it into head, gastric region and stolon. This system first confers upon the cells position dependent quantifiable properties called positional values. Based on this longitudinal pattern circular subsystems elaborate the sites of radial structures such as tentacles (Müller, 1982, 1984; Berking, 1984). Radial patterns are periodic ones as are the colonial patterns. These are characterized by minimum distances between the polyps that emerge on the elongating stolons (Plickert, 1985) and between the branching sites of the stolon pipings (Müller and Plickert, 1982).

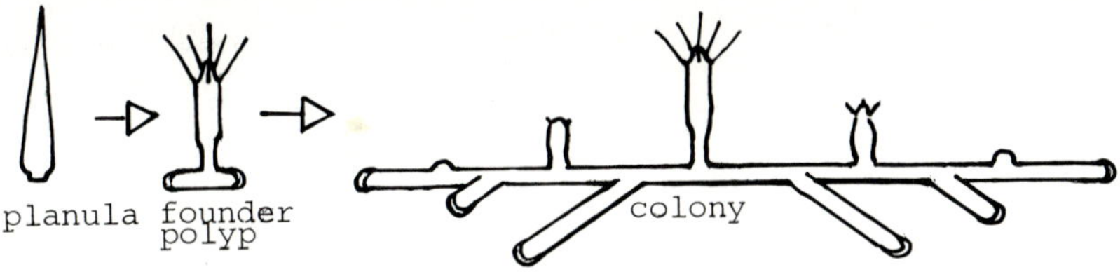

Periodic colonial patterns are controlled by inhibitory signals: When in Hydractinia echinata or in Eirene viridula the polyps are disconnected from the stolons, new polyp buds appear earlier and at only half the normal distance. Removal of only the heads of the polyps has the same effect. The heads are sources of inhibitory signals that are transmitted along the stolons where the signal strength eventually falls below a threshold value. Efflux of the inhibitor into the external medium appears to contribute to the attenuation of the signal strength since inhibitors accumulate in the medium and its frequent renewal causes a shortening of the budding distances as well. On the other hand, transplantation of additional heads onto a founder polyp, or addition of medium enriched with inhibitor, lead to an increase of the budding distances (Plickert et al., 1985).

N-methyl-picolinic acid (homarine) may represent an endogenous inhibitory signal: From hydroids some few low molecular weight compounds which specifically inhibit morphogenetic processes can be extracted. The most prominent compound has been identified to be the pyridine N-methyl-picolinic acid (homarine). In marine Coelenterates such as Hydractinia homarine is present in quite large quantities (up to 10^{-3} M/L in polyps and 10^{-2} M/L in planulae). Dependent on time and dose (10^{-6} to 10^{-5} M) of application, homarine stops metamorphosis, interferes with proportioning in patterning, and inhibits bud specification (Berking, 1985). Equally effective are the chemically related compounds nicotinamide and niacin-methyl betain (trigonelline).

Vitamin A alters the effective signalling range: Upon a pulse-type treatment with all-trans retinal or retinoic acid (10^{-8} to 10^{-7} M), polyps form more tentacles per unit circumference, they reduce their length, and new polyp buds arise at shorter distances (Müller, 1984). At least two interpretations are possible to account for this apparent reduction of the signalling range: (a) Release or transmission of control signals is reduced, (b) the threshold for inhibition in the target cells is increased. A further effect of vitamin A, acceleration of stolon elongation, may demand a different explanation.

Stimulating factors: Control of morphogenesis is certainly not only accomplished by emission or withdrawal of inhibitory signals. Coordinated behaviour of cells also needs mutual stimulation. Control circuits, therefore, may rely on inhibitory and on stimulatory signals as well.

In the hydroid Hydractinia we have traced low molecular weight factors
(1) that induce the formation of lateral tips on stolons and, hence, of stolon branches (new results),
(2) that promote the emergence of polyp buds on stolons which previously were disconnected from existing polyps (new results),
(3) that induce double head formation in regeneration, i.e. the development of head structures at both ends of an isolated gastric region (ref. in Müller, 1982, 1984).

All these factors, however, are still to be chemically identified.

The stimulating activity of phorbol esters: In cell biology phorbol esters and vitamin A are often used as antogonistic tools to stimulate either proliferation or terminal differentiation. Phorbol esters are supposed to intervene in signal transducing pathways that transfer the message of certain growth factors or hormones from receptors into intracellular compartiments (Nishizuka, 1984). In Hydractinia phorbol esters have multiple and dramatic effects:

Phorbol esters induce metamorphosis of planulae into primary polyps. In this respect, they replace or bypass an inducing lipid which, in the natural habitat, is released by environmental bacteria of the genus Alteromonas. The inducing capacity of various phorbol esters is correlated with their tumor-promoting activity. The effectiveness is strongly amplified by Li^{+} and Cs^{+}, or by removal of Mg^{++} (Müller, 1985).

In some aspects phorbol esters act antagonistically to vitamin A: When treated in early metamorphosis, the resulting primary polyps develop fewer tentacles, the polyps become taller whereas elongation of the stolons is retarded (new result).

Phorbol esters stimulate polyp budding on isolated stolons (new result).

Phorbol esters induce double head formation. A short term incubation with TPA (10^{-7} M, 2 h) causes isolated midgastric regions to develop, days after treatment, heads at both ends and often further ones in the middle (Müller, 1985).

In Hydra phorbol esters (TPA 10^{-11} M) stimulate the differentiation of nerve cells out of precursor cells. This effect is also a long term effect and is preceeded by a phase of depression: During the presence of the drug, no new nerve cells are produced (Berking, new result).

In formal terms, tumor-promoting phorbol esters cause an increase in the positional value, that is an increase of the tissues capacity to form and to induce head structures. Phorbol esters confer upon the tissue the capacity to acquire the characteristics of an organizer.

REFERENCES

Berking S (1984). Metamorphosis of *Hydractinia echinata*. Insights into pattern formation in Hydroids. Roux's Arch Dev Biol 193: 370.

Berking S (1985). Is homarine a morphogen in the marine hydroid *Hydractinia*? Roux's Arch Dev Biol, in press.

Müller WA (1982). Intercalation and pattern regulation in Hydroids. Differentiation 22: 141.

Müller WA (1984). Retinoids and pattern formation in a hydroid. J Embryol exp Morph 81: 253.

Müller WA (1985). Tumor-promoting phorbol esters induce metamorphosis and multiple head formation in the hydroid *Hydractinia*. Differentiation, in press.

Müller WA, Plickert G (1982). Quantitative analysis of an inhibitory gradient field in the hydrozoan stolon. W Roux's Archieves 191: 56.

Nishizuka Y (1984). The role of protein kinase C in cell surface signal transduction and tumour promotion. Nature 308: 693.

Plickert G, Heringer A, Hiller B (1985). Analysis of spacing in a periodic pattern. Dev Biol, in press.

Progress in Developmental Biology, Part A, pages 83–86

PATTERN FORMATION IN *DICTYOSTELIUM* AS ANALYZED FROM MOVEMENT OF ANTERIOR-LIKE CELLS.

Ikuo Takeuchi and Tetsuji Kakutani

Department of Botany, Faculty of Science,
Kyoto University, Kyoto 606, Japan

INTRODUCTION

Development of the cellular slime mold, *D. discoideum* provides a good model for the studies of pattern formation within a developing tissue. An apparently homogenous population of cells aggregates to form a slug-shaped tissue which has a polarity in the direction of movement and the pattern of cell differentiation. The anterior and posterior parts of the slug comprize prestalk (ca. 13% of total) and prespore (ca. 77%) cells respectively, which under normal circumstances, eventually become stalk and spore cells of a fruiting body. Besides prespores, Sternfeld and David (1981, 1982) remarked the presence of cells (ca. 10%) having the same staining characteristics as prestalks within the posterior part of the slug and called them anterior-like cells: they are located in the prespore region in an intact slug, but when the prestalk region is cut off some of them move forward and contribute to the regeneration of the prestalk region within the isolate.

To characterize anterior-like cells, we examined their movement during migration of a slug and found that they do not constitute an independent population, but are constantly exchangeable with prestalk cells.

EXCHANGE BETWEEN PRESTALK AND ANTERIOR-LIKE CELLS

When vitally-stained cells are allowed to aggregate and form slugs, prestalk and anterior-like cells are stained while prespore cells are unstained. To examine whether or not prestalk and anterior-like cells mutually exchange during slug migration, the prestalk region of a slug stained with Nile blue sulfate was grafted onto the prespore region of a neutral red-stained slug. After several hours of migration, the prestalk/prespore boundary of the composite slug became purple. The purple region gradually expanded until the whole slug became purple after ca. 24 h migration. This is consistent with previous observations that a considerable mixing occurs between anterior and posterior cells of a slug during migration (Bonner, 1957; Francis and O'Day, 1971; Müller and Hohl, 1973). To determine the rate of exchange between prestalk and anterior-like cells, the number of red cells in the prestalk region of the slugs were counted after various hours of grafting (Table 1).

TABLE 1. Invasion of anterior-like cells into the prestalk region

Time after grafting (h)	% of original anterior-like cells in the prestalk region	
	Observed	Calculated*
2.5	5	1
4	13	13
6	17	24
7	25	27
14	36	40

* Calculated from the frequency distribution of velocities of anterior-like cells, on the assumption that the invasion begins after 2.4 h grafting.

MOVEMENT OF ANTERIOR-LIKE CELLS

To characterize the movement of anterior-like cells, we traced neutral red-stained anterior-like cells within a slug migrating in an agar capillary with a rectangular section. The agar capillary allowed slugs to undergo no changes in shape and direction of movement and provided good optical conditions. The axial component of mean velocity during 315 sec was determined with 94 cells in 4 slugs and normalized by simultaneously measured slug velocities. The frequency distribution of the velocities of the anterior-like cells gave the mean value of 0.92 with a quite large standard deviation of 0.22.

When a single anterior-like cell was tracked for a period of 20-60 min, the cell underwent considerable changes in velocity from time to time. The temporal variation of the mean velocities during 315 sec was determined with 4 different cells. The mean of the standard deviations amounted to 0.16, which is comparable to the standard deviation of the velocity among different cells (0.22). This indicates that the latter is attributable, at least in part, to temporal changes in velocity of the cells. Anterior-like cells are thus characterized by the randomness of their movement.

Whether or not the exchange between prestalk and anterior-like cells is solely accounted for by random movement of the latter was examined by calculating the exchange rate from the frequency distribution of velocities of anterior-like cells. Such calculation was made, based on the assumption that the exchange begins after a lag period of 2.4 h. As shown in Table 1, the calculated values agree well with the observed.

ANTERIOR-LIKE CELLS WITHIN PRESPORE ISOLATES

When the prespore portion is isolated from a slug, some anterior-like cells move to the front and contribute to regeneration of the prestalk region, while the rest remains in the posterior region of the regenerated slug (Sternfeld and David, 1981). The evidence was also presented that a diffusible substance secreted from the prestalk region prevents anterior-like cells from moving forward within an intact slug. We examined the movement of anterior-like cells within prespore isolates after 20-180 min isolation and found that the frequency distribution of their velocities (0.94 ± 0.25) was essentially the same as that within intact slugs (0.92 ± 0.22). This indicates that the movement of anterior-like cells is independent of the presence of the prestalk region.

If one assumes that the number of cells leaving the prestalk region to become anterior-like cells is proportional to the number of prestalk cells, it is possible to estimate the decrease of anterior-like cells within the prespore region of the regenerating isolates. The estimated change agreed well with the observed value. This indicates that the behavior of anterior-like cells within prespore isolates can be explained as a process of restoration of the equilibrium between prestalk and anterior-like cells, which was once upset by removal of the prestalk region.

CONCLUSION

The above results showed that anterior-like cells are in equilibrium with prestalk cells within a slug and both cells are constantly exchanging each other during slug migration. Anterior-like cells are characterized by their random movement and they markedly change in velocity from time to time. A high level of randomness of their movement results in a high rate of exchange between prestalk and anterior-like cells. The behavior of anterior-like cells within prespore isolates is resulted from the upset of the equilibrium owing to removal of the prestalk region. These indicate that, as far as the mode of movement is concerned, anterior-like cells do not constitute the third type of cells but belong to the same population as prestalk cells within a slug.

REFERENCES

Bonner JT (1957). A theory of the control of differentiation in the cellular slime molds. Quart Rev Biol 32: 232-246.

Francis D, O'Day DH (1971). Sorting out in pseudoplasmodia of *Dictyostelium discoideum*. J Exp Zool 176: 265-272.

Müller U, Hohl HR (1973). Pattern formation in *Dictyostelium discoideum*: Temporal and spatial distribution of prespore vacuoles. Differentiation 1: 267-276.

Sternfeld J, David CN (1981). Cell sorting during pattern formation in *Dictyostelium*. Differentiation 20: 10-21.

Sternfeld J, David CN (1982). Fate and regulation of anterior-like cells in *Dictyostelium* slugs. Dev Biol 93: 111-118.

Progress in Developmental Biology, Part A, pages 87–91

EXTRACELLULAR cAMP IN Dictyostelium discoideum TISSUE

R. van Driel, A.P. Otte, M.J.E. Plomp and J.C. Arents
Laboratory of Biochemistry, Univ. of Amsterdam, P.O. Box 20151, 1000HD Amsterdam, The Netherlands

INTRODUCTION

Starvation of D. discoideum amoebae initiates a precisely defined series of developmental events. Within a few hours the cells become adhesive. Guided by chemotactic signals, pulses of extracellular cAMP, the cells come together and form a mobile, slug-like, multicellular organism, comprising up to 10^5 cells. Via a series of morphological changes a fruiting body is formed consisting of a slender column of stalk cells, carrying a globular mass of spores (Loomis, 1982). Presumptive stalk (prestalk) cells and presumptive spore (prespore) cells are found in the slug stage that precedes fruiting body formation. Both cell types are arranged in a well-defined, simple, axial pattern.

Extracellular cAMP plays a crucial role in regulation of D. discoideum development. It simultaneously acts as a chemoattractant, coordinating morphogenetic cell movement, and as a regulator of cell differentiation. Both functions are mediated by cell surface receptors that are highly specific for cAMP (Van Haastert and Kien, 1983). These cAMP functions have been well-studied for cells in the aggregation phase of development. However, much less is known about the role of extracellular cAMP in the multicellular stage. Several studies have shown that a high concentration of exogenous cAMP (0.1-1 mM) permits differentiation to stalk or spore cells under conditions in vitro, i.e. in the absence of normal morphogenesis (Brookman et al., 1982; Kay, 1982). Recently, Schaap and Van Driel (1985) found that the expression of developmental markers that are typical for the multicellular state, can be induced by quite physiological extracellular cAMP concentrations (50% induction at about 0.3 μM cAMP), mediated by the cell surface cAMP receptor. This observation made it pertinent to ask: what is the local extracellular cAMP concentration in D. discoideum tissue. Since no methods are available to measure extracellular cAMP in tissue, we have chosen an indirect approach. Here we report an analysis of the ability of prestalk and prespore cells to produce, detect and hydrolyse extracellular cAMP.

RESULTS AND DISCUSSION

Migrating slugs were enzymatically or mechanically dissociated in single prestalk and prespore cells. The cell types were subsequently separated by Percoll density gradient centrifugation according to Tsang and Bradbury (1981) with minor modifications. The isolated cell preparations were used to determine, (i) adenylate cyclase activity, (ii) cAMP-signal relay capacity, (iii) cell- surface and cytosolic phosphodiesterase activity, and (iv) cell-surface cAMP receptors of prestalk and prespore cells. Details of experimental procedures will be given elsewhere. Results are summarised in Table 1 and Figure 1. Similar measurements

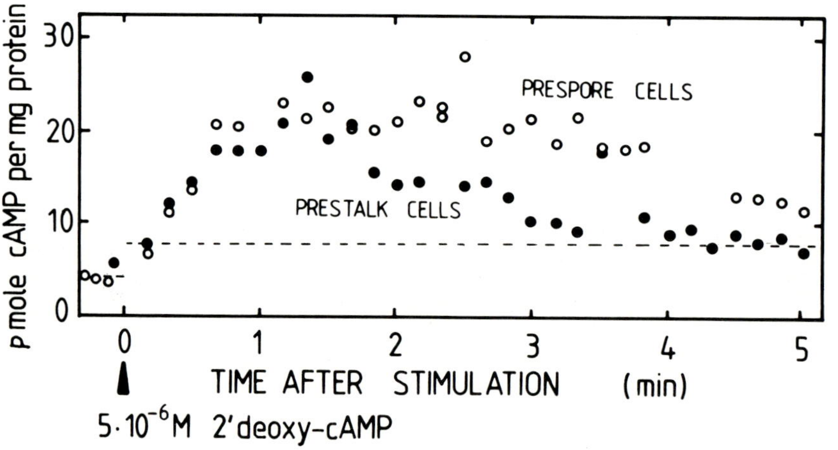

Figure 1. Relay of cAMP signal by prestalk and prespore cells. A suspension of prestalk (•) or prespore (o) cells was stimulated with 5 μM 2'-deoxy-cAMP. The cAMP-induced cAMP production was measured.

TABLE 1. Production, detection and hydrolysis of extracellular cAMP by prestalk and prespore cells isolated from migrating slugs

	Prestalk cells	Prespore cells
Adenylate cyclase	23 ± 10 (n=4)	30 ± 7 (n=4)*
Cell surface phosphodiesterase activity	9.4 ± 4.9(n=2)	3.0 ± 0.5(n=4)*
Cell surface cAMP binding		
measured at 100 nM cAMP	30 (n=2)	7 (n=2)‡
measured at 300 nM cAMP	66 (n=2)	17 (n=2)‡

*) $pmol \cdot min^{-1} \cdot mg^{-1}$; ‡) fmol cAMP bound per 10^7 cells

(except cAMP signal relay) have also been carried out with cells derived from the culmination stage, i.e. just before terminal differentiation. Essentially the same results were obtained. Obviously, no major changes in extracellular cAMP metabolism occurred during the slug-culmination transition.

Our data show that prestalk cells and prespore cells have similar capacities to propagate cAMP signals. Not only do they have similar adenylate cyclase activities, but if stimulated with exogenous cAMP, the two cell types also show the same cAMP response (Figure 1, Table 1). This signal-relay behaviour is similar to that of aggregating D. discoideum cells. At that developmental stage it has been shown that most of the cAMP produced is secreted (Dinauer et al., 1980). It is therefore likely that the slug cells too secrete a major fraction of the newly produced cAMP.

From Figure 1 it can be seen that prestalk cells hydrolyse extracellular cAMP more rapidly that prespore cells. This is consistent with the observation that prestalk cells carry several-fold more cAMP-phosphodiesterase on their cell surface than prespore cells (Table 1), and supports the idea that the relayed cAMP signal is secreted. In contrast, intracellular phosphodiesterase activities are the same in both cell types (not shown).

If prestalk cells and prespore cells produce the same

amount of extracellular cAMP but degrade the cAMP signal at different rates, the average concentration of extracellular cAMP in the prespore tissue will be higher than between the prestalk cells. Such a gradient of extracellular cAMP is expected to be instrumental in maintaining the prestalk/prespore cell pattern in the aggregate. The following two observations support this hypothesis. Firstly, Schaap and Van Driel (1985) showed that extracellular cAMP induces the expression of at least two independent developmental markers in a dose-dependent manner, via the cell surface cAMP receptor. Secondly, Weyer and Durston (1985) showed that elevated levels of extracellular cAMP inhibit prespore to prestalk cell redifferentiation, a transition that occurs readily in a suspension of isolated prespore cells in the absence of cAMP. Conversely, metabolites of cAMP inhibit the reverse transition. Such metabolites are expected to accumulate at sites where cAMP is rapidly hydrolysed, i.e. in the prestalk tissue. Obviously, an increased extracellular cAMP concentration in the prespore tissue would stabilise the prespore differentiated state.

Intact prestalk cells were found to bind 4-fold more cAMP than prespore cells (Table 1). The physiological background of this observation is not clear, because it is not known whether the "extra" receptors on prestalk cells are involved in chemotaxis or regulation of differentiation, or both. It has been suggested that prestalk cells are chemotactically more active than prespore cells (Matsukuma and Durston, 1979), supporting the former possibility.

REFERENCES

Brookman JJ, Town CD, Jermyn KA, Kay RR (1982). Developmental regulation of a stalk cell differentiation-inducing factor in D. discoideum. Developm Biol 91: 191.

Dinauer MC, MacKay SA, Devreotes PN (1980). Cyclic 3',5'-AMP relay in D. discoideum. The relationship of cAMP synthesis and secretion during cAMP signaling response. J Cell Biol 86: 537.

Kay RR (1982). cAMP and spore differentiation in D. discoideum. Proc Natl Acad Sci USA 79: 3228.

Loomis WF (1982). The Development of Dictyostelium discoideum. New York: Academic Press.

Matsukuma S, Durston AJ (1979). Chemotactic sorting in D. discoideum. J Embryol Exp Morph 30: 243.

Schaap P, Van Driel R (1985). The induction of postaggregative differentiation in D. discoideum by cAMP. Evidence for the involvement of the cell surface cAMP receptor. Exp Cell Res 159: 388

Tsang A, Bradbury JM (1981). Separation and properties of prestalk and prespore cells of D. discoideum. Exp Cell Res 132: 433.

Van Haastert PJM, Kien E (1983). Binding of cAMP derivatives to D. discoideum cells. Activation mechanisms of the cell surface cAMP receptor. J Biol Chem 258: 9636.

Weyer CJ and Durston AJ (1985). Influence of cyclic AMP and hydrolysis products on cell type regulation in D. discoideum. J Embryol Exp Morphol, in press.

Progress in Developmental Biology, Part A, pages 93–96

CAMP-DEPENDENT PROTEIN KINASE DURING DEVELOPMENT OF D. DISCOIDEUM.

B.H. Leichtling, K.L. Schaller*, T.C. Chambers and H.V. Rickenberg*
National Jewish Hospital and Research Center; also *The University of Colorado Health Sciences Center, Denver, Colorado

The cellular slime mold Dictyostelium discoideum needs little introduction since it has become a widely used model for the study of cellular differentiation and development. Our interest is in the function of cAMP and of cAMP-dependent protein kinase (cAMP-d PK) in these processes; the cyclic nucleotide plays the dual role of chemoattractant and of presumptive second messenger; a possible causal role in the bifurcation between spore and stalk cell development is not excluded. Our and other worker's earlier studies showed that cAMP-d PK strongly resembled the mammalian enzyme and that the domains on the catalytic and regulatory subunits required for their interaction have been preserved to the extent that the Dictyostelium and bovine regulatory subunits inhibit the activity of the heterologous catalytic subunits in a manner reversible by cAMP (Majerfeld et al., 1984; de Gunzburg et al., 1984; Rutherford et al., 1984; Schoen et al., 1984). The synthesis of the D. discoideum cAMP-d PK is developmentally regulated (Fig. 1) (Leichtling et al., 1984); a maximal increase of the level of both subunits was attained just before the appearance of prespore- and prestalk-specific mRNAs (Barklis et al., 1983; Mehdy et al., 1983). Furthermore, we found that at the slug stage of development prespore cells have about 5 times as much cAMP-d PK as do prestalk cells (Schaller et al., 1984). We assume provisionally that cAMP exerts its effect on the synthesis of developmentally regulated proteins at the level of transcription and that this effect is mediated by the cAMP-d PK.

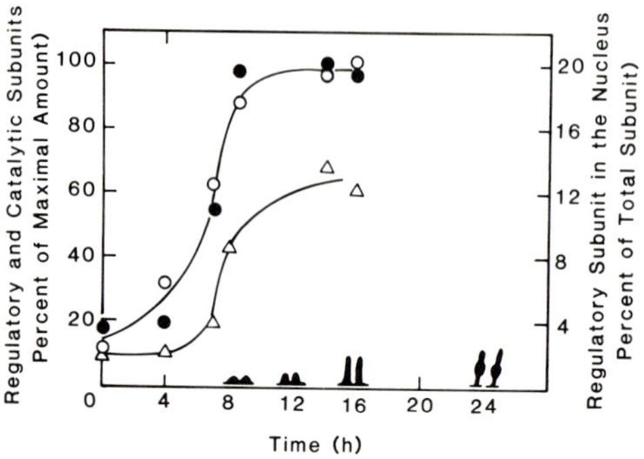

Fig. 1. The level of total soluble regulatory (●) and catalytic subunits (O) and of nuclear regulatory subunits (x) during development. The drawings depict the relevant morphological stages.

There are two possibilities in the context of a role of the cAMP-d PK in transcription; one is that relevant proteins, phosphorylated in the cytoplasm, are translocated into the nucleus; the second is that the cAMP-d PK exerts its effect directly in the nucleus. We find that, as was observed earlier in the case of mammalian cells (Jungmann et al., 1979), a translocation of the cAMP-d PK subunits into the Dictyostelium nucleus occurs and report here on potential nuclear substrates of the enzyme.

Fig. 1 and Table 1 show that the translocation of the two subunits from the cytoplasm to the nucleus is developmentally regulated; in vegetative amoebae 2-2.5% of the total subunits are in the nucleus; this proportion goes up to about 12% at the Tipped Aggregate stage. The maintenance of this level of subunits in the nucleus depends on the integrity of the Tipped Aggregate structure since its disruption to single cells leads to a loss of nuclear cAMP-d PK; the loss can be reversed by exogenous cAMP (Table 1).

We monitored nuclei for substrates of the cAMP-d PK by comparing the occurrence of phosphoproteins formed in isolated nuclei ("in situ") and in nuclear extracts, supplemented with the catalytic subunit ("in vitro") with phosphoproteins formed in in vivo. We find that the addition of purified Dictyostelium or bovine catalytic subunit to nuclear extracts from all developmental stages (Fig. 2) and from the two cell types greatly enhances the phosphorylation of a very basic protein of molecular weight 38,000 which is phosphorylated also in isolated nuclei as well as in vivo. There is evidence (data not shown) that there are several cell type-specific nuclear proteins which are substrates of the cAMP-d PK.

THE EFFECT OF THE DISRUPTION OF TIPPED AGGREGATES AND OF cAMP ON THE LEVELS OF REGULATORY AND CATALYTIC SUBUNITS IN THE NUCLEUS

Condition	Percent of total soluble cAMP-d PK subunits in the nucleus	
	catalytic subunit	regulatory subunit
Vegetative Amoebae	2.1	2.5
Tipped Aggregates	10.5 ± 2.2	11.2 ± 1.9
Disrupted Tipped Aggregates		
-cAMP	5.6 ± 1.1	4.5 ± 1.2
+cAMP at 1,3,4.5 hr	9.5 ± 2.0	9.6 ± 1.6
+cAMP at 4.5 hr only	12.0 ± 2.8	12.3 ± 2.7

Table 1. Amoebae from disrupted Tipped Aggregates were maintained in single-cell suspension and after 5.0 h of incubation the amoebae were harvested. Values for the disrupted Tipped Aggregates are the means of 5 experiments (± standard deviation).

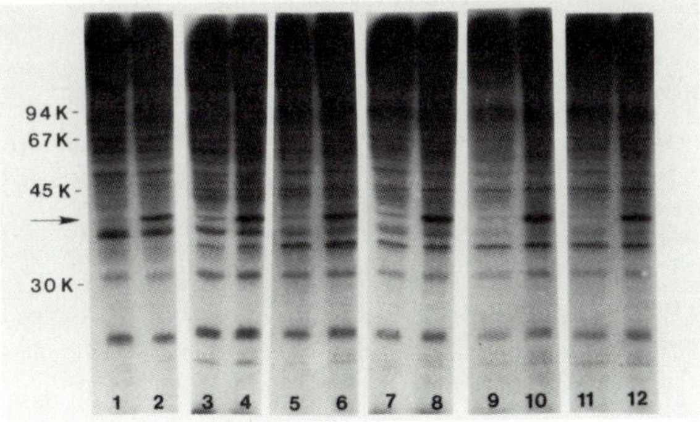

Fig. 2. Phosphorylation of nuclear extracts in the presence (even numbered lanes) and the absence (odd numbered lanes) of added catalytic subunit. Extracts were made from vegetative amoebae (1,2) and amoebae from the Tight Aggregate (3,4), Tipped Aggregate (5,6), Finger (7,8), Slug (9,10) and Mid-Culmination (11,12) stages of development.

We shall next follow the in vivo state of phosphorylation of the 38,000 molecular weight protein throughout development and determine whether the phosphorylated amino acid(s) occur in the same peptide(s) in vivo and in vitro. By the same token the state of phosphorylation of cell type-specific substrates of the cAMP-d PK will be monitored during development. We hope to determine the function, if any, in cell type-specific transcription, of the nuclear proteins which are substrates of the cAMP-d PK.

References

Barklis, E. and Lodish, H.F. (1983). Regulation of Dictyostelium discoideum mRNAs specific for prespore or prestalk cells. Cell 32, 1139-1148.

de Gunzburg, J., Part, D., Guiso, N. and Véron, M. (1984). An unusual cAMP-dependent protein kinase from Dictyostelium discoideum. Biochem. 23, 3805-3812.

Jungmann, R.A., Christensen, M.L. and Derda, D.F. (1979). in "Effects of Drugs on the Cell Nucleus" (H. Busch, S.T. Crooke and V. Daskal, eds.) pp. 507-519. Academic Press, New York.

Leichtling, B.H., Majerfeld, I.H., Spitz, E., Schaller, K.L., Woffendin, C., Kakinuma, S. and Rickenberg, H.V. (1984). A cytosolic cyclic AMP-dependent protein kinase in Dictyostelium discoideum. II. developmental regulation. J. Biol. Chem. 259, 662-668.

Majerfeld, I.H., Leichtling, B.H., Meligeni, J.A., Spitz, E. and Rickenberg, H.V. (1984). A cytosolic cyclic AMP-dependent protein kinase in Dictyostelium discoideum. I. properties. J. Biol. Chem. 259, 654-661.

Mehdy, M., Ratner, D. and Firtel, R. (1983). Induction and modulation of cell-type specific gene expression in Dictyostelium. Cell 32, 763-771.

Rutherford, C.L., Vaughan, R.L., Clouthier, M.J., Ferris, D.K. and Brickey, D.A. (1984). Chromatographic behavior of cyclic AMP dependent protein kinase and its subunits from Dictyostelium discoideum. Biochem. 23, 4611-4617.

Schaller, K.L., Leichtling, B.H., Majerfeld, I.H., Woffendin, C., Spitz, E., Kakinuma, S. and Rickenberg, H.V. (1984). Differential cellular distribution of cAMP-dependent protein kinase during development of Dictyostelium discoideum. Proc. Natl. Acad. Sci. USA 81, 2127-2131.

Schoen, C., Arents, J.C. and Van Driel, R. (1984). Isolation and properties of cyclic AMP-dependent protein kinase from Dictyostelium discoideum. Biochim. Biophys. Acta. 784, 1-8.

Progress in Developmental Biology, Part A, pages 97–100

MECHANICAL AND CELLULAR ASPECTS OF GASTRULATION IN THE SEA URCHIN LYTECHINUS VARIEGATUS

John B. Morrill and Sherry D. Doty

Division of Natural Sciences, New College-U.S.F.
Sarasota, Florida 33580

INTRODUCTION

Theoretical models (Odell, et al., 1981; Ettensohn, 1984) of primary and secondary invagination of the endodermal epithelium and overall morphogenesis of the sea urchin gastrula have been based primarily on observations of live and fixed and sectioned embryos viewed with the light (LM) and transmission electron microscope. To date two dimensional models of endodermal invagination provide only a planar view of the apparent infolding of the endodermal epithelium. Invagination and overall morphogenesis during gastrulation are accompanied by differential increases in cell numbers (Agrell, 1953), changes in the character of extracellular materials or "jelly" in the blastocoel and developing archenteron (Monné et al., 1951) and the non-epithelial behavior and proliferation of cells at the tip of the archenteron (Amemiya et al., 1982). Extracellular physico-mechanical forces and their regulation may play important roles in morphogenesis of the gastrula and the endodermal wall of the archenteron. Furthermore, there may be principle differences in gastrulation processes between different types of sea urchin embryos (Amemiya et al., 1982).

Here we outline several physico-mechanical aspects of gastrulation and scanning electron microscopical (SEM) views of primary invagination in *L. variegatus*.

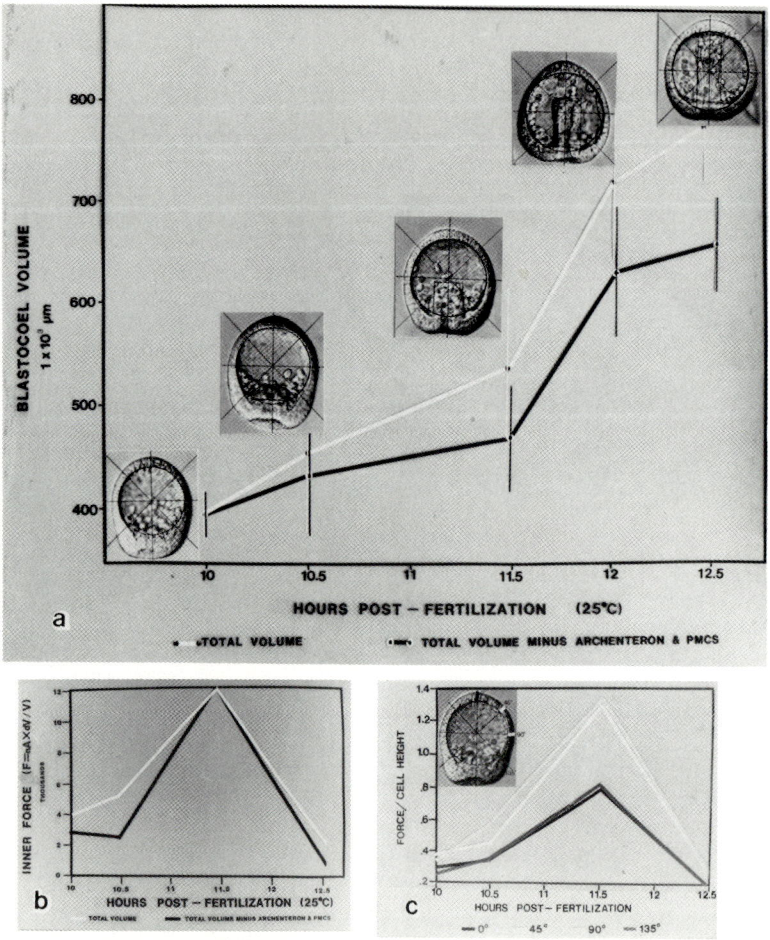

Figure 1. Increase in volume of blastocoel (a), changes in the force of hydrostatic pressure on the ectodermal wall (b) and changes in the force on different regions of the ectodermal wall measured by changes in cell height (c) during gastrulation. In (b) the force on the ectodermal wall is equal to the blastocoel's area times the change in its volume with a constant elastic modulus for the ectodermal cells divided by the blastocoel volume. In (c) the relative force on the ectodermal wall is inversely proportional to the heights of ectodermal cells 0, 45, 90 and 135° from the animal pole.

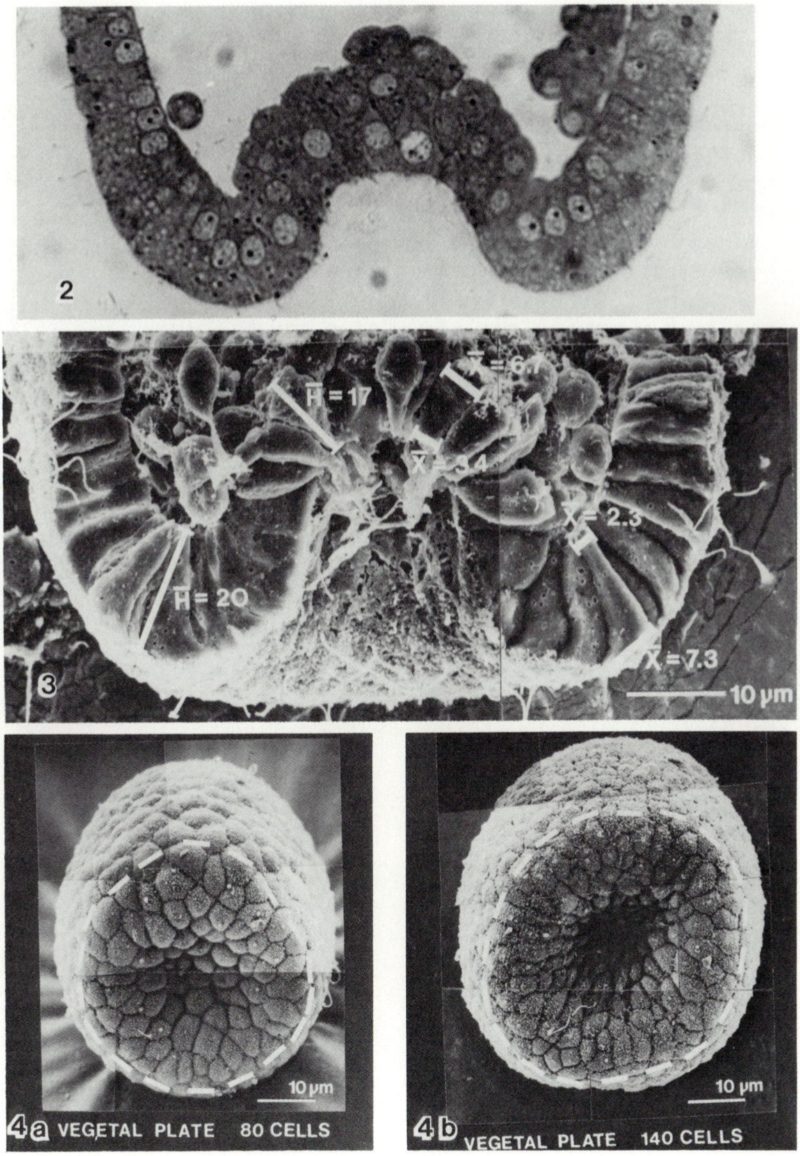

Figures 2-4. Vegetal plate region of early gastrula. Fig. 2. LM of one micron section. Fig. 3. SEM photomontage of dry fractured embryo. Fig. 4. SEMs of deciliated, dehyalinated embryos.

RESULTS AND DISCUSSION

During gastrulation the total volume of the blastocoel increases (Fig. 1a). The increase is small during primary invagination and then marked from the onset of secondary invagination (11.5 hr) until the archenteron is three fourths across the blastocoel. This increase in volume exerts a force on the ectodermal wall that peaks at the onset of secondary invagination and then decreases during later phases of gastrulation (Fig. 1b). Concurrent with the rise and fall of blastocoelic pressure there are differential changes in the heights of ectodermal cells (Fig. 1c) that indicate the relative force on the apical plate and vegetal regions is measureably less than on the lateral walls. These ectodermal cell height changes parallel the relative distribution of mitoses observed in Psammechinus miliaris gastrulae by Agrell (1953).

Primary invagination in L. variegatus involves an infolding of the presumptive endomeres (Figs. 2,3), active cell migration of small micromeres at the tip of the archenteron and their neighboring endomeres (Fig. 3) and proliferation of cells in the vegetal plate (Fig. 4). SEM of dry-fractured embryos (Fig. 3) and the exterior of dehyalinated embryos show the shapes and arrangements of cells and indicate the gastropore has bilateral symmetry.

REFERENCES

Agrell I (1953). A mitotic gradient in the sea urchin embryo during gastrulation. Arkiv Zool 6:213-217.

Amemiya S, Akasaka K, Terayama H (1982). Scanning electron microscopy of gastrulation in a sea urchin (Anthocidaris crassispina). J Embryol exp Morph 67:27-35.

Ettensohn CA (1984). Primary invagination of the vegetal plate during sea urchin gastrulation. Amer Zool 24:571-588.

Monné L, Hårde S (1951). On the formation of the blastocoel and similar embryonic cavities. Arkiv Zool 1:463-469.

Odell GM, Oster G, Alberch P, Burnside B (1981). The mechanical basis of morphogenesis. I. Epithelial folding and invagination. Develop Biol 85:446-462.

Progress in Developmental Biology, Part A, pages 101–104

MORPHOGENETIC ACTIVITY PREPATTERN IN EMBRYONIC EPITHELIA

D. SAVIC, B. N. BELINTZEV, L. V. BELOUSSOV, A. G. ZARAISKY

Inst. Mol. Biol. Acad. Sci. USSR, Biol. Faculty of State Univ. of Moscow, Inst. Bioorg. Chem. Acad. Sci. USSR

INTRODUCTION

It is well known that in embryonic epithelial tissues there is a morphological prepattern preceding the invagination (pattern) during gastrulation. This prepattern consists of strictly separated domains of morphologically polarized and unpolarized cells. In order to explain this we are proposing an alternative to Gierer and Meinhardt's "chemical" model (1972). Our "physical" model is based on three properties of cells and cell interactions: 1) each cell in a tissue can change its form spontaneously, the deformation being the "all or none" kind of transition (Petrov et al., 1984); 2) cell polarization is transmitted from cell to cell via contacts (contact cell polarization - CCP); 3) local cell polarization generates elastic tensions throughout the tissue which in turn stop the expansion of the polarization "front", thereby determining the proportion of polarized and unpolarized compartments, making it size-invariant (property of the embryo not explained by the G.-M. model).

Many experiments give proof for the existence of two types of forces in embryonic tissues: 1) "active" forces that use metabolic energy to change the shape of a cell (mechano-chemical reactions) and 2) "passive" elastic forces that arise as a whole tissue response to the active forces, the latter being 10-20 times faster.

Beloussov et al. (1975) showed that no CCP occurs in the presence of cyanids or cytochalazin B at 0°C but that there is an instantaneous relaxation of tensions in a tissue explant with a previously polarized domain. These deformations are stage and region specific indicating the existence of a characteristic topology of elastic tension fields that change only during a few key stages of development. This is obviously related to the posterior differentiation of the embryonic tissue (Beloussov, 1978, 1980).

FORMAL MODEL

These experimental data allowed us to construct the following formal model. The cytoskeletal movements and contact cell interactions (the local control of the CCP process) phenomenologically described. A dynamic parameter P is introduced whose value is the measure of cell polarization perpendicular to the sheet plane. The individual cell metabolic activity is simulated by a bistable function f(P):

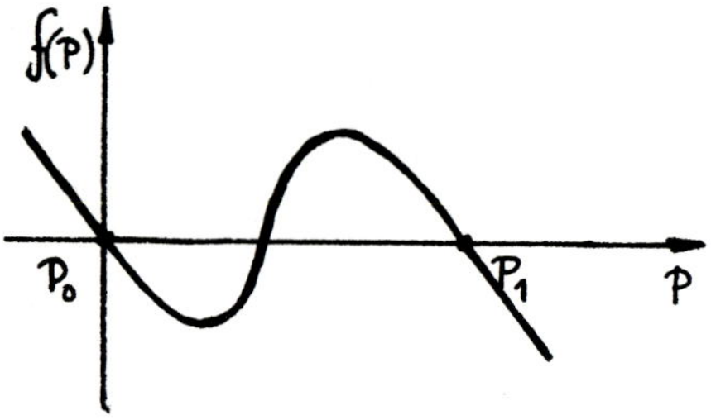

P_0 - morphologically isotropic state, P_1 - polarized state, both being STABLE STEADY states of a cell.

Contact polarization transmission can be approximated by a quasi diffusion term $D\Delta P$ (we are considering a smooth distribution of P in a cell ensemble).

Including the effect of elastic forces we come to the dynamic equation for CCP:

$$\frac{\partial P}{\partial t} = f(P) + D\Delta P - k\delta_{ii}$$

The term $k\delta_{ii}$ is the trace of elastic tension tensor (it is assumed that there is no torsion). The sign is minus because the tangential tensions inhibit perpendicular CP. The dependence of δ_{ii} on active forces (which are in turn linearly dependent on P) we determine from the condition of mechanical equilibrium:

$$\frac{\partial \sigma_{ik}}{\partial x_{ik}} + \frac{\varepsilon \partial p \delta_{ik}}{\partial x_{ik}} = 0$$

We may consider the system as mechanically equilibrated because the elastic reaction is very fast in comparison to active forces.

Phenomenological coefficients ε, k, D are to be determined experimentally.

There is no polarizing influence from the outside and the boundaries are fixed (the stability of the total volume is regulated by turgor). Including these conditions in the calculations we get the following behavior of the system (a one dimensional model is presented here):

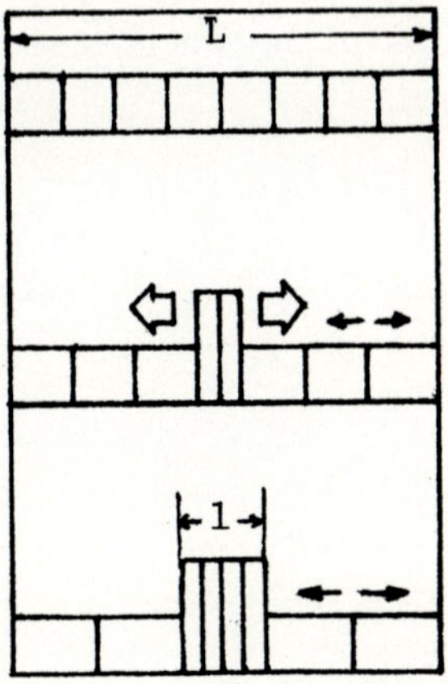

a) polarization = 0
strain = 0

externally induced local
cell polarization

b) polarization front expands
strain small

c) polarization front stopped
strain large

1/L = const: PATTERN SIZE-INVARIANCE

The three basic properties named above are sufficient to provide the selforganization of size-invariant stable prepattern (c) as a consequence of a small fluctuation (b) that took place in a homogeneous sheet of isotropic cells (a).

REFERENCES

Belintzev BN (1983). Dissipative structures and the problem of biological pattern formation. Uspehi fiz. nauk 141(1): 55

-101 (Rus).
Belintzev BN, Beloussov LV, Zaraisky AG (1985). The model of epithelial morphogenesis basing on the elastic forces and contact polarization of cells. Ontogenez 16(1): 5-13 (Rus).
Beloussov LV (1978). Formation and cell structure of tension lines in amphibian embryos. Ontogenez 9(2): 124-130 (Rus).
Beloussov LV (1980). The role of the tensile fields and contact cell polarization in the morphogenesis of amphibian axial rudiments. Roux's Arch Develop Biol 188: 1-7.
Beloussov LV, Dorfman JG, Cherdantsev VG (1975). Mechanical stresses and morphological patterns in amphibian embryos. J Embryol Exptl Morphol 34: 559-574.
Gierer A, Meinhardt H (1972). A theory of biological pattern formation. Kybernetik 12: 30-39.
Petrov KV, Beloussov LV (1984). The kinetics of contact cell polarization in the induced tissues of amphibian embryos. Ontogenez 15(6): 643-648 (Rus).
Savic D, Belintzev BN (1985). Mechanizm of space pattern in epithelial morphogenesis. In press.

Progress in Developmental Biology, Part A, pages 105–108

BIOCHEMICAL AND ULTRASTRUCTURAL STUDIES ON VITAMIN A INDUCED PROXIMALIZATION OF LIMB REGENERATION IN AXOLOTL.

Krishan K. Sharma and Hermann J. Anton

Department of Zoology, University of Rajasthan, Jaipur 302004, India, (K.K.S.), Zoologisches Institut Universität Köln, D5000 Köln 41, B.R.D., (H.J.A.).

INTRODUCTION

In regenerating amphibian limbs only the structures normally distal to the level of amputation are regenerated. However vitamin A and its derivatives (retinoids) have been found to cause serial reduplications in the proximodistal axis in amphibian limb regeneration (Niazi and Saxena, 1978; Sharma 1982; Maden, 1982 ; Thoms and Stocum, 1984). We confirmed this dramatic effect further, and found that regenerating limbs of larval urodeles (*Ambystoma mexicanum*, *Triturus alpestris*, *T. vulgaris*, *T. helveticus* and *Salamandra salamandra*) and anuran tadpoles (*Rana breviceps* and *R. temporaria*) exposed to retinol palmitate consistently exhibited proximodistal duplication of limb segments after certain fundamental alterations in the growing blastema. This work briefly describes some biochemical and ultrastructural changes in the forearm blastema of axolotl growing under the influence of retinol palmitate.

EFFECT OF VITAMIN A

In 35 mm size axolotl 15 days treatment with 75 IU/ml vitamin A palmitate after forearm amputation resulted in proximalization of regeneration. Studies were made on DNA amount, cell cycle, proteins and ultrastructure of cells in the untreated and vitamin A treated blastemas on 10th or 15th day after amputation.

DNA Amount and Blastema Cell Cycle: Measurement of the DNA amount in nuclei (by Feulgen cytophotometry) show difference in the distribution of DNA classes in the mesenchyme of untreated and vitamin A treated blastemas (Fig.1). Vitamin A palmitate decreased 15 % mean DNA content and reduced the number of nuclei in the S phase from 37.4 to 17.7 % in the blastemal mesenchyme on 10th day. It seems as if some mesenchyme cells in the vitamin A treated blastema were arrested in the G1 phase being unable to enter into synthetic phase (S) of cell cycle.

Blastema Proteins: We found that vitamin A interferes during the early events of histodifferentiation. To study this effect of vitamin A on protein pattern, soluble proteins of blastema mesenchyme, epidermis and adjacent stump (without skin) tissues were separated on 15% SDS polyacrylamide gels on 15th day after amputation. The amount of proteins of 11.5, 16.5, 41.6 and 78 Kd molecular weight in blastema epidermis, mesenchyme and adjacent stump decreased after 15 days treatment. In the mesenchyme of treated blastema 12.4 and 13.9 Kd bands were faint and 12 and 38 Kd bands were absent (Fig.2). Similar effects of vitamin A palmitate on blastema proteins have been found in the larvae of Salamandra salamandra (Sharma and Anton, unpublished).

Ultrastructure of Blastema Cells: On 15th day, cells of untreated blastema represent prochondral condensation, the first visible sign of cartilage redifferentiation. These differentiating prochondral cells have oval and dense nuclei with very few nuclear pores in the nuclear envelope. In some cases the plasma membranes are closely aligned one against another, however, desmosomes or tight junctional structures have not been seen. The cytoplasm contains numerous mitochondria having dense matrix crowded with rough endoplasmic reticulum (rER). Free ribosomes are scattered in groups (polysomes) throughout the cytoplasm. The orientation of cisternae profile of rER is usually parallel to the nuclear envelope. Extracellular matrix constitutes condensed ground substance and occasionally oriented collagen fibrils (Fig.3). The cell organelles of vitamin A treated blastemas on 15th day differ in several respects from those of untreated blastemas. The nuclear membrane is porus and the cytoplasm contains numerous free ribosomes and comparatively less rER. Flattend Golgi vesicles which are a rather constant feature of differentiated cells

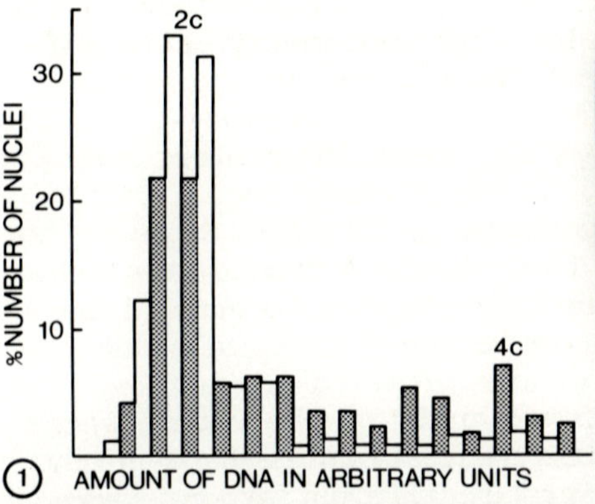

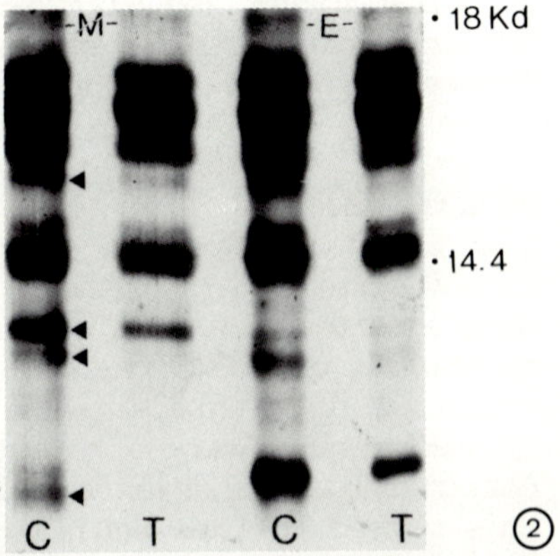

Figure 1. Distribution of DNA classes of nuclei in the mesenchyme of untreated (black staples) and vitamin A treated (open staples) blastemas of axolotl on 10th day. The estimated proportions of the G1, S and G2 phases are: in untreated 48.3% G1, 37.4% S and 14.3% G2. In treated 78.1% G1, 17.7% S and 4.2% G2.

Figure 2. Pattern of soluble proteins separated on 15% SDS-PAG. C = untreated; T = treated; E = epidermis; M = blastema mesenchyme. (Low molecular weight proteins are shown.)

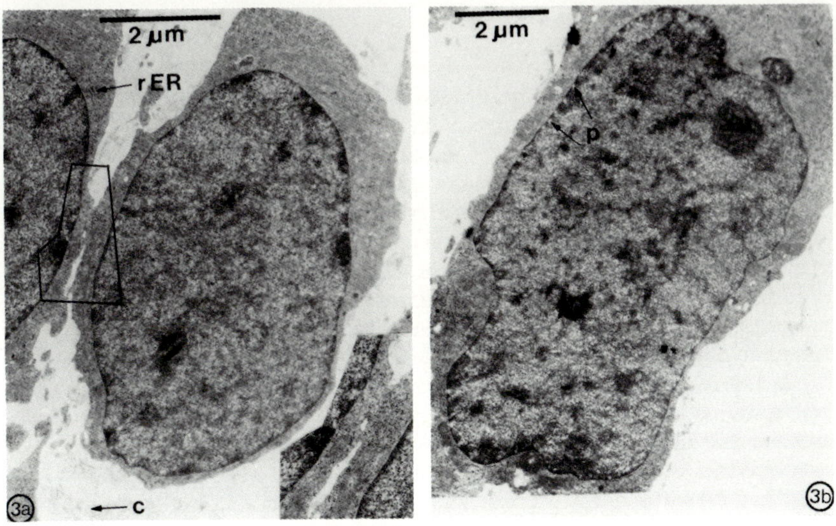

Figure 3. Electronmicrographs of untreated (a) and treated (b) blastema mesenchyme. p = nuclear pores, c = collagen fibrils, rER = rough endoplasmic reticulum.

(Schmidt, 1968), have not been seen in the mesenchymal cells of treated blastemas. Oriented collagen fibrils are almost absent in the extracellular matrix.

CONCLUSION

The results of vitamin A on proximalization are in contrast to its effect on other developing systems. Based on results of the present study we can demonstrate that vitamin A palmitate affects gene expression and alters synthetic machinery of blastemal cells of the regenerate. This resulted in delayed redifferentiation. Perhaps during this period (before beginning of redifferentiation) mesenchymal cells acquire greater morphogenetic potencies to produce serial reduplications of limb elements. Recent reports on the mode of action of vitamin A suggest that a cytoplasmic specific binding protein picks up the retinol and takes it to the cell nucleus where it binds to the chromatin for modulation of gene expression (Ganguly et al., 1980). We are currently investigating the mode of action of retinol palmitate on gene expression during proximalization of regeneration.

REFERENCES

Ganguly J, Rao MRS, Murthy SK and Sarda K (1980). Systemic mode of action of vitamin A. In Munson PL, Glover J, Diczfalusy E, Olson RE (eds): "Vitamin A and Hormones" Vol.38, New York: Academic Press, pp 1-55.

Maden M (1982) Vitamin A and pattern formation in the regenerating limb. Nature (London) 295:672-675.

Niazi IA, Saxena S (1978). Abnormal hindlimb regeneration in tadpoles of the toad, Bufo andersonii, exposed to excess vitamin A. Folia Biol (Krakow) 26:3-11.

Schmidt AJ (1968). "Cellular Biology of Vertebrate Regeneration and Repair". Chicago: The University of Chicago Press, pp 21-117.

Sharma KK (1982) Investigations on limb regeneration in tadpoles and froglets of the anuran Rana breviceps Schneider treated with vitamin A or electrically stimulated. Ph.D Thesis, Univ. of Rajasthan, Jaipur, India.

Thoms SD, Stocum DL (1984) Retinoic acid-induced pattern duplication in regenerating urodele limbs. Dev Biol 103:319-328.

Interactions in the Developing Nervous System

Progress in Developmental Biology, Part A, pages 111–113

DEVELOPMENTAL STUDIES WITH A SPECIFIC CELL SURFACE MARKER FOR LIMBIC SYSTEM NEURONS

Heather Horton and Pat Levitt

Department of Anatomy, Medical College of Pennsylvania, Philadelphia, Pennsylvania 19129

Specific surface markers distributed on related cells may serve as recognition cues to link developing neurons into functional systems. Such a marker should meet specific criteria to be considered as a system specific marker and therefore a candidate for a role in system assembly in brain. These criteria include the following: 1) The antigen should be regionally distributed, expressed only in well-defined areas of the brain; 2) the distribution should reflect both an anatomical and functional relationship among the neurons that carry the antigen; 3) if the molecule serves an an address label, it should be present at the cell surface; 4) a specific marker with a role in system assembly should exhibit a spatiotemporal pattern of expression that correlates with events such as synapse formation; 5) such a molecule should be conserved phylogenetically among homologous structures, particularly in a highly conserved system such as the limbic system. We have produced a monoclonal antibody against rat hippocampal membranes which specifically labels a membrane protein on neurons of the limbic system (Levitt, 1984). The distribution of the Limbic System Associated Membrane Protein (LAMP) meets the first three criteria noted above. In order to test for a functional role of LAMP in limbic system development, immunocytochemical mapping was performed to determine the correlation of the spatiotemporal expression of LAMP with specific developmental events such as axon growth and synapse formation. In addition, the effects of anti-LAMP antibody injection into newborn rats were evaluated to determine whether LAMP binding alters the organization of the mature limbic system.

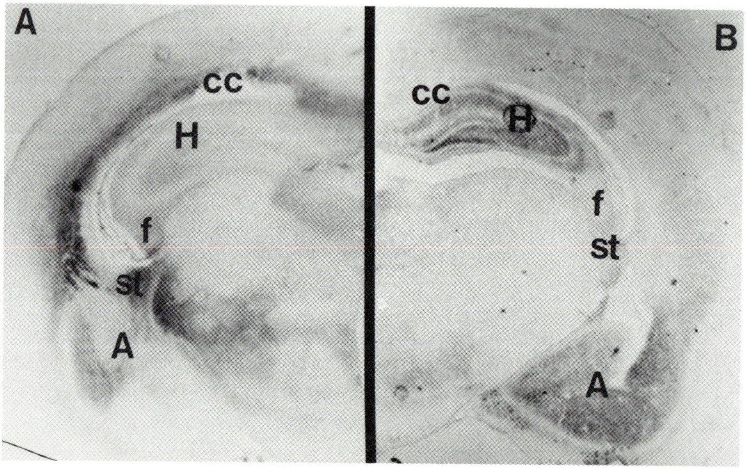

Figure 1. 10 um sections through the rostral hippocampi of Postnatal day 1 (A; x30) and 27 (B; x22) rats. At P1, staining is present in fiber tracts including the fornix (f), stria terminalis (st), and corpus callosum (c), which are no longer immunopositive by P 27. Staining of the hippocampus (H) and amygdala (A) at P 27 is much more pronounced than at P1.

Immunocytochemical mapping using the anti-LAMP-antibody was performed on postnatal Sprague-Dawley albino rats from birth through two months. Important differences in the distribution of LAMP between birth and adulthood were evident. In the adult rat, staining is restricted to cortical and subcortical gray matter of the limbic system. This pattern is evident at birth, although the hippocampus and other areas which are most immunopositive in the adult are only moderately stained in the newborn rat. In addition to immunoreactivity in the gray matter of the limbic system, staining in neonatal rats is also evident in limbic-related fiber tracts, including the fornix and stria terminalis. During development, the intensity of fiber tract staining decreases, and that of the limbic nuclei increases. By the end of the first week, the staining of the hippocampus and the fornix are of equal intensity. Staining of fiber tracts disappears completely by the end of the third postnatal week. The intensity of staining in limbic areas increases until the fourth week, at which point both the distribution and intensity of staining are identical to that of the adult. All areas that are not immunoreactive in the adult are never stained during development, indicating that only limbic associated structures express this unique surface protein.

During development, the data indicates that LAMP is present only in areas which interconnect to form the limbic system. The transient appearance of LAMP in fiber tracts may indicate that LAMP has a role in the formation of limbic system connections. During this important developmental period, LAMP is present both pre- and post-synaptically which would suggest the possibility of homophilic (LAMP-LAMP) binding as is seen with N-CAM (Hoffman, 1983).

The existing data indicate a role for LAMP in the development of the neural connections in the limbic system, functioning as one of many system-specific molecules guiding the development of the nervous system. To test this more directly, anti-LAMP antibody (40ug) was injected intracisternally into postnatal (P) day 1 and 3 rats. At one month, stereological analysis revealed an alteration in the hippocampus. A statistically significant 22% decrease in Ammon's Horn (containing pyramidal cells) was measured, whereas the dentate gyrus (containing granule cells) was unchanged. These results demonstrate the specificity of the treatment, because although both areas are immunoreactive in the adult, only Ammon's Horn cells contain LAMP at the time of injection. The isolation of LAMP and further functional tests in vitro will provide more direct evidence of the role that this system-specific molecule plays in development. This work was supported by March of Dimes Research Grant 1-919, a National Down Syndrome Award, and a Sloan Foundation Fellowship.

REFERENCES

Levitt P (1984). A monoclonal antibody to limbic system neurons. Science 223:299-301.

Hoffman S, Edelman GM (1983). Kinetics of homophilic binding by E and A forms of the neural cell adhesion molecule. Proc Nat Acad Sci 80:5762-5766.

Progress in Developmental Biology, Part A, pages 115–118

INDUCTION OF NEURAL CONNECTIONS IN F9 EMBRYONAL CARCINOMA CULTURES

Jorma Wartiovaara and Leena Rechardt

Departments of Electron Microscopy and Medical Biology, (J.W.), University of Helsinki, 00280 Helsinki, and Biomedical Sciences, (L.R.), University of Tampere, 33101 Tampere, Finland

INTRODUCTION

Early developmental events can to a varying degree be mimicked in vitro by embryonal carcinoma (EC) cells. Endodermal differentiation is the first developmental change in the inner cell mass cells which in the mouse embryo are the normal counterpart of the EC cells. This differentiation stage has been one of the most studied ones with the teratocarcinoma model (cf. Hogan et al., 1983). Accumulating evidence indicate that many EC cell lines including F9 cells have the capacity to differentiate also into neural direction (cf. Wartiovaara et al., 1984a; Wartiovaara and Rechardt, 1985). This differentiation, in contrast to endodermal differentiation, seems to require a more prolonged culture in the presence of retinoic acid and cyclic AMP and seems to be enhanced by the addition of nerve growth factor (NGF; Liesi et al., 1983). Under these conditions the F9 cells seem capable of differentiating into adrenergic type of neurons without appearance of glial type cells.

By using the above conditions we have investigated on the ultrastructural level the development of neural characteristics and especially the formation of intercellular connections in the differentiating F9 cells.

RESULTS

When F9 cells are cultivated in the presence of retinoic acid and dibutyryl cyclic AMP for longer periods of time the embryonal carcinoma cells differentiate into multipolar neural type cells which form extensive networks through cell processes (Fig. 1; Wartiovaara et al., 1984a, b). In the presence of NGF such cells show immunoreactivity to tyrosine hydroxylase and are capable of L-dopa uptake (Liesi et al., 1983). In thin sections the cell processes contain in addition to longitudinally arranged microtubules and mitochondria also synaptic like vesicles in large numbers (Fig. 2; Wartiovaara et al., 1984c). The localization of specific acetylcholin esterase to such neural like cells has also been established (Rechardt et al., 1984).

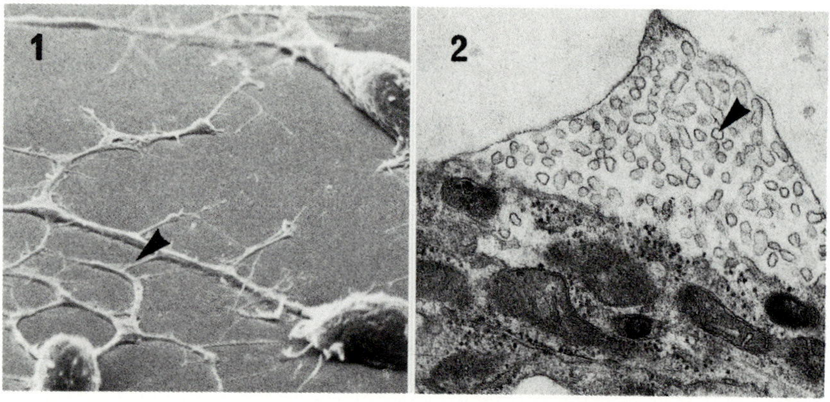

Fig. 1. Scanning electron micrograph of F9 embryonal carcinoma cells cultivated for 14 days with 10^{-7} M retinoic acid and 10^{-3} M dibutyryl cyclic AMP. Note formation of numerous contacts between cell processes from different neural type cells. x 3000.

Fig. 2. F9 cell process from a culture as in Fig. 1. Note numerous synaptic like vesicles. x 60 000.

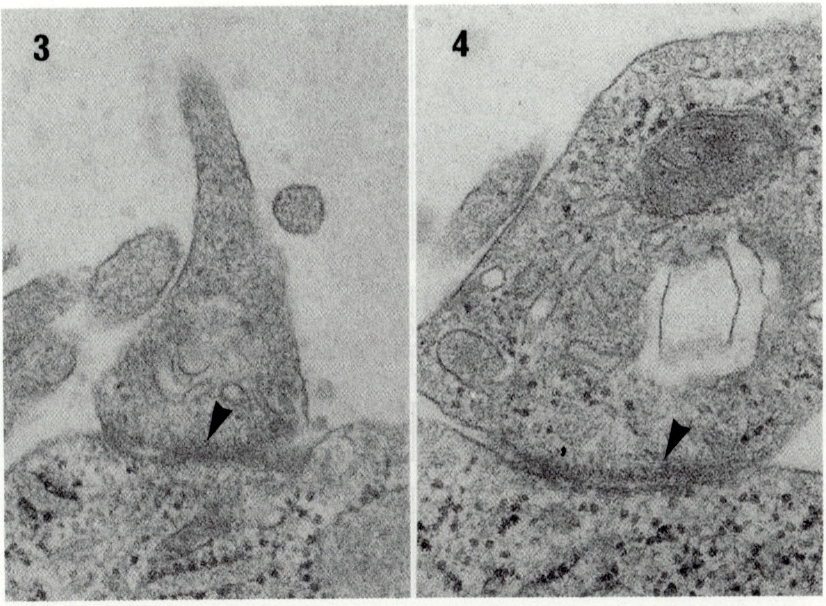

Figs. 3 & 4. F9 cells cultured for 18 days as in Fig. 1. Note specialized membrane junctions between cell process and other cell. x 75 000.

In the cultures numerous contact sites between cells and processes were seen and often specialized membrane junctions had developed (Figs. 3 & 4). Nexus type structures could also be found (cf. also Wartiovaara et al., 1984c). To what extent synaptic type contacts develop between cells in the cultures is currently under study.

The results suggest that the differentiation of F9 cells into neuronal cells with the formation of specialized cell contacts could probably be employed for the study of synaptogenesis and the formation of exitable neuronal networks in vitro.

ACKNOWLEDGEMENTS

This work was supported by grants from the Sigrid Jusélius Foundation and the Finnish National Research Council for Medical Sciences.

REFERENCES

Hogan BLM, Barlow DP, Tilly R (1983). F9 teratocarcinoma cell as a model for the differentiation of parietal and visceral endoderm in the mouse embryo. Cancer Surveys 2:115-140.

Liesi P, Rechardt L, Wartiovaara J (1983). Nerve growth factor induces adrenergic neuronal differentiation in F9 teratocarcinoma cells. Nature 306:265-267.

Rechardt L, Lehtinen S, Wartiovaara J (1984). Scanning electron microscopy and x-ray microprobe analysis in detection of acetylcholinesterase in cultured embryonal carcinoma cells. J Histochem Cytochem 32:1154-1158.

Wartiovaara J, Liesi P, Hervonen H, Rechardt L (1984a). Neurodifferentiation in cultures of F9 teratocarcinoma cells. In Duprat AM, Kato AC, Weber M (eds): "The Role of Cell Interactions in Early Neurogenesis", New York: Plenum Press, p 145-154.

Wartiovaara J, Liesi P, Rechardt L (1984b). Expression of laminin and fibronectin in endodermal and neural differentiation of F9 embryonal carcinoma cells. In Hinchliffe R, Kemp R (eds): "Matrices and Cell Differentiation", New York: Alan R Liss, p 233-247.

Wartiovaara J, Liesi P, Rechardt L (1984c). Neuronal differentiation in F9 embryonal carcinoma cells. Cell Differ 15:125-128.

Wartiovaara J, Rechardt L (1985). Neural differentiation in embryonal carcinoma cells. In Lash JW, Saxén L (eds): "Developmental Mechanisms: Normal and Abnormal", New York: Alan R Liss, p 3-13.

Progress in Developmental Biology, Part A, pages 119–122

SELECTED OPTIC LOBE EXTRACTS STIMULATING RETINAL NEURITE OUTGROWTH

Nestor G. Carri[†] and Ted Ebendal[*]

[†]IMBICE, OC 403, 1900 La Plata, Argentina,
[*]Department of Zoology, Uppsala University,
Box 561, S-751 22 Uppsala, Sweden

INTRODUCTION

Interactions between target areas and their innervating neurons play an important role during development. The hypothesis that the target stimulates growth of invading neurites by releasing neuronotrophic factors is attractive but little is known concerning such a mechanism in e.g. the visual system. In order to study trophic factors in the CNS we have developed a short-term bioassay measuring neurite outgrowth from chicken embryonic neural retina (Carri and Ebendal, 1981a, 1981b). Our previous experiments demonstrated that macromolecules extractable from brain areas support the growth of retinal neurites in culture (Carri and Ebendal, 1983a) and that this stimulation was even greater in target areas not receiving their normal innervation (Carri and Ebendal, 1983b). We also demonstrated that brain extracts exert their trophic stimulation in a dose-response manner (Ebendal *et al.*, 1985). Current work is focused on stimulative effect of selected regions of the optic lobe on explanted retina. In particular, we wanted to test whether the target for a certain set of growing visual neurites exerted a specific influence on the retinal neurons normally projecting to this part of the tectum. The results obtained so far are summarized here.

MATERIALS AND METHODS

Explants of neural retina from chicken embryos (White Leghorn, stages 28-29) were cultured on a collagen matrix substratum prepared according to Elsdale and Bard (1972) covered by Eagle's Basal Medium with 10% fetal calf serum and 10% horse serum (Fig. 1). They were incubated at 37.5^{o}C, 5% CO_2 and 80% relative humidity. After four days, maximum neurite length was measured using an inverted microscope equipped with phase contrast optics (x 125 final magnific-ation). The medium was supplemented with tissue extracts prepared from the anterior and posterior parts of the ED 18 optic lobe (the middle part being discarded to avoid possible overlap). The extract was prepared by homogenizing 1 g wet weight of tissue in 5 ml of Eagle's basal medium. The 100 000 x g supernatant was dialyzed against the culture medium, sterilized by filtration and tested at a series of dilutions ranging from 100 to 0.16 mg of tissue per ml.

Figure 1. Diagram showing the experimental design. (a) Head of a 6-day-old chicken embryo. Two types of explants were taken from the neural retina of the right eye: Nasal (N) and temporal (T). (b) Dorsal and (c) lateral aspect of the brain of an 18-day-old chicken embryo. The left optic lobe shows the two extracted parts (shadowed): the anterior (A) and posterior (P). Tissue extracts prepared from A and B were tested on nasal and temporal retina. All experiments were performed as blind test.

Figure 2. Dark field micrograph showing outgrowth from a retinal explant after 2 days in culture. In this experiment nasal retina was stimulated by posterior optic lobe extract (12.5 mg tissue per ml of medium).
Calibration bar 0.3 mm.

1

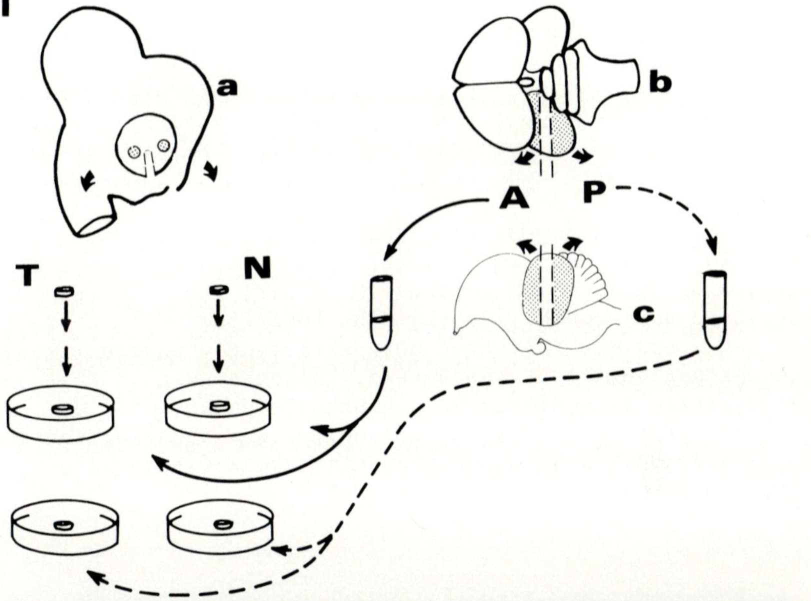

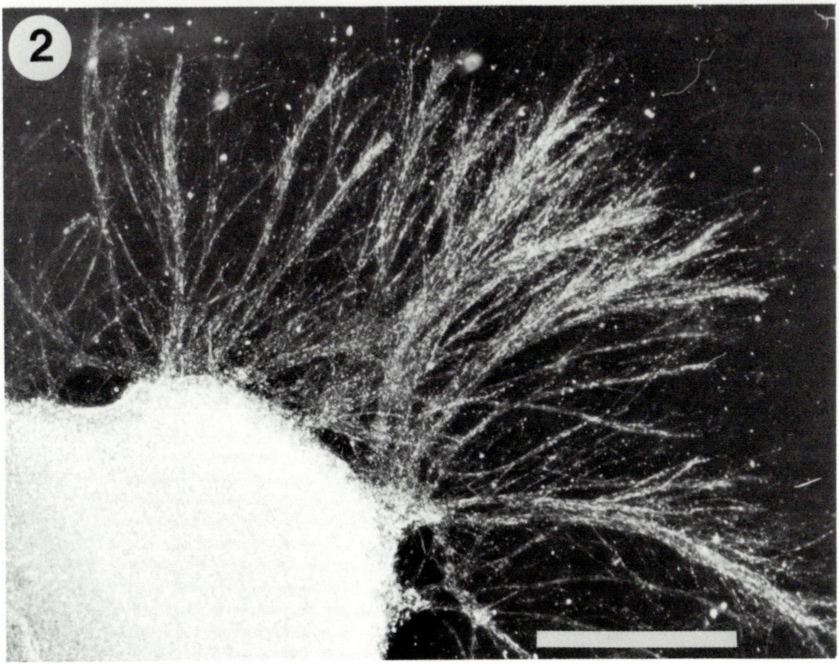

RESULTS AND CONCLUSION

The optic lobe supernatants improved neurite outgrowth from the retina in a dose-response fashion. Nasal retina was stimulated by both extracts. The best results were obtained with the posterior optic lobe. Temporal retina was also stimulated by both extracts but in this case best results were obtained at high concentration of the anterior optic lobe.

Our results demonstrate that extractable macromolecular material from the normal target areas can stimulate the growth of innervating neurons. Moreover, the results indicate the possibility that the topographically correct target areas may influence more efficiently neurons from the proper region of the retina. The data suggest that the stimulation is mediated by a neuronotrophic factor or factors present in the developing optic tectum but the molecular mechanism underlying the observed responses remains to be studied.

REFERENCES

Carri NG, Ebendal T (1981a). Optic lobe extract stimulating neurite extension in embryonic retinal explants *in vitro*. Soc Neurosci Abstr 7:547.

Carri NG, Ebendal T (1981b). Optic lobe extract stimulating retinal outgrowth in organotypic culture. J Cell Biol 91:83a.

Carri NG, Ebendal T (1983a). Organotypic cultures of neural retina: Neurite outgrowth stimulated by brain extracts. Develop Brain Res 6:219-229.

Carri NG, Ebendal T (1983b). Optic lobe from anophthalmic chick embryos stimulates retinal neurite outgrowth in organotypic cultures. Fidia Res Series 1:158-159.

Ebendal T, Norrgren G, Carri NG, Belew M (1985). Nerve growth factors in peripheral tissue in CNS. In Sir John Eccles and Dimitrijevic MR (eds.): "Recent Achievements in Restorative Neurology", Vol 1, Basel: Karger (in press).

Elsdale T, Bard J (1972). Collagen substrata for studies on cell behavior. J Cell Biol 54:626-637.

Progress in Developmental Biology, Part A, pages 123–126

NEURONAL DEVELOPMENT IN *DROSOPHILA* EMBRYONIC CULTURES

Nicole Bournias-Vardiabasis, Paul S. Salvaterra and Tina P. Nair
Division of Cytogenetics, City of Hope Medical Center (N.B.V) and Neurosciences (P.M.S., T.P.N.), City of Hope Beckman Research Institute, Duarte, CA 91010.

INTRODUCTION

Early gastrula stage embryonic cells of *Drosophila* can be cultured under conditions where they form differentiated neurons (Seecof et al 1971, 1973; Shields et al 1975). These cultures offer a unique biological preparation to study neurogenesis. In the past Dewhurst and Seecof (1975) have shown evidence that the enzymes involved in acetylcholine (Ach) metabolism in *Drosophila* are expressed *in vitro*. Here we report on the normal temporal expression of choline acetyltransferase (ChAT), acetylcholinesterase (AchE) and [^{125}I] α-bungarotoxin binding in *Drosophila* embryonic cultures along with the expression of a neuron-specific cell surface marker recognized by horseradish peroxidase (HRP).

RESULTS

Figure 1 shows results for the expression of the Ach metabolizing enzymes and [^{125}I] α-bungarotoxin binding sites under normal conditions. The *in vitro* expression of ChAT and AchE is identical to that seen *in vivo*. The appearance of ChAT, the biosynthetic enzyme for Ach production, is first detected in 5 hr old cells. There is an initial phase of low level expression followed by a rapid rise in ChAT activity, shortly after the clusters of neurons make contact with each other. ChAT activity reaches a plateau level in 36 to 48 hr old cells. AchE activity can be detected several hours before ChAT and also shows a period of low level expression followed by a rapid increase and plateaus around 36-48 hr. [^{125}I] α-bungarotoxin binding appears in about 4 hr old cells and approaches maximum levels by about 36 hr. AchE has been also localized histochemically to the neurons and their processes. Only approximately 50% of neuron clusters stain while there is no staining of myotubes, fat and chitin secreting cells (at 48 hr).

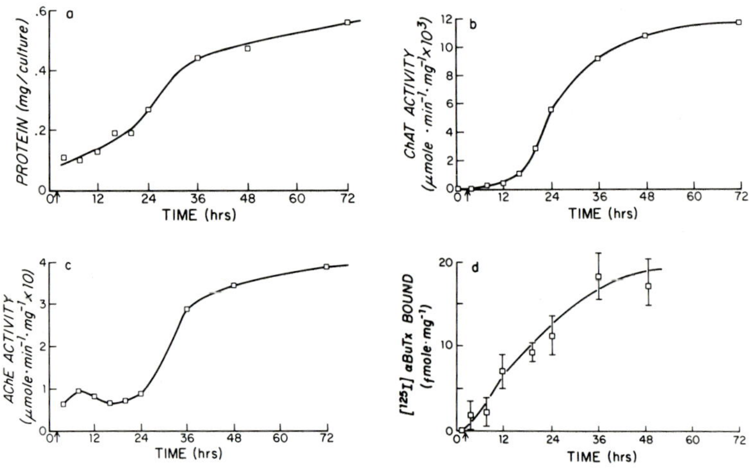

Figure 1. Normal in vitro expression of a) protein, b) ChAT activity, c) AchE activity and d) [^{125}I] α-bungarotoxin binding in Drosophila primary embryo cell cultures. Eggs were collected for 1-2 hr and put into culture at 3.5 hr (arrows). The standard error of mean is indicated for toxin binding measurements. All other parameters have SEM less than the height of the symbols.

The normal in vitro expression pattern for ChAT and AchE can be altered by adding various cholinergic drugs to the culture medium during cell differentiation, d-Tubocurarine, scopolamine, carbachol and BW284C51 (a specific AchE inhibitor) significantly inhibit normal ChAT and AchE expression levels (Figure 2). Medium conditioned by 7 day old cultures can also result in lower levels of ChAT and AchE. These results strongly suggest that neurotransmitter metabolism can be influenced by environmental conditions.

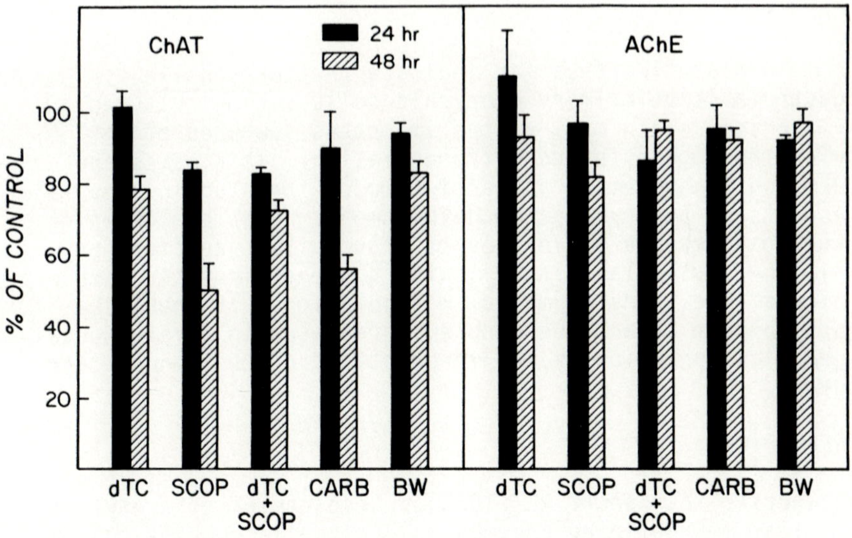

Figure 2. Time dependence of decreased enzyme expression following exposure to cholinergic drugs. All drugs were added to fresh cultures following a 30 min cell attachment period at a concentration of 1X10E-5M except CABB (1X10E-4M). Values represent the mean of triplicate assays + the SEM. The drugs are: dTC = d-tubocurarine, CARB = carbamyl choline, SCOP = scopolamine, BW = BW284c51.

Recently Jan and Jan (1982) have described the appearance of an insect neuron-specific cell surface marker which can be recognized by antibodies raised against the soluble enzyme horseradish peroxidase (HRP). We have studied the appearance of this marker *in vitro* as a function of the developmental age of *Drosophila* embryonic cells. Some positive cells can be detected as early as 5 hr, when no overt morphological differentiation is apparent. As the cells differentiate, the staining pattern is limited to the small neuronal clusters and their axons. These results indicate that the specific marker recognized by anti-HRP antibodies is expressed in gastrula stage cells and its function might be to serve as a marker for undifferentiated cells destined to become neurons.

Our studies thus far indicate that *Drosophila* gastrula stage primary embryonic cells can be cultured *in vitro* and retain many of the important features of *in vitro* neurogenesis and differentiation. This makes the *Drosophila* system an attractive model for studying several aspects of nervous system development. Also by combining the considerable advantages of *Drosophila* (genetics, physiology, molecular biology) with those of *in vitro* techniques (i.e., environmental manipulations) it should be possible to attack the problems presented in developmental genetics, principally the interactions of environment and genes.

REFERENCES

Dewhurst, SA, Seecof RL (1975). Development of acetylcholine metabolizing enzymes in *Drosophila* embryos and in cultures of *Drosophila* cells. Comp Biochem Physiol 50C:53-58.

Jan LY, Jan YN (1982). Antibodies to horseradish peroxidase as specific neuronal markers in *Drosophila* and grasshopper embryos. Proc Natl Acad Sci USA 79:2700-2704.

Seecof RL, Alleaune N, Teplitz RL, Gerson I (1971). Differentiation of neurons and myocytes in cell cultures made from *Drosophila* gastrulae. Exp Cell Res 69:161-173.

Seecof RL, Donardy JJ, Teplitz RL (1973). Differentiation of *Drosophila* neuroblasts to form ganglion-like clusters of neurons *in vitro*. Cell Differ 2:143-149.

Schields G, Dubendorfer A, Sang JH (1975). Differentiation *in vitro* of larval cell types from early embryonic cells of *Drosophila melanogaster*. J Embryol Exp Morph 33:159-175.

Progress in Developmental Biology, Part A, pages 127–130

GLIOGENESIS: THE IMPORTANCE OF APPROPRIATE CELL ASSOCIATIONS

Paul J. Linser

C. V. Whitney Laboratory and the Department of Anatomy of the University of Florida, St. Augustine, Florida 32086

Previously, it was shown that the mature chicken optic tectum (OT) possesses at least three distinct classes of neuroglia: S-100 + oligodendroglia; GFA/glutamine synthetase (GS) + fibrous astroglia; GS/carbonic anhydrase-C (CA-C) + protoplasmic astroglia (Linser, 1985a). To study the intrinsic capacity of embryonic OT cells to differentiate into these classes of glia, we first examined the normal developmental pattern of glial maturation. Immunohistochemical techniques as well as a sensitive dot-radioimmunoassay were employed to localize and quantitate the glial antigens described above. In addition, myelin basic protein (MBP) was also analyzed as a secondary marker for oligodendroglial development. Briefly, we found that all markers showed marked developmental patterns of expression. GS was first detected in ependymal cells and their vertical processes on day 9 of embryonic development. None of the other glial markers were detectable at this time. Antibodies to MBP began to decorate axon fascicles by day 12 and by day 16 the cell bodies of oligodendroglia could be discerned by immunolocalization of S-100. Also on day 16, GFA was first detected in processes of typical fibrous astrocytes. Immunostaining for CA-C in the GS-containing protoplasmic astroglia was not noted until just before hatching at 21 days of development. Immunostaining for all of the antigens increased gradually following the time of their first appearance. The immuno-histochemical findings were mimicked by quantitative analyses.

To examine the role of cell interactions in glial

phenotype maturation, the levels and localizations of the panel of glial markers was examined in two tissue culture systems as a function of extrapolated developmental time in vitro. The first culture system was rotation mediated suspension culture of dissociated day 7 embryonic OT cells in which the OT cells aggregated forming 3 dimensional clusters of cells. The second culturing method involved plating day 7 OT cells in adherent monolayers.

In both culturing systems utilized, no evidence of oligodendroglial development was noted even following culturing for periods of time in excess of 3 weeks (i.e. well beyond the normal time of oligodendroglial development in vivo). Appreciable levels of S-100 and MBP could not be detected in aggregates or monolayers either by RIA or immunohistochemistry. Furthermore, there was little if any evidence of glial-specific accumulation of CA-C in these cultures. However, development of GS and/or GFA positive astrocytes was marked in both culturing systems. In aggregates, GS expression, measured either by RIA or by activity assay, paralleled the normal developmental pattern and even exceeded tissue levels of GS. Furthermore, GFA was detectable in aggregates earlier in development than in the intact tissue. Again, as with GS, the levels of GFA achieved by aggregates was significantly higher than that of the intact tissue of equivalent developmental age (Fig. 1). Double-label immunohistochemical analyses showed that most GS-positive cells were also positive for GFA. However, significant numbers of cells which labelled for only one or the other of the two astrocyte markers were also encountered.

When embryonic OT cells are plated in adherent monolayer on glass or tissue culture plastic, the cells initially form numerous small clumps. Over a period of 3-4 days, the clumps attach and flat cells migrate out of the clumps and proliferate to form an epithelial-like sheet. Round cells then extend axon-like processes over the surface of the flat cell layer (Adler et al, 1979). Thus, such cultures eventually consist of small multicellular aggregates, interconnected networks of neuron-like cells and flat cells both juxtaposed to neuron-like cells and in comparative isolation from the neuronal cells. Quantitative analyses showed that the levels of GFA in such cultures was higher than equivalent age whole tissue at all stages examined although somewhat lower than in the aggregate cultures. GS levels however were very low relative to both whole tissue and aggregates (Fig. 1).

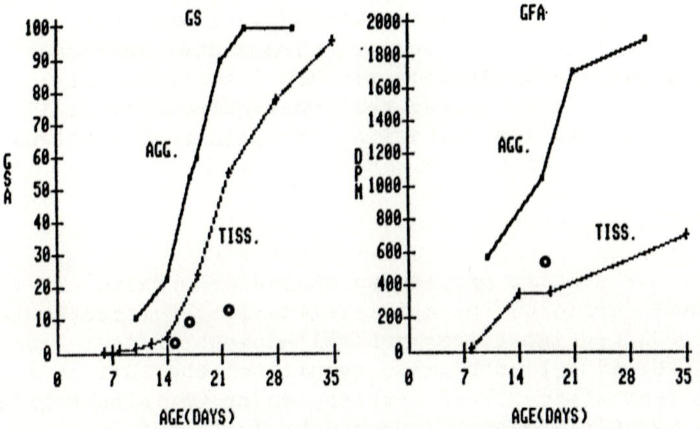

Figure 1. Levels of GS and GFA in tissue (tiss.), aggregate (agg.) and monolayer cultures (**O**) as a function of developmental age. GS enzyme activity (GSA) and the relative amount of GFA (DPMs via RIA) were normalized against total protein.

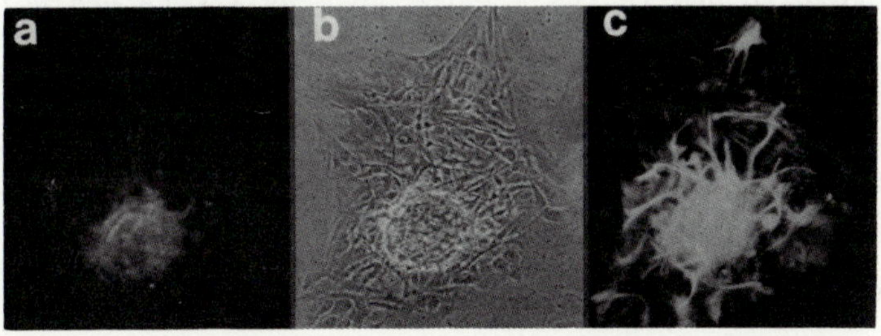

Figure 2. Double-label immunoflourescence localization of GS (a) and GFA (c) in OT-7 cells following 10 days in culture. Note variably intense yet generalized staining for GFA in flat cells (c) but presence of GS (a) only in the flat cells immediately subjacent to the cluster of neurons seen in the lower-center (b).

Examination of adherent "monolayer" cultures by immunohistochemistry revealed that most of the flat cells could be specifically immunostained with antibodies to GFA. The intensity of labelling for GFA varied over a broad range. It was generally true that flat cells in apparent contact with neurons or neuronal processes or within small aggregates labelled more intensely than those in relative isolation from neurons. Frequent exceptions to this were noted however. GS was also localized in flat cells in these cultures. However, in contrast to GFA, GS was only detectable in flat cells that were immediately adjacent to neuronal cells, cell clusters or axon-like processes. Double-label analyses showed that most GS-positive cells were also GFA positive but again, exceptions were noted. (Fig. 2).

DISCUSSION

Previous studies have shown that GS expression in the Muller glial cells of the chick retina is regulated by contact-mediated input from retinal neurons (Linser and Moscona, 79; Linser, 85b). The results of the current study indicate that similarly the expression of GS in astroglial cells of the OT is also influenced by juxtapositional relationships between the glia and neuronal cells. Furthermore, expression of GFA in OT astrocytes seems to be enhanced by proximity of the glia to neurons, although the stringency of the neuronal-glial relationship in this case does not appear to be as extreme as with GS expression. These results add to the growing body of literature which indicates that many phenotypic characteristics of neuroglial cells are subject to regulatory influences from neuronal and neuroblast cells. In addition, these results also raise several questions concerning the failure of oligodendroglial development in vitro and with regard to the lack of glial CA-C expression in cultures of embryonic OT cells.

REFERENCES

1. Adler, R., Manthorpe, M., and Varon, S. (1979) Separation of neuronal and nonneuronal cells in monolayer cultures from chick embryo optic lobe. Develop. Biol. 69: 424-435.
2. Linser, P. J. and Moscona, A. A. (1979) Induction of glutamine synthetase in embryonic neural retina: localization in Muller fibers and dependance on cell interactions. Proc. Natl. Acad. Sci. USA 76: 6476-6480.
3. Linser, P. J. (1985a) Multiple marker analysis in the avian visual cortex: three biochemically distinct classes of neuroglia and carbonic anhydrase-containing neurons. J. Neuroscience (in press).
4. Linser, P. J. (1985b). Lectin and antibody mediated interference with retinoblast differentiation in vitro. Inv. Ophthal. (Supplement) 26: 335.

Progress in Developmental Biology, Part A, pages 131–133

ASPARAGINE-LINKED GLYCOPROTEINS ON THE CELL SURFACE AS RECEPTORS RESPONSIVE TO THE NEURAL-INDUCING STIMULUS

Kiyoko Yamazaki-Yamamoto[1], Kenzo Takata[2] and Noriko Takahashi[3]
[1]Research Institute of Environmental Medicine, [2]Radioisotope Center, Nagoya University, Nagoya 464 and [3]Department of Biochemistry, Nagoya City University Medical School, Nagoya 467, Japan

In spite of intensive study using embryological and biochemical procedures, the molecular basis of the neural induction still remains ambiguous. To approach this problem, we have employed lectins as the probes due to reasons that their modes of action have been well elucidated biochemically and some of them have the neural-inducing ability (Takata et al., 1981).

The presumptive ectoderm of early gastrulae of the Japanese newt, *Cynops pyrrhogaster*, differentiated into the neural tissue by Con A or UEA I (*Ulex europaeus* agglutinin I) treatment (200-300 g/ml, 3 hr, 20° C), whereas it developed into the atypic epidermis when cultured alone in Steinberg's balanced salt solution. The autoradiograms of the presumptive ectoderm incubated with ^{3}H-Con A revealed that Con A molecules were much more bound to the interior surface than the exterior surface of the ectoderm, and a part of the bound Con A to the cell surface was successively incorporated into the cytoplasm and the nucleus. By using Con A-Sepharose beads, on the other hand, it was found that Con A molecules immobilized on Sepharose preserved the neural-inducing ability. From these results, we concluded that Con A works primarily in the plasma membrane but not in the cytoplasm. Our results agree well with those of Tiedemann and Born (1978) who used the neural-inducing factor isolated from chick embryos. The primary importance of plasma membrane has been also pointed out by Duprat's group (1982) who employed lectins in combination with the organizer in the induction experiments. The neural-inducing effect of Con A was also demonstrated in *Xenopus* presumptive ectoderm (Grunz, 1985).

50% of Con A bound to the ectoderm explant could be removed by washing explants with 2 mM mannoside and this treatment gave rise to failure of neural induction. This implies that retention of a certain amount of Con A in the cell surface for a given period such as 6 hr is prerequisite for evoking the neural induction. Some membrane mechanism thus might be involved in the induction machinery, i.e., Con A-binding to certain membrane constituents, perhaps glycoproteins, would activate the machinery.

Electron microscopic examination demonstrated tremendous architectural changes of the cell surface to which lectins, such as Con A and DBA (Dolichos biflorus agglutinin) for examples, were bound (Yamazaki-Yamamoto et al., 1981). Recently, scanning electron microscopic examination by using colloidal gold-conjugated Con A confirmed that the receptors for Con A molecules were mainly localyzed on the interior (blastocoelic) surface of the presumptive ectoderm consisting of one-cell layer. This is in good agreement with the results of autoradiograms of the ectoderm explant labelled with ^{3}H-Con A. Electron micrographs of the ectoderm treated with ferritin-conjugated DBA, on the other hand, revealed that the DBA-binding sites were abundant in the exterior surface but quite sparse in the interior. The observed mode of distribution of the respective lectin in the explant is of interest in view of the facts that the inductive interaction takes place between the organizer and the interior surface of the ectoderm, and that DBA is one of lectins which are unable to evoke the neural induction.

Almond glycopeptidase is a specific enzyme which can remove the mannose-type oligosaccharides intact from the asparagine residue of peptide chain. The presumptive ectoderm became unresponsive to Con A when pre-treated by almond glycopeptidase (5 mU/ml, 4 hr, 26° C) and differentiated into the atypic epidermis (Takata et al., 1984), suggesting the involvement of glycoproteins in the neural competence. We confirmed, in fact, that the competent period of the ectoderm to Con A was same as that to the organizer. The oligosaccharide fractions obtained from the competent ectoderm by the glycopeptidase were fractionated through Bio-Gel P-4. The oligosaccharides from the cell surface were entirely different from those from the cytoplasm. Most recently, by means of a high performance liquid chromatography we have obtained an indication of existence of unique high mannose-type oligosaccharides in the cell surface.

Results so far obtained led us to the conclusion that a certain glycoprotein with the particular type of oligosaccharide(s) in the interior surface of the competent ectoderm plays a primary role in responding to the neural-inducing signal and its appearance on the cell surface may give rise to the genesis of the neural competence. Con A-competent ectoderm system would be a promising experimental model for analyzing the molecular basis of the neural induction.

REFERENCES

Duprat AM, Gualandris L, Rouge P (1981). Neural induction and the structure of the target cell surface. J Embryol Exp Morphol 70:171-187.

Grunz H (1985). Effect of Concanavalin A and vegetalizing factor on the outer and inner ectoderm layers of early astrulae of Xenopus laevis after treatment with Cytochalasin B. Cell Differentiation 16:83-92.

Takata K, Yamazaki-Yamamoto K, Ozawa R (1981). Use of lectins as probes for analyzing embryonic induction. Wilhelm Roux's Archives 190:92-96.

Takata K, Yamazaki-Yamamoto K, Ishii I, Takahashi N (1984). Glycoproteins responsive to the neural-inducing effect of Concanavalin A in Cynops presumptive ectoderm. Cell Differentiation 14:25-31.

Tiedemann H, Born J (1978). Biological activity of vegetalizing and neuralizing inducing factors after binding to BAC-cellulose and CNBr-Sepharose. Wilhelm Roux's Archives 184:285-299.

Yamazaki-Yamamoto K, Ozawa R, Takata K, Kitoh J (1981). Cell surface changes of the presumptive ectoderm following neural-inducing treatment by Concanavalin A. Wilhelm Roux's Archives 190:313-319.

Progress in Developmental Biology, Part A, pages 135–138

EMBRYONIC INDUCTION IN AMPHIBIANS

Horst Grunz and Lothar Tacke
Department of Zoophysiology,FB 9(Biologie),
University GHS of Essen, 4300 Essen 1, FRG

There exist sufficient experimental data on amphibian development, which support the view that the determination of early embryonic cells is triggered by various morphogenetic factors protein in nature. During cleavage the mesodermal marginal zone is established by the interaction of vegetal and animal blastomeres (Nieuwkoop,1969).During gastrulation the chordamesoderm stimulates the overlaying neuroectoderm to differentiate into the central nervous system. Because of the complexity of the processes, responsible for embryonic induction, determination and differentiation, especially the events at the molecular level are still a matter of speculation. On the other hand it must be pointed out that our knowledge about those events has substantially increased by recent results of different laboratories. The vegetalizing factor (a protein with a molecular weight of about 13 000 daltons,Tiedemann,1982) induces in competent ectoderm endodermal and mesodermal derivatives (Grunz, 1983). Similar factor(s) could be responsible for the determination of the mesodermal zone of the embryo during cleavage. While the vegetalizing factor apparently must be internalized, neuralizing factor(s) are thought to interact with specific receptors on the plasmamembrane of the target cells (neuroectoderm).By recent experiments a factor with neural inducing activity could be isolated from early neurulae (Tiedemann, 1984). The mechanism of transport of inducing factor between inducing chordamesoderm and reacting neuroectoderm is still unclear. The migration (Saxen, 1961) of the factor(s) could take place in the zones of close cell-to-cell contact (Grunz and Staubach,1979). Preliminary results together with Dr.Tacke in my laboratory indicate that the tranfer of inducing factors depends on close cell-to-cell contact between inducing tissue and target cells. Further electronmicroscopic analysis are in progress to decide this question. There is good evidence that neuralizing factors interact with specific receptors on the plasmamembrane(Born et al., 1980). The view that receptors on the plasmamembrane are part of the chain in neural induction is supported by experiments with the plant lectine Concanavalin A. Con A stimulates competent ectoderm to differentiate into neural derivatives (Takata et al., 1981, Grunz, 1984,1985). Con A does not loose its neural inducing activity after immobilization by binding to Sepharose beads (Takata et al., 1984).These results indicate that Con A really must interact with receptors (α-D-mannoside or α-D-glucoside-residues) on the plasmamembrane and its biological activity does not depend on its internalization. By extensive experiments with gold-Con A together with Dr.Tacke in my laboratory we can show that the inducing activity depends on the amount of Con A bound to the surface of the ectodermal cells. These results, which also inform about the fate of the gold-Con A-receptor-complex(patching, receptor-mediated endocytosis and degradation in lysosomes) will be published in detail elsewhere. While informations are available that neuralizing factors are interacting with specific receptors on the plasmamembrane, the further chain of events within the target cells leading to neuralization are obscure. However, so far we can exclude ions and cyclic nucleotides as factors initiating neural induction (Grunz and Tiedemann, 1977; Grunz, 1985; Siegel et al.,1985). Treatment of ectoderm of Triturus alpestris and Xenopus laevis with cyclic nucleotides did not cause neuralization. Also the incubation of ectoderm with calciumionophore A 23187 does not stimulate Xenopus ectoderm to form neural tissue.

On the other hand our data indicate that the differentiation pattern can be changed by simultaneous treatment of ectoderm with calciumionophore and inducing factors (Grunz,1985). Pattern formation apparently depends on secondary cell interactions (cited by Tiedemann,1982;Grunz,1979; Minuth and Grunz, 1980). By the increase of free intracellular calcium after treatment of ectoderm with A 23187 the intercellular communication via gap junctions and in turn normal pattern formation could be inhibited (Grunz, 1985).However it is unlikely that the primary signals of neuralizing and vegetalizing factors are transmitted via gap junctions, because the relative big size of such factors does not allow their passage through gap junctions. Warner et al.,1984 reported that injected antibodies against gap junction protein severely disturbs cell communication. Also the "opposite" effect to the disruption of cell communication, i.e. cell fusion, could give further valuable informations about cell differentiation and gen regulation. We have started experiments together with Dr.Schweiger and Dr. Koop,Ladenburg/Heidelberg to follow the fate of cells of different germ layers after electrofusion (Fig.1-3).

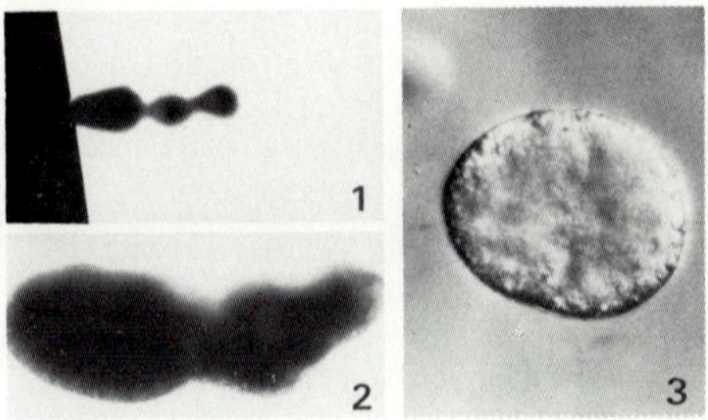

Electrofusion of amphibian cells.Fig.1 shows 3 ectodermal cells of the early gastrula immediately after fusion.The same cells after 10 min.(Fig.2).Fig.3 Hybridcell formed by fusion of two ectodermal cells.

REFERENCES

Born J, Grunz H, Tiedemann H, Tiedemann H (1980) Biological activity of the vegetalizing factor: Decrease after coupling to polysaccharide matrix and enzymatic recovery of active factor. Wilhelm Roux's Archives 189: 47-56.

Grunz H (1983) Change in the differentiation pattern of Xenopus laevis ectoderm by variation of the incubation time and concentration of vegetalizing factor. Wilhelm Roux's Archives 192: 130-137.

Grunz H (1984) Early embryonic induction: The ectodermal target cells.In Duprat A M, Kato M, Weber M (eds): "The role of cell interactions in early neurogenesis", Plenum Press, Ser.A, Life Sciences, Vol.77, pp. 21-38

Grunz H (1985) Effect of Concanavalin A and vegetalizing factor on the outer and inner ectoderm layers of early gastrulae of Xenopus laevis after treatment with Cytochalasin B. Cell Differentiation 16: 83-92.

Grunz H (1985) Information transfer during embryonic induction in amphibians (submitted,JEEM)

Grunz H, Staubach J (1979) Cell contacts between the chordamesoderm and the overlying neuroectoderm (presumptive central nervous system) during the period of primary embryonic induction in amphibians. Differentiation 14: 59-65.

Minuth M, Grunz H (1980) The formation of mesodermal derivatives after induction with vegetalizing factor depends on secondary cell interactions. Cell Differentiation 9: 229-238.

Nieuwkoop P (1969) The formation of mesoderm in urodelean amphibians. I. Induction by the endoderm. Wilhelm Roux's Archives 162: 341-373.

Saxen L (1961) Transfilter neural induction of amphibian ectoderm. Dev. Biol. 3: 140-152.

Siegel G, Grunz H, Grundmann U, Tiedemann H, Tiedemann H (1985) Embryonic induction and cation concentrations in amphibian embryos (in press, Cell Differentiation)

Takata K, Yamamoto KY, Ozawa R (1981) Use of lectins as probes for analyzing embryonic induction. Wilhelm Roux's Archives 190: 92-96.

Takata K, Yamamoto KY, Takahashi N (1984) A molecular aspect of neural induction in CYNOPS presumptive ectoderm treated with lectins.In Duprat A M, Kato M, Weber M (eds): "The role of cell interactions in early neurogenesis", Plenum Press, Ser.A, Life Sciences Vol.77, 83-88

Tiedemann H (1982) Signals of cell determination in embryogenesis.In Jaenicke L (ed): "Biochemistry of Differentiation and Morphogenesis", Springer Verlag, Berlin, Heidelberg, pp 275-287.

Tiedemann H (1984) Neural embryonic induction. In Duprat A M, Kato M, Weber M (eds): "The role of cell interactions in early neurogenesis", Plenum Press, Life Sciences, Vol.77, 89-105.

Warner AE, Guthrie SC, Gilula NB (1984) Antibodies to gap-junctional protein selectively disrupt junctional communication in the early amphibian embryo. Nature 311: 127-131.

Progress in Developmental Biology, Part A, pages 139–141

TROPHIC AND REGENERATIVE EFFECTS OF NEURAL TUBE IMPLANTS IN THE CHICK LIMB

Betty F. Sisken and Ira Fowler
Wenner Gren Lab and Department of Anatomy
University of Kentucky, Lexington, KY 40506

INTRODUCTION

Limb regeneration in chick embryos does not occur after removal of the apical ectodermal ridge. We have reported, however, that implantation of neural tube (6 segments) from 2 day embryos into the amputated bud of the 4 day chick embryo induces regeneration of middle and distal segments in 30% of the embryos (Fowler and Sisken, 1982). Substitution of somites, notochord or heart from 2-7 day embryos for the neural tube does not induce regrowth. The responses obtained in any group are designated as : Type I (no regeneration), Type II (tissue regeneration) and Type III (limb regeneration).

The effects of the amputation and implantation were assessed nine days later. Morphometric analyses of the following structures were performed on histological sections of control and experimental embryos; the peripheral load (defined as the area of the limb skeleton), dorsal root ganglia, numbers and area of nerve fibers in the proximal segment, area of the neural implant and numbers of lateral motor column neurons. Except for the area of the implant, all measurements were made on both amputated and unamputated sides of each embryo to normalize variation between embryos. Data from these measurements (ratios of amputated: unamputated) were analyzed and subjected to a Dunnett's many to one t statistic at the p=.05 level of significance.

RESULTS

The peripheral load area (% A/U) in Type I and III embryos were significantly greater than that of control embryos. In Type I embryos, this response was represented exclusively as an increase in growth of the humerus (Sisken et al, 1985). The area of the dorsal root ganglia was increased significantly only in Type III embryos and this response correlated positively with the peripheral load. Although no differences were found in the numbers of nerve fibers in the distal portion of the proximal segment in control and experimental embryos, the area occupied by peripheral nerve fibers/area of amputated limb was significantly increased in Type I and III embryos (Sisken et al, 1984).

The total area of neural implant in Type I response embryos was compared to that of Type III embryos, and the cellular composition of the implants were identified. Regeneration-inducing implants are larger and contain relatively more larger size neurons; in addition, the neuropil of these implants is more mature. Finally, the numbers of neurons in the lateral motor column (%A/U) were found to be unaffected by amputation in the presence of the neural implant. Whereas the ratio in control embryos dropped to 68%, the ratios in Types I and III remained close to 100% (Sisken et al, 1985). It appears that the neural implants have a dual function: the stimulation of new limb segment growth (Type III) or humeral growth (Type I), and the preservation of host motor neurons from amputation-induced cell death.

The regenerative response obtained with implanted neural tube supports Singer's hypothesis (Singer, 1952) that successful regeneration can be induced by augmentation of the nerve supply to the limb. The induction of growth of new segments found in Type III response embryos and the preservation of motor neurons may be the result of neurotrophic factors produced by the neural implant. An anatomical route for the transmission of molecules from the implant to the host CNS has been detected in embryos sacrificed 3 days post-implantation; similar routes (nerve fibers) have been found emanating from the implant to the limb musculature. Both routes may play important roles in producing the observed responses. Supported by NIH19798-02.

REFERENCES

Fowler, I. and Sisken, B.F. (1982) Effect of augmentation of nerve supply upon limb regeneration in the chick embryo. J. Exp. Zool. 221:49-59.

Singer, M. (1952). The influence of the nerve in regeneration of the amphibian extremity. Quat. rev. Biol. 27: 169-200.

Sisken, B.F., Fowler, I., Barr, E.J. and Kryscio, R.J. (1984). The threshold quantity of nerve required to induce limb regeneration in the chick embryo. J. Neurosc. Res. 12: 623-632.

Sisken, B.F., Fowler, I., Matson, G., Thudium, P. and Barr, E. (1985). Responses of host motor and sensory neurons to a neural tube implant in amputated chick limbs. Dev. Br. Res. 19:245- 252.

Progress in Developmental Biology, Part A, pages 143–146

CELL CONTACTS BETWEEN INDUCING TISSUE AND TARGETS CELLS DURING NEURAL INDUCTION

Lothar Tacke and Horst Grunz
Department of Zoophysiology,FB 9(Biologie),
University GHS of Essen, 4300 Essen 1, FRG

It is still unclear, if cell contacts, the transfer of free diffusible inducing factors or both between inducing chordamesoderm and reacting ectoderm are responsible for the induction of the central nervous system during early development of amphibian embryos. For all these possibilities there exist experimental indications. Although the results of the in vitro transfilter experiments of Toivonen and Saxen indicate the free diffusion of neuralizing factor(s), the necessity of cell contacts between inducing and reacting tissue in normogenesis can not be excluded.
Therefore we checked the interface between inducing cells of the upper blastopore lip and the reacting cells of competent ectoderm in the electron microscope, to analyze the formation of cell contacts and morphological features, characteristic for this interface. In the sandwich-method we combined upper blastopore lip from early gastrulae of _Xenopus laevis_ or _Triturus vulgaris_ with two animal caps of _X. laevis_. We could observe cell contacts between ectodermal cells and the cells of the inner layer of the upper blastopore lip.

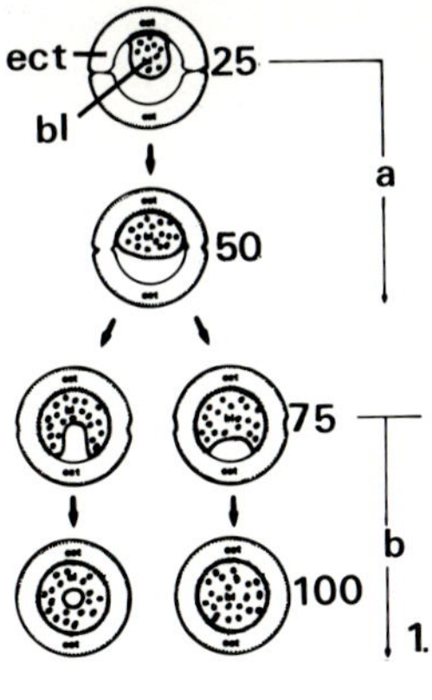

Fig.1: Percentage of numbers of cells from the upper blastopore lip (bl), which get in contact to the ectodermal cells of the sandwich-halves (ect). Cells of the superficial cell layer of the upper blastopore lip, which can be recognized by their pigment granules in the cell periphery, show no contact to the ectodermal cells. A lumen remains within the

sandwich between the inner layers of one of the ectodermal caps and the superficial layer of the upper blastopore lip. When at least 75% of the inner surface cells of the inducing tissue have formed contacts to the surrounding ectodermal cells, the further progression of cell contacts takes place in two different ways: 1.Since the pigmented cells of the superficial layer of the blastopore lip have not yet established cell contacts to the ectoderm cells a lumen still persists. 2.The superficial pigmented layer of blastopore lip undergoes cell surface changes followed by their contact to ectoderm cells. In this case no lumen persists within the cell mass (a=differentiation into atypical epidermis;b=75% of cells in close contact results in differentiation into neural structures).

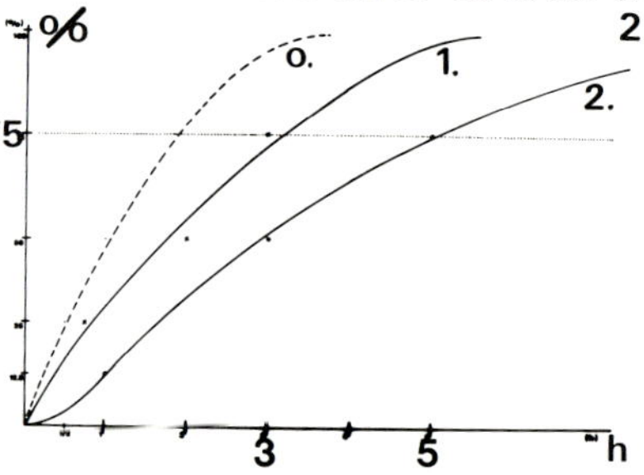

Fig.2.Percentage of contacted cells of upper blastopore lip (see fig.1) in relation to the time of culture: 0. A piece of ectoderm (X. laevis) was combined with two sandwich-halves. 1. Upper blastopore lip of X.laevis takes up contact more rapidly than 2. upper blastoporal lip of T.vulgaris. In the control series (explants cultured for up to 5 days) a neuralizing effect can be observed after 3h in 'Xenopus-sandwiches'(1.), and at least after 5h in 'Triturus-sandwiches'(2.).This is obviously correlated in both cases to the same percentage of cells in close contact (75%).In early stages of contact formation filopodia and blebs are formed by the ectodermal target cells. Both types of cytoplasmatic protrusions bridge the space between inducing and reacting cells and can be easily distinguished not only by their different shape but also by their cytoplasmatic components. Within the filopodia there can be observed mitochondria, lipid droplets, yolk platelets etc.. On the other hand there are only ribosomes, glycogen and endoplasmatic reticulum within the blebs (Fig.3).

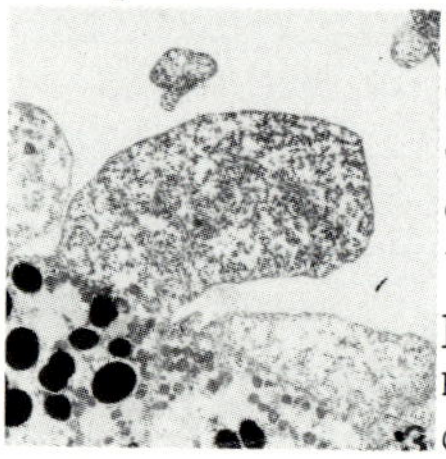

Similar blebs have been described in Cynops-ectoderm (Komazaki,1982), Ambystoma-ectoderm (Stanisstreet and Smith, 1978) and even in the chicken (Fabian and Eyal-Giladi, 1981). Their function remains unclear. They possibly play a role during morphogenetic movements (Komazaki,1982) and/or early cellular differentiation (Fabian and Eyal-

Giladi,1981). The filaments and vesicles within the peduncle of the blebs are similar to those known from telophase in cell division. Similar to those they may possibly removed from the ectoderm cells by shedding. In any case the ectodermal blebs and filopodia are good features to distinguish between reacting and inducing tissue in the early steps of cell-to-cell contact. In a later stage, when great parts of cell surfaces are in close contact with each other, it is quite difficult to recognize the ectodermal-mesodermal interface. So we used the xenoplastic combination of X.laevis-ectoderm and upper blastopore lip of T.vulgaris. The cells of both species can be distinguished by their different pigment granules. It is likely that neuralizing factors interact with specific receptors on the membranes of the target cells (Born et al., 1980). Furthermore the experiments with gold-labelled Con A (the lectine has neural inducing activity), show that the Con A-receptor-complex is internalized by coated pits (Tacke and Grunz,1986). Those coated pits could be observed in small amounts at the inner cell surfaces of the superficial ectoderm layer and on cell membranes of the inner layers of the inducing tissue. These pits could neither be found in the membranes of blebs and filopodia nor at the outer cell surfaces of the superficial ectoderm layer. The highest amount of coated pits could be observed at cell membranes of the inner ectoderm layer, i.e. in those target cells,which come in direct contact to the inducing chordamesoderm in normogenesis. On the other hand there exist conspicious exocytotic vesicles, which often appear accumulated on surfaces of inducing cells and also sometimes on ectodermal cells

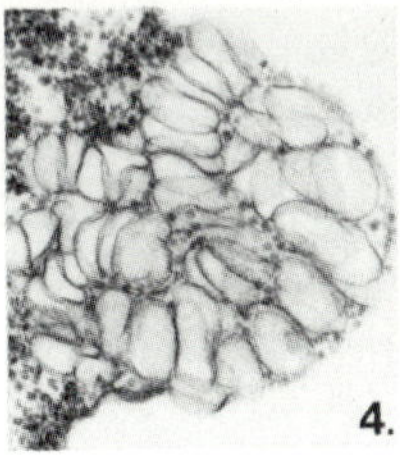

4.

(Fig.4). It needs further investigation, to find out, whether these vesicles take part in the primary steps of neural induction. On the other hand they may only represent structures of rapid membrane growth, as Bluemink et al., 1983 assumed for similar vesicles in neuroblastoma cells.

Telophase bridges with mid-bodies have been described in mammalian secondary spermatocytes, in cnidoblasts of Hydra (Fawcett,1961) and in chick blastoderm (Buck et al.,1975). Over these intercellular bridges cell communication is possible. Similar to chick blastoderm the bridges found in Xenopus ectoderm connect two daughter cells only, which may be situated in a relative big distance from each other (Fig.5).

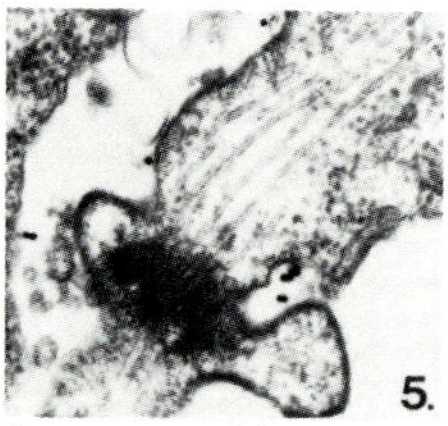
5.

Fig.5 Via those bridges cell communication couldtake place during secondary cell interactions, prerequiste processes for specific pattern formation.

Our results support the view that a close cell-to-cell contact is necessary for neural induction in amphibians. Furthermore there are indications that a certain number of inducing cells must come into contact with a certain number of ectodermal target cells to evoke neural induction.

REFERENCES

Bluemink J G, van Maurik P A M, Tertoolen L G J, van der Saag P T, de Laat S W (1983) Ultrastructural aspects of rapid plasma membrane growth in mitotic neuroblastoma cells. Eur.J.Cell Biol. 32: 7-16.

Born J, Grunz H, Tiedemann H, Tiedemann H (1980) Biological activity of the vegetalizing factor: Decrease after coupling to polysaccharide matrix and enzymatic recovery of active factor. Wilhelm Roux's Archives 189: 47-56.

Buck R C, Ohara P T, Daniels W H (1975) Intercellular bridges of chick blastoderm studied by scanning and transmission electron microscopy. Experientia 32: 505-507.

Fabian B, Eyal-Giladi H (1981) A SEM study of cell shedding during the formation of the area pellucida in the chick embryo. J.Embryol.exp.Morph. 64: 11-22.

Fawcett D W (1961) Intercellular bridges. Exp.Cell Res. suppl.8: 175-187

Komazaki S (1982) Morphological changes of the inner surface of the blastocoelic wall before and during gastrulation on the newt, Cynops pyrrhogaster. Develop.Growth and Differ. 24: 491-499.

Stanisstreet M, Smith J L (1978) Scanning electron microscopy of cells isolated from amphibian early embryos. J.Embryol.exp.Morph. 48: 215-223.

Tacke L, Grunz H (1986) Electron microscope study of the binding of Con A-gold to superficial and inner ectoderm layers of Xenopus laevis in correlation to the neural inducing activity of this lectine. Wilhelm Roux's Archives (submitted)

Sex Determination

Progress in Developmental Biology, Part A, pages 149–158

ON THE STUDY AND CONTROL OF SEX DIFFERENTIATION

Stephen Wachtel[1,2], Dean Nakamura[1] and W.A. King[3]

[1]Center for Reproductive Biology, Spring Creek Ranch, 871 Ridgeway Loop Road, Memphis, TN 38119; [2]University of Tennessee Center for the Health Sciences, Memphis, TN 38163; and [3]Université de Montréal, St. Hyacinthe, Québec, Canada J2S 7C6

INTRODUCTION

Inasmuch as cells come into contact with one another at the cell surface, it is reasonable to assume that components of the cell surface are involved in the cell-cell interactions that occur during the various processes of embryogenesis and development. The manipulations of immunogenetics are thus especially useful in the study of embryogenesis, because they deal with cell surface antigens.

In general, differentiation is signalled by the appearance of novel cell surface antigens, called *differentiation alloantigens*; and because they are identified by specific antibodies, differentiation alloantigens are amenable to serological analysis. Consider the T cells. The population of T cells consists of the thymocytes and the thymus-derived T-lymphocytes, each of which is characterized by expression of particular differentiation alloantigens. Of the antigens denoted Thyl, TL, Lytl, Lyt23, and Lyt5, for example, only the last (Lyt5) is found on the prothymocyte, the immediate precursor of the thymocyte. TL is found on the thymocyte but not in the other T cells, and Lytl and Lyt23 are differentially expressed in the T cell subsets according to function: Lytl is present in helper lymphocytes but not in cytotoxic suppressor (CS) lymphocytes, and Lyt23 is present in the CS lymphocytes but not in helper lymphocytes (Table 1). In contrast Lyt5 is present in all T cells but it is not found in other cells (Boyse and Cantor, 1978).

So the surface phenotype of the prothymocyte is TL^- $Thy1^-$ $Lyt1^-$ $Lyt23^-$ $Lyt5^+$, but when the prothymocyte is induced to become a thymocyte, the surface phenotype changes to TL^+ $Thy1^+$ $Lyt1^+$ $Lyt23^+$ $Lyt5^+$; and later differentiative steps involving appearance of the T cell subsets are signalled by still other surface phenotypes.

INDUCERS

Differentiation of the thymocyte from the prothymocyte is induced by the thymic hormone *thymopoietin*. But appearance of the thymocyte antigens can also be induced by cyclic AMP, bacterial endotoxin, ubiquitin and other agents. Evidently the prothymocyte is committed to a particular pathway of differentiation -- before it assumes residence in the thymus. It follows that induction of the thymocyte is permissive; the program leading to acquisition of the thymocyte phenotype need only be triggered by any of several agents, specific or nonspecific.

INDUCTION OF THE GONAD

In contrast, induction of sex differentiation among non-germinal elements of the primitive gonad seems to be *instructive*: given the proper signal, XX gonadal cells organize a testis. This is nowhere more apparent than in the unambiguous male gonads of certain XX/XY chimeric male mice (McLaren, 1976). Evidently the XX cells are induced to organize testicular structures by an agent that is produced by XY cells (Müller et al, 1982).

Development of the testis is signalled by appearance of the Sertoli cells; these organize the seminiferous tubules, thereby encasing the primordial germ cells, and excluding the mesenchymal elements from which the Leydig cells arise. Evidently Sertoli cells and the granulosa cells of the ovary share a common lineage. Their morphology is similar; they are rich in microsomal aromatase; they act as "nurse cells" for their respective germ cells; and they express a common cell surface antigen that is not found in other cells of the gonad (Table 1) (Ohno, 1979).

TABLE 1. Presumptive Surface Phenotype of Lymphocytes, Erythrocytes, Epidermal and Gonadal Cells of the Mouse

	TL	Thyl	Lytl	Lyt23	Lyt5	Skn1	Skn2	H-2	H-Y	H-Y rec	SG
Prothymocyte	−	−	−	−	+	−	−	+	+	−	−
Thymocyte	+	+	+	+	+	−	−	+	+	−	−
Helper T Cell	−	±	+	−	+	−	−	+	+	−	−
CS T Cell	−	±	−	+	+	−	−	+	+	−	−
Erythrocyte	−	−	−	−	−	−	−	+	−	−	−
Epidermal Cell	−	+	−	−	−	+	+	+	+	−	−
Sertoli Cell	−	−	−	−	−	−	−	+	+	+	+
Granulosa Cell	−	−	−	−	−	−	−	+	−	+ (?)	+

TL, the thymus-leukemia antigen, is found in some mice but not others. Thyl is found in T-cells and skin and is thus part of at least two differentiative programs (Boyse and Cantor, 1978). Skn1 and Skn2 are selectively expressed in the neuro-ectodermal derivatives: skin and brain. H-2 and H-Y are not properly categorized as differentiation antigens because they are expressed ubiquitously. The Sertoli-granulosa-cell lineage antigen (SG) and H-Y receptor are gonad-specific markers; see text.

MORPHOGENESIS AND THE CELL SURFACE

When separated from one another, differentiated cells of a primitive or developing organ or tissue retain the potential for reaggregation and reconstitution of structures characteristic of the organ or tissue from which they were derived. A classic example involves the marine sponge *Microciona prolifera*. Wilson (1908) cut the tissues of the sponge into small pieces and dispersed the individual cells by passing the tissues through a sieve; when they came into contact with one another on glass slides, the cells organized functional miniature sponges.

Similar observations have been reported in the warm-blooded vertebrates. Moscona (1961) obtained single cell suspensions from embryos of the chicken and mouse, and placed them in rotating flasks, to promote aggregation. The cells aggregated with respect to tissue -- not species. When limb-bud mesoblasts from chick and mouse were co-cultured, together with cells from chick liver, the aggregates contained well-organized areas of chimeric mouse-chick cartilage with cells from each species interspersed; the third-party liver cells were not incorporated, and organized themselves separately.

H-Y ANTIGEN AS INDUCER

The mammalian indifferent embryonic gonad arises as a stratification of the coelomic epithelium near the mesonephric kidney. After the primordial germ cells complete their migration from the yolk entoderm, the Sertoli cells appear. As noted above the Sertoli cells form the testicular tubules and encase the germ cells. These events mark the first stages of testicular organogenesis. It may be inferred that the inducer of testicular organogenesis is phylogenetically conservative; that it is associated with the cell surface; that it may be released from the cell surface; and that it is involved in differentiation of the Sertoli cells. The inducer of male secondary sex differentiation (androgen), operates by engagement of a specific receptor; by analogy it may be supposed that the inducer of male gonadal development also engages a specific receptor.

Sera from male-grafted female mice contain antibodies that identify H-Y antigen. Those antibodies recognize H-Y or a crossreactive molecule in the heterogametic (XY) sex in species representing each of the major vertebrate classes. On the basis of widespread phylogenetic persistence, it was proposed that H-Y is the inducer of the heterogametic gonad, the testis in mammals and the ovary in birds (Wachtel, 1983).

Phylogenetically conservative H-Y is an excellent candidate for primary sex determinant. By definition, the molecule is a component of the cell surface; there are reasons for believing that it is released by Sertoli cells or their precursors; there is evidence that the soluble moiety engages a gonad-specific receptor.

Apparently H-Y depends for its residence in the plasma membrane on association with cell surface components of the major histocompatibility complex (the HLA system of man; H-2 of the mouse; see Table 1). Cultured Daudi cells, which lack HLA, secrete H-Y into the medium. When exposed to Daudi-secreted H-Y, cells of the presumptive bovine ovary organize a testis in whole organ culture (Ohno et al, 1979). Dispersed cells of the perinatal rat ovary likewise organize testicular structures in rotatory culture, in the presence of soluble H-Y from the rat testis; and that effect is blocked by the addition of monoclonal H-Y antibody (Müller and Urban, 1981).

Monoclonal H-Y antibody binds soluble H-Y from cultured TM-4 cells, of presumptive Sertoli cell origin (Brunner et al, 1984); the reaction is obviated by absorption of the antibody with male cells, but not female cells.

As for the H-Y receptor, Nakamura et al (1984) found that soluble H-Y of the mouse testis is bound in gonadal cells of the male chicken (ZZ), but not in cells of the blood, bursa, kidney, liver or spleen. Similar results were obtained by Müller et al (1978) in the rat. In that study, soluble H-Y was bound in cells of the XX ovary, but not in cells of the somatic tissues. The implication is that reaction of H-Y and its receptor is a common event in organogenesis of the heterogametic gonad, which may be testis or ovary, depending on the species.

H-Y NEGATIVE MALES?

There are reports that challenge the presumptive sex-determining role of H-Y antigen. Recently McLaren et al (1984) described male development in sex-reversed XX male mice ostensibly lacking H-Y antigen. The mice were descendents of a single female carrying Sxr (X^{Sxr}, a Y-X crossover) and the T(X;16)16H X-autosome translocation (T16H), which is preferentially activated in this system. Of eleven T16H/X^{Sxr} females studied, ten were H-Y^+; only one "consistently proved to be negative for H-Y" when tested by the method of cell-mediated cytotoxicity. Inasmuch as the H-Y^- phenotype was heritable in males, the authors proposed that an alteration had occurred within the Sxr segment, whereby male determination had been separated from expression of the H-Y transplantation antigen. The finding has important implications for the study of sex determination, for occurrence of an H-Y negative male would certainly defeat the proposition that H-Y is an inducer of male gonadal differentiation. But the situation warrants scrutiny, for the following reasons.

First, cell-mediated cytotoxicity of H-Y incompatible target cells is notorious for dependency on H-2 haplotype of effector and target cells (Hurme et al, 1977) and there are numerous instances (Simpson and Gordon, 1977; von Boehmer et al, 1977; Hurme et al, 1978) in which H-Y^+ cells are not recognized by cytotoxic T cells for lack of proper association of H-Y and H-2K/D. Mutation within H-2 could accordingly preclude lysis of male cells despite presence of H-Y (it is worth emphasizing that H-2 genes are remarkable for frequent mutation and that H-2 is among the most polymorphic loci in the murine genome). Second, unambiguous typing for H-Y might have been accomplished by inoculating cells from the propositus and male descendents into C57BL females with subsequent test-grafting of C57BL male skin, but this was not done. Third, mutant enzymes retain antigenicity yet lose function in certain cases of enzyme deficiency. We have said that serological H-Y is part of a molecule that binds a gonadal receptor, and loss of receptor-binding activity has already been described in a molecule that retained antigenicity (Nagai et al, 1980). Thus H-Y antigenicity and receptor-binding activity are separable by mutation and apparent absence of H-Y could be due to mutation of the antigenic moiety. Fourth, as pointed out by the authors, absence of H-Y on account of mosaicism is not likely to be inherited. Yet it would seem wise to evaluate H-Y phenotype in the gonads of the mice in question by serological methods. This is a critical point and one that bears emphasis (Ohno, in press). The fact is that when T16H/X^{Sxr} mice were produced, some, but not all, developed as females. Thus selective inactivation of the intact X carrying Sxr did not always extend the full length of the chromosome, to include Sxr. Given a mosaic field of T16H/X^{Sxr} cells in which inactivation of the Sxr segment is variable, male or female development could occur, depending on the frequency of X^{Sxr} cells inactivated in the gonads. The gonads of XX/XY chimeric mice produced by blastocyst fusion tend to become testes even when the ratio of XX:XY cells is 80:20; and when XX ovarian cells are cultured with XY testicular cells, testicular reaggregation occurs in all cases in which the proportion of XY cells exceeds 20 percent (Urban et al, 1981). The inevitable conclusion is that the gonads of the T16H/X^{Sxr} male mice that were typed H-Y^- were populated by a residuum of H-Y^+ cells. In fact the conclusion of McLaren et al (1984) is based on an assumption. They tested spleen cells. The gonads could not be tested in these experiments, and it is not likely that the method used would detect a minor population of H-Y^+ cells in the spleen; the killer-to-target ratio required in the cytotoxic method of the authors is 30:1.

H-Y AND EMBRYO TRANSFER

Presence of H-Y in the developing male embryo is of interest in relation to the bovine embryo transfer industry. Some 75,000 cattle were born in the USA and Canada in 1984 as a consequence of embryo transfer; this represents 150,000 embryos implanted. Indications are that those numbers will increase substantially in the years ahead.

Superovulation and embryo transfer are performed to increase the breeding efficiency of valuable stock. According to the standard method, a donor cow is injected with gonadotropin to induce superovulation, and is artificially inseminated 5 days later. The fertilized eggs, which are free-floating in the uterus (for 28 days), are flushed with saline solution about a week after artificial insemination and are implanted surgically or nonsurgically into the uterine horns of surrogate mothers; alternatively the eggs may be frozen and stored indefinitely in liquid nitrogen.

From zero to fifty eggs may be obtained in a setting from a single donor (the average is from five to seven) and the procedure can be repeated about thrice annually. Thus a cow could produce multiple offspring in a given year -- without having to carry any of them (Seidel, 1981).

TABLE 2. Identification of Gender in Bovine Embryos with Monoclonal H-Y Antibody

Experiment	Embryo	H-Y Phenotype	H-Y Sex	Sex Chromosomes
2	Blastocyst	-	Female	XX (?)
3	Blastocyst	+	Male	XY (?)
4	Blastocyst	+	Male	XX
	Morula	+	Male	XY
12	Blastocyst	+	Male	XX
13	Blastocyst	-	Female	XX
	Blastocyst	+	Male	Y (?)
	Blastocyst	+	Male	XY (?)
	Morula	-	Female	XX
	Blastocyst	+	Male	Y (?)
	Blastocyst	-	Female	Y (??)*
	Blastocyst	+	Male	XY (?)

Metaphase spreads were obtained for 18 embryos; both sex chromosomes or a Y could be identified in 12. A single X was detected in five (not shown) of which 3 were typed $H\text{-}Y^+$ and two, $H\text{-}Y^-$. *(??) Very questionable.

In general cattle breeders prefer one sex or the other; dairymen prefer heifer calves for example. So the question is raised whether H-Y antibody could be used for selective implantation of male or female during embryo transfer. In fact H-Y is present in bovine morulae and blastocysts, corresponding to the stages at which embryo transfer is practicable. When we applied monoclonal H-Y antibody and FITC-conjugated goat antimouse Ig in early fluorescent assays with bovine embryos, we obtained three classes: H-Y^+, H-Y^- and questionable, in a ratio of 3:3:1. Those results were evaluated by cytogenetic analysis and by visual inspection of transferred embryos, at 70 days of gestation or at birth. Of 20 embryos treated in preliminary trials, 17 were found to be correctly identified with regard to sex (Wachtel, 1984). Those numbers have not held up under the conditions of a field test involving transportation over substantial distances, but more recent data collected in our new laboratory have given reason for believing that H-Y serology may well provide the means for identification of gender in bovine embryo transfer (Table 2). It might be added that the fluorescent method does not appear to traumatize the embryo, and there is thus the opportunity and the incentive for renewed testing and development.

REFERENCES

von Boehmer H, Fatham CG, Haas W (1977). H-2 gene complementation in cytotoxic T cell responses of female against male cells. Eur J Immunol 7: 443-447.

Boyse EA, Cantor H (1978). Immunogenetic aspects of biologic communication: a hypothesis of evolution by program duplication. Birth Defects Orig Art Ser 14: 249-283.

Brunner M, Moreira-Filho CA, Wachtel G, Wachtel S (1984). On the secretion of H-Y antigen. Cell 37: 615-619.

Hurme M, Hetherington CM, Chandler PR, Gordon RD, Simpson E (1977). Cytotoxic T-cell responses to H-Y: *Ir* genes and associative antigens map in H-2. Immunogenetics 5: 453-459.

Hurme M, Hetherington CM, Chandler PR, Simpson E (1978). Cytotoxic T-cell responses to H-Y: mapping of the *Ir* genes. J Exp Med 147: 758-767.

McLaren A (1976). "Mammalian Chimaeras." Cambridge University Press.

McLaren A, Simpson E, Tomonari K, Chandler P, Hogg H (1984). Male sexual differentiation in mice lacking H-Y antigen. Nature 12: 552-555.

Müller U, Urban E (1981). Reaggregation of rat gonadal cells in vitro: experiments on the function of H-Y antigen. Cytogenet Cell Genet 31: 104-107.

Müller U, Aschmoneit I, Zenzes MT, Wolf U (1978). Binding studies of H-Y antigen in rat tissues: indications for a gonad specific receptor. Hum Genet 43: 151-157.

Müller U, Singh L, Grund S, Jones KW (1982). Ovarian cells participate in the formation of tubular structures in mouse/rat heterosexual gonadal cocultures: a direct demonstration by in situ hybridization. Differentiation 22: 136-138.

Moscona AA (1961). Rotation-mediated histogenic aggregation of dissociated cells: a quantifiable approach to cell interactions in vitro. Exp Cell Res 22: 455-475.

Nagai Y, Iwata H, Stapleton DD, Smith RC, Ohno S (1980). Testis-organizing H-Y antigen of man may lose its receptor-binding activity while retaining antigenic determinants. *In* Steinberger A, Steinberger E (eds): "Testicular Development, Structure, Function," New York: Raven Press, pp 41-47.

Nakamura D, Wachtel SS, Gilmour D (1984). Gonad-specific uptake of H-Y antigen in the chicken. J Reprod Immunol 6: 11-17.

Ohno S (1979). "Major Sex Determining Genes." Springer-Verlag, Berlin-Heidelberg-New York.

Ohno S (in press). The Y-linked testis determining gene and H-Y plasma membrane antigen gene: are they one and the same? Endocrinol Rev

Ohno S, Nagai Y, Ciccarese S, Iwata H (1979). Testis-organizing H-Y antigen and the primary sex-determining mechanism of mammals. Rec Prog Hormone Res 35: 449-476.

Seidel GE Jr (1981). Superovulation and embryo transfer in cattle. Science 211: 351-358.

Simpson E, Gordon RD (1977). Responsiveness to H-Y antigen: Ir gene complementation and target cell specificity. Immunol Rev 35: 59-75.

Urban E, Zenzes MT, Müller U, Wolf U (1981). Cell reorganization in vitro of heterosexual gonadal cocultures. Differentiation 18: 161-168.

Wachtel SS (1983). "H-Y Antigen and the Biology of Sex Determination." Grune and Stratton, New York.

Wachtel SS (1984). H-Y antigen in the study of sex determination and control of sex ratio. Theriogenology 21: 18-28.

Wilson HV (1908). On some phenomena of coalescence and regeneration in sponges. J Exp Zool 5: 245-258.

Progress in Developmental Biology, Part A, pages 159-162

TEMPERATURE DEPENDENT SEX DETERMINATION IN THE OLIVE RIDLEY TURTLE

P. Mohanty-Hejmadi and Marie T. Dimond

Department of Zoology, Utkal University, Bhubaneswar 751004, INDIA, (P.M-H.), and Department of Biology, Trinity College, Washington, DC 20017, USA (M.T.D.)

INTRODUCTION

Temperature influences sexual differentiation in turtles. Incubation temperature during a critical series of stages (Bull, 1980; Bull and Vogt, 1979, 1981; Morreale et al., 1982; Pieau and Dorrizzi, 1981; Spotila et al., 1983; Yntema, 1979; Yntema and Mrosovsky, 1980, 1982) determines the sex. Males are produced at cooler and females at warmer temperatures with both sexes developing at an intermediate temperature. So far four of the seven species of marine turtles have been found to conform to the same general pattern: the loggerhead Caretta caretta (Yntema and Mrosovsky, 1980); the green turtle Chelonia mydas (Miller and Limpus, 1981; Morreale et al., 1982); the hawksbill Eretmochelys imbricata (Dalrymple et al., 1985); and the olive ridley Lepidochelys olivacea (Dimond and Mohanty-Hejmadi, 1983; McCoy et al., 1983; Mohanty-Hejmadi et al., 1985).

MATERIALS AND METHODS

Clutches of eggs were removed from nests at Gahirmatha, Bhitar Kanika Sanctuary, Orissa, and transported to the laboratory at Utkal University. The eggs were placed between two layers of cotton and incubated at three temperature levels: cold (27-28C); room (fluctuating around 29-30C); and hot (31-32C). In addition, eggs were transferred reciprocally between hot and cold at the critical period of development. Embryos were staged according to Yntema's (1968) series. Sex was determined by morphological characteristics of the gonads followed by histological examination. Karyotypes were prepared from the spleen of hatchlings.

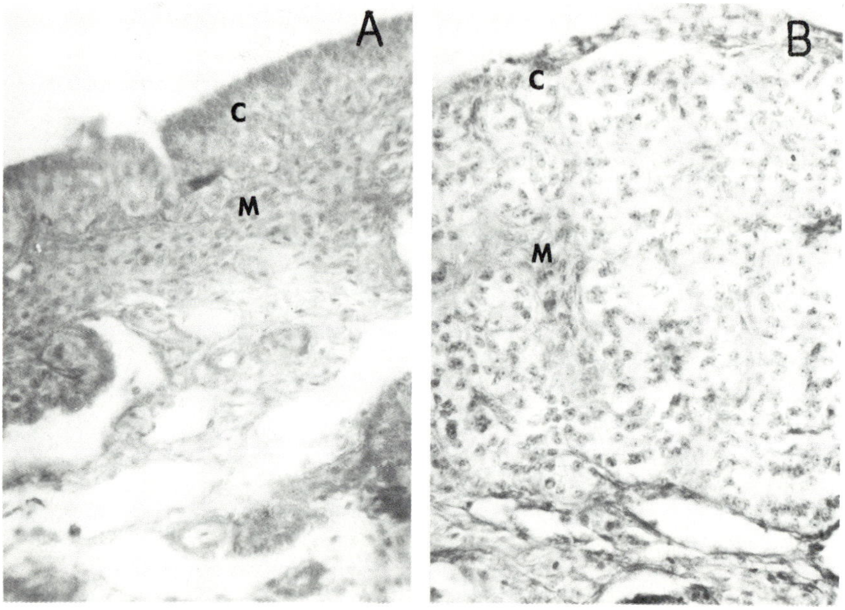

Figure 1. Photomicrograph of *Lepidochelys olivacea* hatchling ovary (A) and testis (B) showing characteristics of cortex (C) and medulla (M).

TABLE 1. Effect of One-shift Transfer on Sex of Olive Ridley Turtle, *Lepidochelys olivacea*

TREATMENT	STAGE SHIFTED (N)	TOTAL HATCHED (DAYS)	PHENOTYPE ♀	PHENOTYPE ♂	INTERSEX (%)
HOT -- COLD	20-21 (12)	10 (63)	5	5	100
COLD --HOT	15-16 (10)	5 (57.5)	4	1	80

RESULTS AND DISCUSSION

Temperature of incubation affected the duration of development, the rate of development and the sex. Females only were produced at 30C and above, and males only were produced at 28C and below (Fig. 1). Both males and females were produced at 29.5C indicating that this probably represents the pivotal temperature. The results confirm earlier observations (Dimond and Mohanty-Hejmadi, 1983; McCoy et al., 1983; Mohanty-Hejmadi et al., 1985) that in the olive ridley cool and warm temperatures induce the differentiation of male and female gonads respectively. In hot to cold shift eggs, development was prolonged and resulted in equal numbers of phenotypic males and females (TABLE 1). Histologically, however, all were intersexes with both cortex and medulla well developed. In the reciprocal experiment, 80% developed into phenotypic females which were intersexes histologically, and 20% into males both phenotypically and histologically. Results indicate that a temperature shift during critical stages of development affects differentiation of the gonads.

Karyotypic analyses indicate that the diploid number in males is 56 with bimodality consisting of 24 macro- and 32 microchromosomes. No heteromorphic pair of sex chromosomes could be discerned. The karyotype pattern conforms to the pattern reported for other chelonids (Bhunya and Mohanty-Hejmadi, 1985).

REFERENCES

Bhunya SP, Mohanty-Hejmadi P (1985). Morphometric analysis of somatic chromosomes of a chelonid turtle Lepidochelys olivacea. In preparation.

Bull JJ (1980). Sex determination in reptiles. Quart Rev Biol 55: 3-21.

Bull, JJ, Vogt RC (1979). Temperature dependent sex determination in turtles. Science 206: 1186-1188.

Bull, JJ, Vogt RC (1981). Temperature-sensitive periods of sex determination in emydid turtles. J Exp Zool 218: 435-440.

Dalrymple GH, Hampp JC, Wellins J (1985). Male-biased sex ratio in a cold nest of a Hawksbill sea turtle (Eretmochelys imbricata). J Herpetol 19: 158-159.

Dimond MT, Mohanty-Hejmadi P (1983). Incubation temperature and sex differentiation in a sea turtle. Amer Zool 23: 1017.

McCoy CJ, Vogt RC, Censky EJ (1983). Temperature-controlled

sex determination in the sea turtle *Lepidochelys olivacea*. J Herpetol 17: 404-406.

Miller JD, Limpus CJ (1981). Incubation period and sexual differentiation in the green turtle *Chelonia mydas*. Proc Melbourne Herpetol Symp, Zool Board of Victoria: 66-73.

Mohanty-Hejmadi P, Behera M, Dimond MT (1985). Temperature dependent sex differentiation in the olive ridley *Lepidochelys olivacea* and its implications for conservation. Proc Symp on Endangered Marine Animals and Marine Parks. The Marine Biol Assoc India, Cochin, India.

Morreale SJ, Ruiz GJ, Spotila JR, Standora EA (1982). Temperature-dependent sex determination: current practices threaten conservation of sea turtles. Science 216: 1245-1247.

Pieau C, Dorizzi M (1981). Determination of temperature sensitive stages for sexual differentiation of the gonads in embryos of the turtle, *Emys orbicularis*. J Morph 170: 373-382.

Spotila JR, Standora EA, Morreale SJ, Ruiz GJ, Puccia C. (1983). Methodology for the study of temperature related phenomena affecting sea turtle eggs. US Fish and Wildlife Service, Endangered Species Report 11, 51p.

Yntema CL (1968). A series of stages in the embryonic development of *Chelydra serpentina*. J Morph 125: 219-252.

Yntema CL (1979). Temperature levels and periods of sex determination during incubation of eggs of *Chelydra serpentina*. J Morph 159: 17-28.

Yntema CL, Mrosovsky N (1980). Sexual differentiation in hatchling loggerheads (*Caretta caretta*) incubated at different controlled temperatures. Herpetologica 36: 33-36.

Yntema CL, Mrosovsky N (1982). Critical periods and pivotal temperatures for sexual differentiation in loggerhead sea turtles. Can J Zool 60: 1012-1016.

Progress in Developmental Biology, Part A, pages 163-169

TEMPERATURE AND ESTROGEN DEPENDENT CHANGES OF SEX PHENOTYPE AND H-Y ANTIGEN EXPRESSION IN GONADS OF A NEWT

Patrick Zaborski

CNRS, Centre de Cytologie Experimentale, and Laboratoire de Biologie du Développement (Univ. of Paris VI), 67 rue Maurice Gunsbourg, F-94200 Ivry-Sur-Seine, France

INTRODUCTION

The serologically defined H-Y antigen has been detected in all vertebrate classes (review: Wachtel, 1983). Since it is a heterogametic sex specific factor, it has been proposed to be the primary inducer of gonad differentiation of the heterogametic sex in vertebrates (Wachtel *et al.*, 1975). Although numerous data in favour of this hypothesis have been found, some opposing views have been presented (Silvers *et al.*, 1982).

A prerequisite to the study of the role of the H-Y antigen in gonads is to define the conditions of its expression in relation with the well known action of the epigenetic factors that can influence gonad differentiation (review: Foote, 1964). Such an investigation has already been undertaken in different laboratories using nonmammals (review: Zaborski, 1985). However, much work remains to be done, especially in the species in which sex differentiation can be disturbed under the influence of sex hormones or environmental conditions. Among these species, *Pleurodeles waltlii* (Amphibia, urodela), a ZZ-male/ZW-female newt, provides a useful model in which the sex is already differentiated at metamorphosis but ZZ larvae can be feminized by estrogens (Gallien, 1962), and ZW larvae can be masculinized by heat; the genetic proof of the heat-dependent sex inversion has been well demonstrated (Dournon and Houillon, 1984). Moreover, the female heterogametic sex is H-Y positive (Zaborski, 1979b), and the H-Y expression together with the sex phenotype of the gonads can be inversed by estrogen treatment (Zaborski and Andrieux, 1980; and in preparation). Thus *Pleurodeles* was used to study the possible effect of environmental factors on gonadal H-Y expression. Larvae and adults were heat treated in the context of known hormonal influence on gonad differentiation. In the present work it is shown that high temperatures can act specifically on H-Y expression in gonads.

MATERIAL AND METHODS

Experimental Procedures

A genetically female (ZW) unisexual progeny was used; it was produced in our laboratory by mating a normal ZZ male with a WW female (Collenot, 1975). The resulting ZW spawnings were reared at room temperature (20-22°C) from the egg stage until the beginning of the heat treatment. At 25 days of age (about 40 days before the first histologic signs of sex differentiation of the gonads), larvae were divided in three groups. The first one was reared at room temperature to yield control individuals(ZW, 20°C). The second group was maintened at 32±0.5°C until the end of metamorphosis (about 3 weeks after the beginning of heat treatment), and thereafter at room temperature for the following 5 months (ZW,32°C) The third group was treated by both estradiol benzoate (BzE2 500 ng/ml aquarium water) and heat (32±0.5°C) until the end of metamorphosis, and then shifted to room temperature without BzE2 for the following 5 months (ZW, 32°C+BzE2).

For treatment of adults, four groups of ZW females were constituted. The first one served as control females at room temperature (♀, 20-25°C). The second group was heat treated at 32±0.5°C for 3 weeks before sacrifice (♀, 32°C). The third group contained females that were hypophysectomized and immediately heat treated at 32±0.5°C for the subsequent 3 weeks before sacrifice (♀, HX, 32°C). The forth group consisted of hypophysectomized females maintained at room temperature during 3 weeks before sacrifice (♀, HX, 20°C).

The identification of the sexual phenotype of the gonads was done at the time of sacrifice. The number of sacrificed and subsequently H-Y typed individuals is indicated in the RESULTS.

Detection of H-Y Antigen

For each sacrificed individual, washed blood cells were used as nongonadal cells from blood samples withdrawn from the arterial bulb; washed gonadal cells were isolated from manually dissociated gonads in cold 0.7% NaCl. The H-Y typing of each catergory of cells in each individual was run in parallel by immunoabsorption of portions of a specific mouse H-Y antiserum. The H-Y antiserum was produced in C57 BL/6 female mice immunized with spleen cells from syngeneic male mice. The residual anti-H-Y activity was estimated using the immunobacterial technique on mouse sperm (H-Y positive target) previously described (Zaborski, 1979a), slightly modified (Zaborski *et al.*, 1982). During the immunoabsorption procedure, H-Y positive gonadal or nongonadal cells would remove H-Y antibodies from the H-Y antiserum and, therefore, the absorbed serum would be characterized by a low percentage of positively labeled mouse sperm. The converse would be found when H-Y negative cells are used. Statistical significances were anlyzed by the Wilcoxon-Mann-Whitney-test.

RESULTS

Treatments in Larvae (Figure 1)

Larvae of ZW genotype developed as females when reared at room temperature, and their ovaries typed H-Y positive (low residual anti-H-Y activity). On the contrary, all the ZW larvae reared at 32±0.5°C became phenotypic males with testes that were H-Y negative, as in normal ZZ male controls, but their blood cells absorbed anti-H-Y antibodies as efficiently as blood cells of normal ZW females . The difference between H-Y expression in testes and blood cells of these sex-inversed animals was highly significant (P = 0.00001), while the same comparison showed no difference in control females issued from the *ZW, 20°C* larvae group. The testis cells of the sex-inversed animals also presented the same degree of H-Y expression as did testis cells of the normal ZZ males. In contrast, the ZW larvae reared at 32±0.5°C in aquarium water containing BzE2 developed as phenotypic females, as did ZW larvae reared at room temperature, and the blood cells as well as the ovary cells typed H-Y positive as in control females. The difference of H-Y expression between the 'ZW testes' and the 'ZW ovaries' was about 60%.

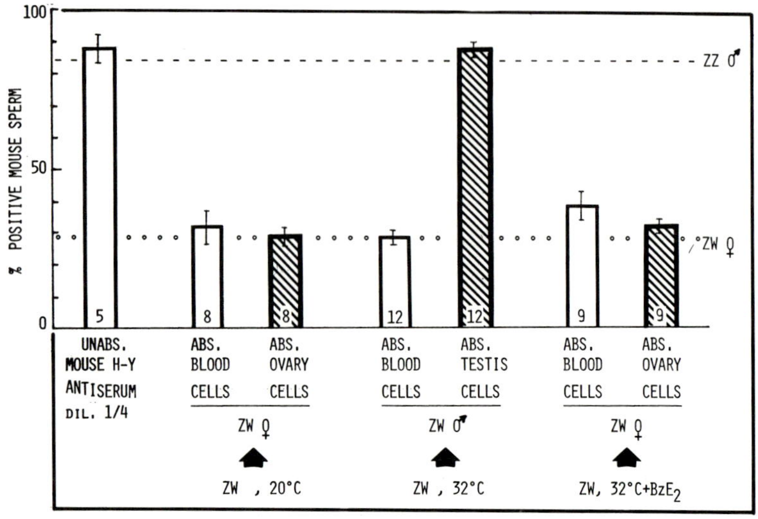

Figure 1. H-Y antigen expression in metamorphosed newts after rearing at different temperatures during the larvae stage. The *hatched bars* (gonads) or *open bars* (blood or unabsorbed H-Y antiserum) represent mean values of the indicated number of determinations ± SD (one individual by determination); *abs.* denotes the residual activity of H-Y antiserum, diluted 1/4, absorbed with the indicated cells from individuals of the indicated sexual phenotype and genotype issued (*arrows*) from larvae of female (*ZW*) sex genotype reared under different hormonal (*BzE2*: estradiol benzoate) or temperature (*°C*) conditions. *Horizontal dotted lines*, show the mean data of residual anti-H-Y activity after absorption with gonadal cells from normal adult males (*ZZ O*) or females (*ZW ♀*).

Treatments in Adults (Figure 2)

The sex phenotype of adult females maintained at 32±0.5 °C for 3 weeks was not modified and blood as well as ovary cells typed H-Y positive as in control females. Females hypophysectomized for 3 weeks and heat treated during this period did not shown any sign of sex change in ovaries, all of which remained normal; however, ovary cells showed a large difference in their H-Y expression in comparison with the other experimental situations. These ovary cells became H-Y negative, similar to testis cells of control males. The hypophysectomy alone did not modify the sex phenotype nor the H-Y expression of the ovaries in the foth group (♀,*HX*,*20°C*).

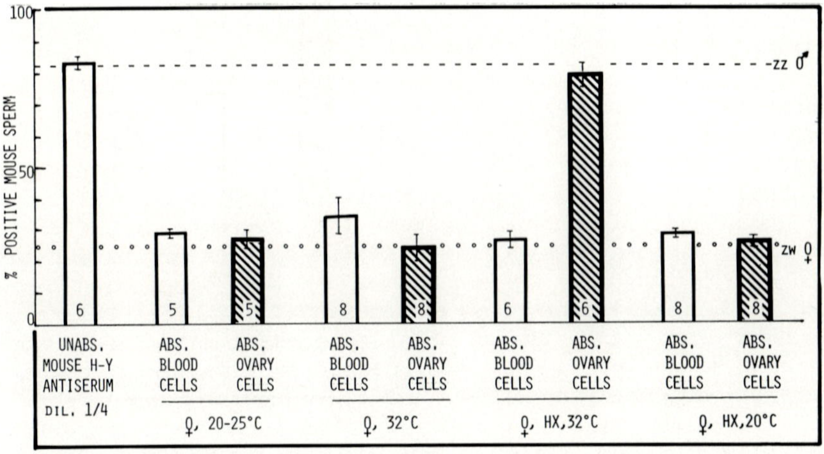

Figure 2. H-Y antigen expression in adult female newts after maintenance at different temperatures. *abs*. denotes the residual activity of H-Y antiserum, diluted 1/4, absorbed with the indicated cells from intact or hypophysectomized (*HX*) females. See figure 1 legend for other details.

DISCUSSION

The present study shows that the masculinizing effects of high temperature on gonad differentiation giving testes in genetic female larvae are inhibited by simultaneous estrogen treatment which induces a ovarian differentiation. It is also shown that the positive H-Y status in gonads depends on the development of ovarian tissues in this species with female heterogamety. This is new experimental evidence for correlation found between the H-Y antigen expression in gonads and histomorphologic features specific of the heterogametic sex.The morphologic consequences of heat treatments on gonad differentiation in *P. waltlii* have been well documented (Dournon and Houillon, 1984); present observations confirm them. Temperature dependent sexual differentiation of the gonads is a widespread phenomenon in reptiles (Bull,1980) and the present findings are consistent with similar findings which have been described in the turtle *Emys orbicularis*.

In this turtle, the expression of H-Y antigen is correlated with ovarian development which is under the control of the egg incubation temperature; the gonadal and blood cells display a different H-Y type when sex-inversion has occured (Zaborski *et al.*, 1982). The present experiments in adult females show that it is possible to obtain only the inversion of H-Y expression in already differentiated ovaries without any histomorphologic change. This observation seems to favor a causal relationship between the absence of H-Y expression and the masculinizing events provoqued in gonadal tissues by high temperatures. However, it also account for an occasional discrepancy between the occurence of H-Y expression and the heterogametic features. This may be explained by the fact that no sex change can take place once a gonad is differentiated. The used dosage for BzE2 in this study counteracts the effects of high temperatures. This dosage is also known to provoque a complete feminization at normal temperature (Gallien, 1962), and to induce H-Y expression in *P. waltlii* (Zaborski and Andrieux, 1980; and in preparation). On the other hand, hypophysectomy, which is known to decrease or inhibit gonadal steroidogenesis, permits the effects of high temperatures on differentiated ovaries by inhibiting the H-Y antigen expression. Thus, the mechanism of action of the high temperature might involve the hormonal pathways by modifying the activity of some steroid regulated events. Even though this might occur in differentiated ovaries, it is not clear whether the action of high temperatures implies a similar mechanism in undifferentiated gonads. Although these results do not directly demonstrate that the H-Y antigen is responsible for the sex-reversal, they do, however, argue for a role of H-Y antigen in some process of gonadogenesis.

REFERENCES

Bull JJ (1980). Sex determination in reptiles. Quart Rev Biol 55:3-21.

Collenot A (1975). Unisexual female offspring in the salamander *Pleurodeles waltlii* Michah. In Reinboth R (ed): "Intersexuality in the Animal Kingdom", Berlin, Heidelberg: Springer-Verlag, p 311.

Dournon C, Houillon Ch (1984). Démonstration génétique de l'inversion fonctionnelle du phénotype sexuel femelle sous l'action de la température d'élevage chez l'Amphibien Urodèle: *Pleurodeles waltlii* Michah. Reprod Nutr Dévelop 24: 361-378.

Foote CL (1964). Intersexuality in amphibians. In Armstong, Marshall (eds): "Intersexuality in Vertebrates Including Man", New York: Academic Press, p 233.

Gallien L (1962). Comparative activity of sexual steroids and genetic constitution in sexual differentiation of amphibian embryos. Gen Comp Endocrinol Supp 1:346-355.

Silvers WK, Gasser DL, Eicher EM (1982). H-Y antigen, serologically detectable male antigen and sex determination. Cell 28:439-440.

Wachtel SS (1983). "H-Y Antigen and the Biology of Sex Determination". New York, Grune and Stratton: p 302.

Wachtel SS, Ohno S, Koo GC, Boyse EA (1975). Possible role for H-Y antigen in the primary determination of sex. Nature 257:235-236.

Zaborski P (1979a). Detection of H-Y antigen on mouse sperm by the use of *Staphylococcus aureus*. Transplantation 27: 348-350.

Zaborski P (1979b). Sur la constance de l'expression de l' antigène H-Y chez le sexe hétérogamétique de quelques Amphibiens et sur la mise en évidence d'un dimorphisme sexuel de l'expression de cet antigène chez l'Amphibien Anoure *Pelodytes punctatus* D . C R Acad Sci 289:1153-1156.

Zaborski P (1985). H-Y antigen in nonmammalian vertebrates. Arch Anat Microsc Morphol Exp 74:33-37.

Zaborski P, Dorizzi M, Pieau C (1982). H-Y antigen expression in temperature sex-reversed turtles (*Emys orbicularis*) Differentiation 22:73-78.

Progress in Developmental Biology, Part A, pages 171–174

GONADAL SEX REVERSAL OF FETAL MOUSE OVARIES FOLLOWING TRANSPLANTATION INTO ADULT MICE

Teruko Taketo and Horacio Merchant-Larios

The Population Council, New York, New York 10021
and Instituto de Investigaciones Biomedicas,
Universidad Nacional Autonoma de Mexico, Mexico

INTRODUCTION

The sex of an individual is genetically determined at the time of fertilization by the combination of sex chromosomes in mammals. Morphological sex differences first become evident during gonadal organization, i.e., differentiation of gonadal primordia into testes or ovaries. During normal development of a fetus, the gonadal sex coincides with the genetic sex. However, it has been reported that fetal ovaries develop testicular structures when transplanted into various sites of adult host animals (Buyse, 1935; Moore and Price, 1942; Torrey, 1950; Taketo et al., 1984). To clarify the mechanism of gonadal sex reversal following transplantation, we studied in the mouse (1) the influence of sex and hormonal condition of hosts on the development of ovarian grafts and (2) the sequence of the appearance of testicular components in ovarian grafts.

RESULTS

Fetal ovaries with mesonephroi attached were dissected from mouse fetuses on the 12th day of gestation (Rockefeller Univ. NCS[S] strain), and transplanted beneath the kidney capsules of adult mice of the same strain. On the 14th day after transplantation, many ovarian grafts developed testicular structures in addition to ovarian structures (ovotestes). The frequency of ovotestis development (Table 1) was low in female hosts and slightly increased by castration of the host. It was higher in male hosts than in females, and not affected by castration. Testosterone implants increased the frequency of ovotestis development, whereas 17β-estradiol implants decreased it. These results suggest that sex hormones play an important role during gonadal sex differentiation. However, sex steroids appear not to be the sole factor responsible for the induction of testicular development in ovarian grafts. We found that when fetal ovaries were separated from mesonephroi before transplantation, they developed testicular structures equally in female (79 %, n=18) and male (88 %, n=19) hosts. These results suggest that the mesonephros also plays a key role in sex determination of ovarian grafts, possibly by protecting ovarian grafts from masculinizing stimuli of the hosts in a female environment.

Fetal ovaries separated from mesonephroi were transplanted into male mice, and examined between the 4th and 66th day after transplantation. Only ovarian components were found in all grafts until the 7th day. By the 11th to 14th day, follicles, seminiferous cords, and transitional

Table 1. Influence of The Hormonal Condition of Hosts on The Development of Ovarian Grafts

Sex of hosts	Treatment of hosts	No. of exp.	Ovotestis development (%)
Female	Untreated	40	19
	Castrated[a)]	90	30
	Castrated + E_2 implants[b)]	37	17
	Castrated + T implants[b)]	34	89
Male	Untreated	25	55
	Castrated[a)]	48	49
	Castrated + E_2 implants[b)]	38	19

a) Castration was performed 5 to 8 weeks before transplantation.

b) Pellets containing 2.6 mg of 17β-estradiol (E_2) or testosterone (T) were implanted into s.c. position of hosts one week before transplantation.

structures had developed (Fig. 1). The transitional structure was composed of pregranulosa cells, Sertoli cells, a third type of cells with intermediate characteristics between Sertoli and pregranulosa cells, and occasional oocytes, which were enclosed by common basal lamina (Fig. 2). Leydig cells or peritubular myoid cells were not seen in this area. These cell types were present while oocytes were absent in the testicular area. On the 33rd and 66th day, ovarian and testicular structures were seen whereas the transitional structures had disappeared. These observations suggest that pregranulosa cells of fetal ovaries can differentiate into Sertoli cells, which are responsible for the organization of the seminiferous cords, degeneration of oocytes, and differentiation of other testicular somatic cell types.

The present study suggests that Sertoli cell differentiation leads the testicular development of ovarian grafts, and this process is in part under the control of sex hormones.

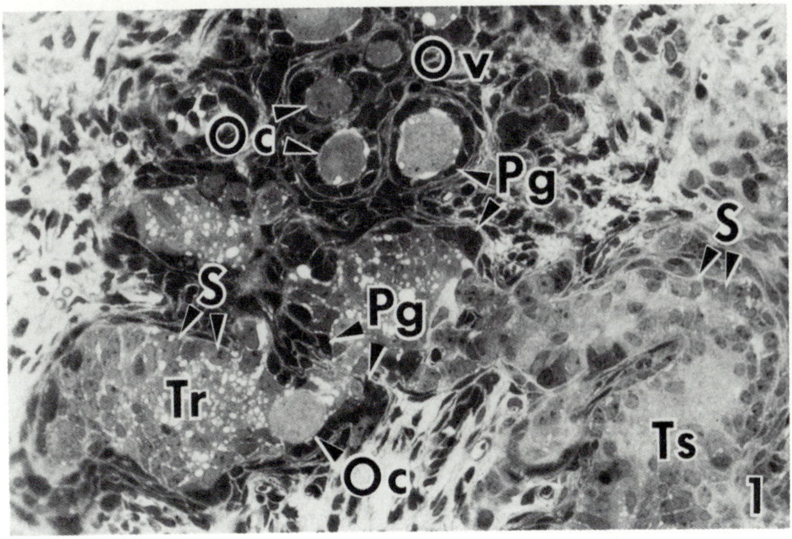

Figure 1. A fetal ovary on the 14th day after transplantation into an adult male mouse. In the follicles of an ovarian area (Ov), several oocytes (Oc) are surrounded by pregranulosa cells (Pg). The cords in a transitional area (Tr) contain pregranulosa cells, an oocyte, and Sertoli cells (S). Only Sertoli cells are seen in the seminiferous cords of a testicular area (Ts). Magnification, X 190.

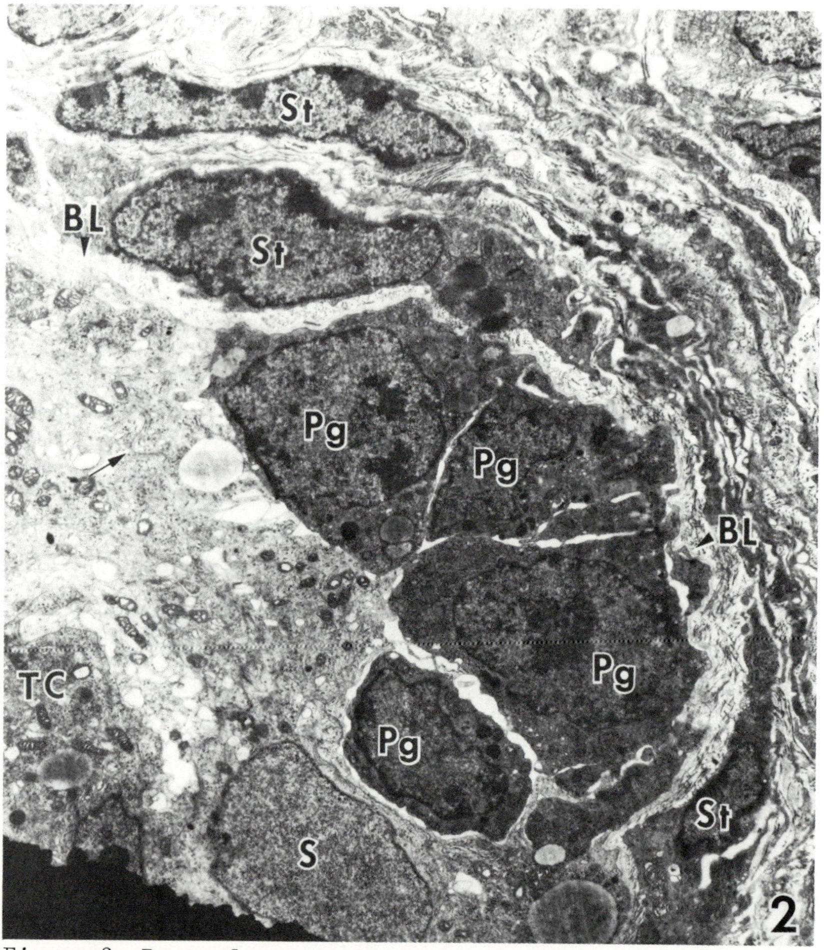

Figure 2. Part of a transitional area of an ovotestis. Pre-granulosa cells (Pg), Sertoli cells (S), and transitional cells (TC) are enveloped by common basal lamina (BL). An interSertoli junctional complex (arrow) has been formed. Undifferentiated stromal cells (St) are in parallel with the basal lamina. Magnification, X 4100.

REFERENCES

Buyse A (1935) J Exp Zool. 70: 1.

Moore CR, Price D (1942) J Exp Zool. 90: 229.

Torrey TW (1950) J Exp Zool. 115: 37.

Taketo T, Merchant-Larios H, Koide SS (1984) Proc Soc Exp Biol Med. 176: 148.

Progress in Developmental Biology, Part A, pages 175–178

GERM CELLS CONTROL SEX DETERMINATION IN HYDRA.

C. Lynne Littlefield

Developmental Biology Center, University of California, Irvine, California 92717.

Sexual differentiation in hydra is remarkably simple in that males and females possess no secondary sexual characteristics. Gametes arise from the differentiation of stem cells, called interstitial cells, which also give rise to somatic cells, including nerve cells and nematocytes. Testis and egg formation result from the accumulation of gamete-precursor interstitial cells in the outer ectodermal tissue layer of the body column of the animal. No cell types other than the interstitial cells appear to be involved in this process. The sexual phenotype expressed, however, does appear to be influenced by the cellular environment. This is illustrated by the following observations.

In all dioecious species of hydra tested to date (*Hydra oligactis*, Brien, 1963; *H. attenuata*, Tardent, 1968; *H. carnea*, personal observation), females become masculinized when brought in to contact with males by grafting. In control grafts (female to female) no masculinization occurs. To determine which cell types in the male are responsible for female sex reversal, chimeras were constructed which consisted of male ectodermal tissue and female interstitial cells and their products (*H. carnea*, Littlefield, 1984; *H. oligactis*, Campbell, 1985). These chimeras were female, indicating that male epithelial tissue has no influence on the sexual phenotype of the animal. Therefore, it was concluded that cells of the male interstitial cell population, or some differentiation product thereof, must be responsible for the masculinization of females in heterosexual grafts.

Several findings have provided new approaches for determining which cells of the male interstitial lineage can cause sex reversal of females in heterosexual grafts of *H. oligactis*. First, a monoclonal antibody, AC2, has been isolated that binds specifically to a subpopulation of interstitial cells, as well as to cells of the entire sperm pathway in normal males (Littlefield et al., 1985). This antibody thus allows one to distinguish sperm-precursor interstitial cells from those that give rise to the somatic cells in asexual animals. Second, animals have been isolated which contain developmentally-restricted interstitial cells that have the ability to differentiate into sperm, but do not produce nerve cells or nematocytes (Littlefield, 1985). Such animals were obtained by treating normal males with hydroxyurea, which lowers the interstitial cell population to approximately 1-2 cells in most animals. Those cells remaining represent a stem cell population that is capable of extensive proliferation, yet restricted to germ cell production (Littlefield and Bode, 1985); these characteristics are suggestive of a separate male germline. Over 95% of the interstitial cells present in these animals (C-clone animals) are AC2+, whereas less than 1% are AC2+ in normal, asexual males maintained at 22°C (Littlefield et al., 1985). The findings indicate a heterogeneity of the interstitial cell population that may be involved in controlling sexual phenotype.

With the isolation of sperm precursor interstitial cells in C-clone animals, it has been possible to examine their role in sex-reversal of females. Five reciprocal grafts were made, as illustrated in Fig. 1, and were left intact for several weeks. During this time the interstitial cells of the 2 halves become intermingled via cell migration between the grafted fragments (Tardent and Morganthaler, 1966; personal observation). This leads to incoporation of C-clone interstitial cells into the female tissue. Following this, the sexual phenotype of the animals was examined. In all cases the animals differentiated testes, indicating that masculinization of the females had occurred. This masculinization is complete, as asexually produced offspring (buds) from these animals have remained male over a 2 year period. As controls, animals from another clone, E-clone, which lack all interstitial cells and their products, were grafted to normal females. Of the five grafts with E-clone animals, all were female, confirming that male epithelial tissue has no effect.

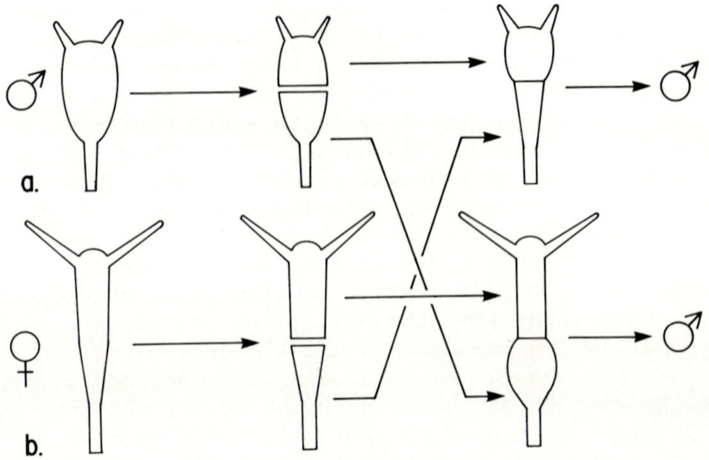

Figure 1. Grafts between a) C-clone animals containing germline-restricted male interstitial cells and b) normal females. The swollen appearance of C-clone animals reflects the inability of nerve-free hydra to osmoregulate (Wanek et al., 1980). Sexuality was induced by lowering culture temperature from 18°to 10°C.

The isolation of sperm-restricted interstitial cells from males indicates a developmental heterogeneity in the interstitial population of males. These cells are also capable of causing masculinization of females when introduced by grafting, and, thus, may exert a dominant influence in determining sex. This dominance is seen in the cessation of egg production concomitant with the onset of sperm differentiation, implying an interaction of the migrating sperm-precursor interstitial cells with the resident female interstitial cells that give rise to eggs. Whether the female interstitial cells undergo a switch from egg to sperm production, or if the invading cells of the male suppress gamete differentiation of the female cells while continuing to produce sperm, is not known.

That the AC2+ interstitial cells play a fundamental role in sex determination is supported by the following observation. Sex-reversed males from a culture of normal males grown at 22°C have recently been isolated (Littlefield, in preparation). Such animals no longer contain AC2+ interstitial cells, yet retain all other products of interstitial cell differentiation, including nerve cells and nematocytes. Thus, these animals appear the same as normal males with one exception; they cease sperm differentiation and begin making eggs. This suggests that in normal males, sperm differentiation requires the presence of AC2+ interstitial cells, and in their absence, egg production can and does occur.

This work was supported by NIH grants HD 08086 and HD 16440 and NSF grant PCM 84-02807.

REFERENCES

Brien, P (1963). Contribution à l'étude de la biologie sexuelle chez les Hydres d'eau douce. Induction gamétique et sexuelle par la méthode des greffes en parabiose. Bull Biol France/Belg 97:213-283.

Campbell, RD (1985). Sex determination in Hydra: Roles of germ cells (interstitial cells) and somatic cells. J Exp Zool 234:451-458.

Littlefield, CL (1984). The interstitial cells control the sexual phenotype of heterosexual chimeras of hydra. Dev Biol 102:426-432.

Littlefield, CL, Dunne, JF, Bode, HR (1985). Spermatogenesis in Hydra oligactis. I. Morphological description and characterization using a monoclonal antibody specific for cells of the spermatogenic pathway. Dev Biol 110:308-320.

Littlefield, CL (1985). Germ cells in Hydra oligactis males: I. Isolation of a subpopulation of interstitial cells that is developmentally restricted to sperm production. Dev Biol (in press).

Littlefield CL, Bode HR (1985). Germ cells in Hydra oligactis males. II. Characterization of a subpopulation of interstitial cells whose differentiation is limited to sperm production. (in preparation)

Tardent, P (1968) Experiments about sex determination in Hydra attenuata Pall. Dev Biol 17:483-511.

Tardent, P, Morganthaler, U (1966). Autoradiographische Untersuchungen zum Problem der Zellwanderungen bei Hydra attenuata (Pall). Rev Suisse Zool 73:468-480.

Wanek, N, Marcum, BA, Lee, H-T, Chow, M, Campbell, RD (1980). Effect of hydrostatic pressure on morphogenesis in nerve-free hydra. J Exp Biol 211:275-280.

Progress in Developmental Biology, Part A, pages 179–182

SEXING OF MAMMALIAN SPERM

Hideo Mohri, Shigeru Oshio and Satoru Kaneko

Department of Biology, University of Tokyo, Meguro-ku, Tokyo 153 (H.M., S.O) and Department of Obstetrics and Gynecology, Keio University School of Medicine, Shinjuku-ku, Tokyo 160, Japan (S.K)

INTRODUCTION

Approximately 200 sex-linked genetic diseases including hemophilia, muscular dystrophy, achromatopsia, and others have been described. Prior selection of sex represents the most accessible and reliable method for preventing these diseases. Various attempts have been made to separate human X- and Y-bearing sperm *in vitro* by centrifugation, electrophoresis, gel filtration and passage through albumin layers, etc. We have recently shown that human X- and Y-bearing sperm could be separated by means of the discontinuous density gradient centrifugation in Percoll, and the free-flow electrophoresis (Kaneko et al., 1984a, b).

MATERIALS AND METHODS

Human semen was obtained by masturbation and was allowed to liquefy for 30 min. Sperm counts were determined with a hemocytometer and sperm motility was estimated subjectively.

For Percoll density gradient centrifugation, isotonic Percoll was diluted with 10 mM HEPES buffered Hanks solution to give 80 to 25% Percoll solution in 5% stepwise, and 0.7 ml each solutions were layered successively in a centrifuge tube. The ejaculated semen (1.0 ml) was loaded on the gradient and centrifuged with a swing-out rotor. After centrifugation, the sperm that distributed on density interfaces and at the bottom were recovered from a pin hole at the bottom of the tube.

For free-flow electrophoresis, the chamber and electrode buffers were composed of 10 mM HEPES-NaOH, 30 mM NaCl, 0.25 M sucrose containing 1 mg/ml human serum albumin, pH 7.4 and 100 mM HEPES-NaOH, 80 mM NaCl, pH 7.4, respectively. The liquefied semen was diluted 2-fold with Hanks solution, pH 7.4, and an aliquot (4.0 ml) was loaded on 4.0 ml of the chamber buffer and centrifuged at 250 x g for 20 min. The precipitated sperm were resuspended in the chamber buffer to the final volume of 1.0 ml. Electrophoresis was performed at 20°C and 85 V/cm electric field. The washed sperm suspension was introduced in the separation chamber (4 x 13 x 0.3 cm, width x hight x depth) of the free-flow electrophoresis apparatus (ACE 710, Hirschman, FRG), and the sperm were fractionated from the bottom nozzles into 20 fractions.

Human Y-bearing sperm were detected by fluorescent staining with quinacrine mustard according to the method of Barlow and Vosa (1970). Sperm with a fluorescent body (F-body) at the center of head were identified as Y-bearing sperm and those without F-body as X-bearing ones. The percentage of Y-bearing sperm in unseparated semen was found to be 46 to 48% in the present study.

RESULTS AND DISCUSSION

In Percoll density gradients, a higher percentage of Y-bearing sperm was found in lighter fractions, and the percentage decreased as the density increased. The sedimentation velocity of X-bearing sperm was faster than that of Y-bearing sperm, despite of little difference between the apparent densities of X- and Y-bearing sperm obtained by equilibrium sedimentation (Kaneko et al., 1984a). The ratio of the two types of cells, however, varied with the length and force of centrifugation. Since it seemed difficult to prepare both types of cells simultaneously in one procedure, the discontinuous density gradient with 12 steps was employed for the selective isolation of X-bearing sperm, which is rather important for sex-linked genetic diseases. Among the conditions examined, centrifugation at 250 x g for 30 min gave the optimum results. Under these conditions, 23.3 ± 6.3% sperm was recovered in the sediment, of which the percentage of Y-bearing sperm was only 6.4 ± 1.8%, i.e. the purity of X-bearing sperm reached approximately 94% (Fig. 1). Almost all the sperm in the sediment was active and forward motile.

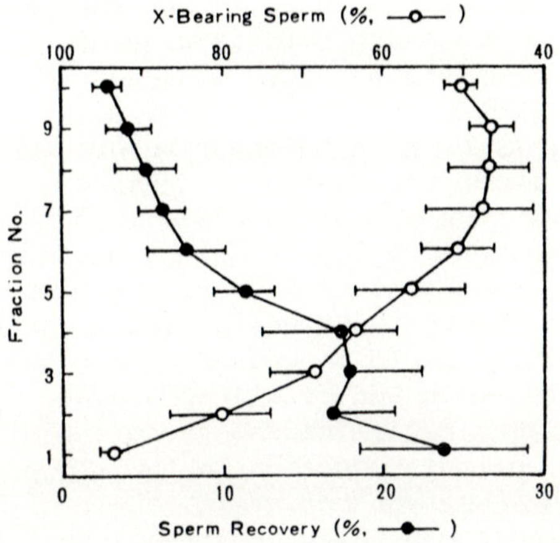

Figure 1. Selective isolation of human X-bearing sperm in Percoll density gradient. X-bearing sperm (%) = 100 - percentage of Y-bearing sperm. Values are mean ± S.D. (n = 6). (From Kaneko et al, 1984a. With the permission of the publisher)

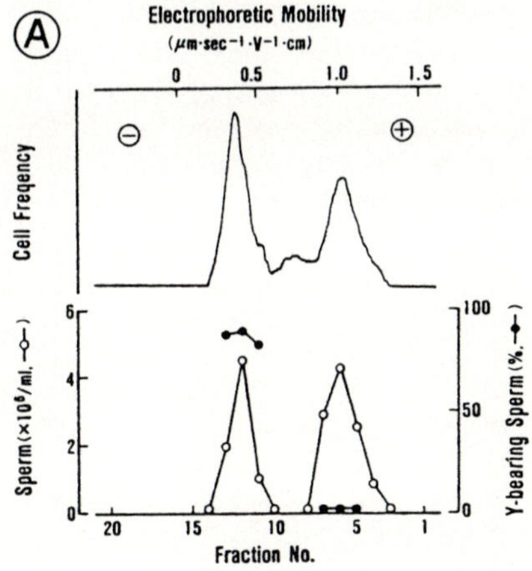

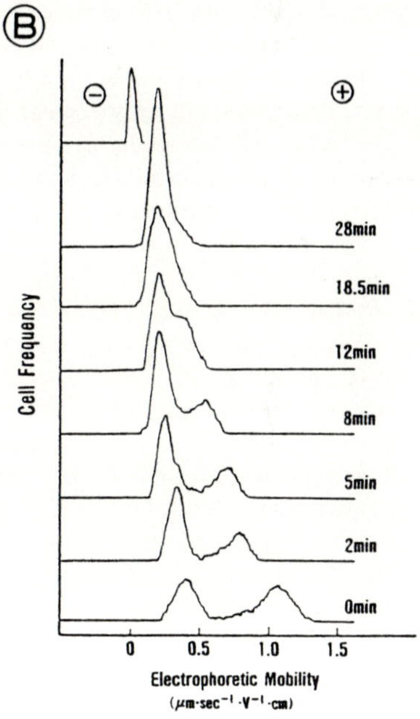

Figure 2. Free-flow electrophoresis of human sperm. A: Upper panel, The profile of densitometric tracing of cell distribution. Lower panel, Sperm counts. B: Effect of neuraminidase treatment of the electrophoretic mobility. The small peak at the top left of the figure shows the elution profile of dextran, a non-migrating substance used as a control. (From Kaneko et al., 1984b. With the permission of the publisher)

Free-flow electrophoresis was also used to separate X- and Y-bearing sperm. When washed sperm were introduced in the free-flow electrophoresis apparatus, they migrated to the anode and the profile of cell distribution showed two main peaks (Fig. 2A). No sperm bearing F-body were observed in the faster peak near the anode whereas the percentage of Y-bearing sperm in the slower peak was 83 to 89%. This result suggests that the net negative charge on the surface of X-bearing sperm is higher than that of Y-bearing sperm, thus allowing the separation of the two types of cells. The movement of cells in an electric field depends on their surface charge, which is usually dominated by glycoproteins containing neuraminic acid. The neuraminidase treatment of the washed sperm induced an obvious reduction in the electrophoretic mobilities of the two types of cells, and the rate of reduction in X-bearing sperm was higher than that of Y-bearing sperm (Fig. 2B). Finally the two peaks joined together and the percentage of Y-bearing sperm in the joint peak was found to be 43 to 46%. It appears that the higher net negative charge on X-bearing sperm was mainly due to a larger content of glycoproteins containing neuraminic acid on their surface relative to that of Y-bearing sperm. No distinct difference in morphology was detected between X- and Y-bearing sperm.

REFERENCES

Barlow P, Vosa CG (1970). The Y-chromosome in human spermatozoa. Nature 226: 961-962.

Kaneko S, Oshio S, Kobayashi T, Mohri H, Iizuka R (1984a). Selective isolation of human X-bearing sperm by differential velocity sedimentation in Percoll density gradients. Biomed Res 5: 187-194.

Kaneko S, Oshio S, Kobayashi T, Iizuka R, Mohri H (1984b). Human X- and Y-bearing sperm differ in cell surface sialic acid content. Biochem Biophys Res Commun 124: 950-955.

Progress in Developmental Biology, Part A, pages 183–186

A TIME CLOCK FOR MEIOSIS INITIATION IS PROGRAMMED IN AVIAN GERM CELLS.

Françoise Dieterlen-Lièvre, Claude Martin and Khadija Hajji

Institut d'Embryologie du C.N.R.S. et du Collège de France 94130 Nogent-sur-Marne, France
Laboratoire de Zoologie, Faculté des Sciences, Agdal Rabat, Maroc.

INTRODUCTION

In female vertebrates, meiosis is initiated during embryonic life and becomes arrested at diplotene, while in males it occurs only at puberty and proceeds continuously. The mechanisms which control the onset of this event at such widely separated periods are largely unknown. Here we present the results of experiments in which quail germ cells were made to colonize the stroma of chick gonads.

To our knowledge, one aspect of the development of the gonads has never been studied previously. It concerns the respective roles of the primordial germ cells (PGCs) and the stroma in determining the time at which meiosis begins in female embryos. In the control female quail embryos PGCs enter meiosis at 10 days of incubation and in female chick at 16 days. This 6-day difference made it possible to determine whether meiosis is timed by an internal PGC clock or by a stromal signal.

MATERIAL AND METHODS

Chimeras were constructed surgically in ovo according to Cl. Martin's procedure (1972). The grafting scheme was devised so that quail PGCs would move into the chick gonadal territory (Fig. 1). The origin of cells could be identified through the quail nucleolar marker (Le Douarin, 1969), recognizable in quiescent GCs (Callebaut, 1973 ; Tachinante, 1974 ; Eyal-Giladi et al., 1981) and until the leptotene stage of the first meiotic prophase.

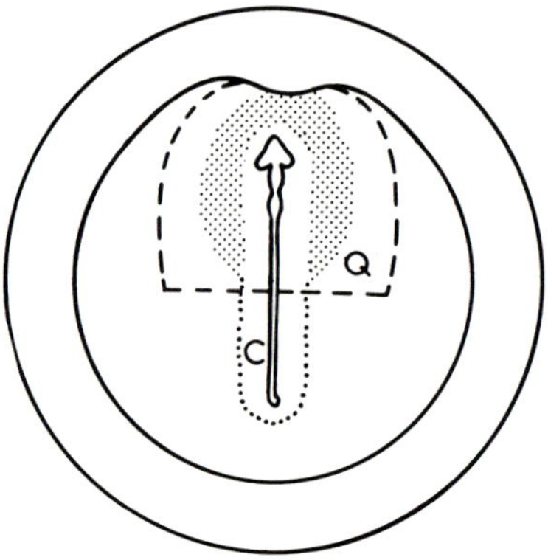

Fig. 1

The genetic sex of the two components was identified using a sex-linked recessive albino mutation available for both quail and chick (kindly provided by Dr. A. Perramon, INRA, Jouy-en-Josas).

At the time of grafting pigmentation had not yet developed and the genetic sex of the blastodiscs could not be ascertained. Thus sex combinations were realized randomly.

RESULTS

In the normal female quail, a few PGCs located in the median part of the left gonad showed preleptotene figures at 10 days of incubation. Twenty-four hours later typical leptotene figures became numerous. In the female chick embryo, preleptotene and leptotene figures appeared at 16 days. In both quail and chick, a fraction of PGCs remain dormant, while the active ones progressed to the diplotene stage within 4 to 5 days (Table 1).

In chimeras, the genetic sex of the stroma always determined the sexual phenotype of the gonads. In the male stroma, PGCs - whether male or female - did not undergo meiosis within the experimental period.

Female quail PGCs in female chick stroma were examined at 11, 12 and 13 days (Table 2). At 11 days very few early preleptotene figures were present making the gonad look like that of the 10-day control quail. Later all chimeric gonads showed meiotic figures, but at 12 days they were preleptotene, corresponding to those in 11-day control female gonads.

UNFOLDING OF MEIOSIS IN THE LEFT OVARY OF NORMAL FEMALE QUAIL AND CHICK EMBRYOS

days of incubation	10	11	12	13	14	15	16	17	18	19	21	23	25
QUAIL	pL	pL +L	pL +L +Z +P	pL +L +Z +P	pL +L +Z +P +D	pL +L +Z +P +D		1st follicles					
CHICK			Interphase				pL +L	pL +L +Z	pL +L +Z +P	pL +L +Z +P			

Two left ovaries were studied at each stage

pL : preleptotene ; L : leptotene ; Z : zygotene ; P : pachytene ; D : diplotene

Table 1

On the other hand male quail PGCs in female chick stroma only entered meiosis at 13 days of incubation. At 12 days preleptotene figures were present in none of the 5 chimeric gonads (Table 2).

EVOLUTION OF QUAIL P G Cs IN THE FEMALE CHICK STROMA

Days of incubation	Stages of meiotic prophase in	
	♀ P G Cs	♂ P G Cs
11	pL (2)	Interphase (4)
12	pL (4)	Interphase (5)
13	pL + L (6)	pL (6)

() number of cases

Table 2

DISCUSSION

Irrespective of their genetic sex, quail PGCs which entered a female chick stroma underwent meiosis during the second week of incubation, at an earlier time than would chick GCs. However, the first signs of the meiotic process were delayed in comparison to the control female quail and occurred at different dates according to whether GCs were female or male. Female quail PGCs followed the quail clock with, a 24-hour delay, while male quail PGCs exhibited a further 48-hour delay (Table 1). Meiosis initiation still occurred earlier than in control female chick gonads. Thus the clock must be programmed both in female and male germ cells.

The experimental intervention is not responsible for the delay in onset of meiosis, since chick-chick chimeras built on the same model initiated meiosis according to the chick clock, whether the GC were genetically female or male (data not shown). Rather it seems that a normal block to the onset of meiosis has to be abrogated, presumably through a signal emitted by the female stroma. If such is the case the delay could be due to the low level of the stromal signal six days prior to chick meiosis.

In conclusion in this avian model it is clear that, provided the PGCs are in the female stroma, meiosis is initiated according to a clock built into the cells themselves. It is probably triggered by a stromal signal which must be given off much earlier than required in the chick species.

REFERENCES

Callebaut, M (1973). La synthèse de l'ADN prémitotique et préméïotique dans les cellules germinales de l'embryon femelle de la caille japonaise (Coturnix coturnix japonica). Experientia 29: 471-472.

Eyal-Giladi, H., Ginsburg, M. & Farborov, A. (1981). Avian primordial germ cells are of epiblastic origin. J. Embryol. exp. Morph. 65: 139-147.

Le Douarin, N.M. (1969). Particularités du noyau interphasique chez la Caille japonaise (Coturnix coturnix japonica). Bull. Biol. Fr. Belg., 103: 435-452.

Martin, C. (1972). Technique d'explantation in ovo de blastodermes d'embryons d'oiseaux. C.R. Soc. Biol., Paris 166: 283-285.

Tachinante, T. (1974). Sur les échanges interspécifiques de cellules germinales entre le Poulet et la Caille, en culture organotypique et en greffes coelomiques. C.R. Acad. Sci. Paris 278: 1895-1898.

Pattern Formation in Plants

Progress in Developmental Biology, Part A, pages 189–193

POLARITY ACQUISITION AND PROGRESSION DURING CARROT AND ALPHA-ALPHA SOMATIC EMBRYO DEVELOPMENT

V.Nuti Ronchi, M.A.Caligo, G.Martini and G.Luccarini
Istituto di Mutagenesi e Differenziamento - CNR

Via Svezia, 10 - 56100 PISA, Italy

The relative anaccessibility of the embryo in plant seeds and the consequent difficulty for biochemical investigation has been overcome with the possibility of performing somatic embryogenesis from plant cultures. In fact, the development of carrot somatic embryos (triggered by the sole omission of hormone from the medium) closely resembles zygotic embryogenesis and the unique features presented by the system are being exploited by various research groups for studies on the biochemistry and the genetics of early embryogenesis (Sung et al., 1984). Unfortunately, not many other plant systems are capable of producing embryos in liquid cultures in such a uniform and fairly synchronised way as carrot cultures.

Our study, which was initially directed towards an identification of the functional phases within the traditional embryogenic globular, heart and torpedo stages, has now been oriented to distinguish differences and similarities in embryo ontogeny in species which produce embryos in culture. The esterase reaction (Gahan, 1981) which reveals polarity through precocious staining of cells committed to stelar differentiation, has been used.

In carrots we have been able to identify two functional phases (Nuti Ronchi, Giuliano, 1984; Caligo et al., 1985) within Stage I (globular): the first phase (<u>isodiametric growth</u>), characterised by non-oriented cell divisions clearly distinguishable in the second phase (<u>polarity acquisition</u>) when polar distribution of esterase activity is evident on the site of the cotyledon emergence. <u>Polarity progression</u>

marks the gradual extension of the esterase activity to the hypocotyl and root in this order as the embryo proceeds from the heart to the torpedo stage, reaching the fourth phase (bipolar growth) when meristems are fully formed. This distinction of functional phases has been confirmed using various substances (a pirimidine analogue ECP, two aminoacids, proline and serine (Nuti Ronchi et al. 1982; Caligo et al. 1985) which deregulate normal development suspending it at different stages. Added at the beginning of embryogenesis, serine arrests it at phase I whereas proline affects polarity acquisition (as does ECP) giving an esterase reaction in the centre of the globular embryo, which grows isodiametrically, differentiating vascular tissue in an ectopic position. When added to different purified fractions of embryos, proline blocks development progressively from cotyledonary traces to hypocotyl and root in this order. A partial recovery of treated embryos in medium lacking the inhibitors resulted in structures lacking the hypocotyl and the root (when the block is early) or only the root. No inhibition was induced when bipolar growth had begun indicating, in this phase, an irreversible determination of the morphogenetic pattern of the embryo.

The possible pathways which can account for the induced deregulation of polarity acquisition and progression in carrot somatic embryos with the use of proline and serine, are under study. Either a morphoregulatory role, through the hydroxyproline and serine rich protein 'estensin', or a modification of the cell osmotic balance and pH, influencing the levels of endogenous auxin (Leguay and Guern, 1977) may be supposed.

Interestingly, many arrested root-less proline induced aberrant forms are phenocopies of the structures differentiated in carrot cultures stressed by various agents: (premature medium change, high cell density, altered temperature atc.) and in ts mutants (Terzi et al. 1982) given the temperature pulse after stage 2. For a comparison with carrot embryogenesis, which so far appears as the best regulated in vitro system, we are studying the pattern of initiation and development of embryos from different strains of Medicago callus cultures, growing in different conditions.

We have studied strain K-11 (From Dr.Novàk, Olomouc,Czechoslovakia) cultured on Bladys medium- 2gr/1 yeast extract 1 mgr/1 2,4-D (K-11 A) - or in MS,-2gr/1 yeast extract, 0,5 mgr/1 2,4-D (K-11 B)- and strain P_1 Adriana (from Dr.

Arcioni, Perugia, Italy) cultured on Ms medium-2mgr/l 2,4-D- Embryogenesis was initiated in agarized hormon-free medium.

Medicago embryos appear to develop through three distinct morphogenetic processes: 1) formation (from single cells or small clusters of cells: strain K-11 A) of an ovoidal structure from which cotyledons develop, esterase reaction indicating the commitment of the cotyledonary traces somewhere more centrally located than in carrots. The ovoidal structure is maintained on the axis below cotyledons and the root is functional only on transfer to fresh medium. 2) Formation (strain K-11 B) of a smooth, green ovoid outgrowth similar to 1) but which enlarges, developing in a central position a vascular bundle system and which, depending on conditions not yet completely clear, may produce, from epidermal cells, either embryos (developing ad described in 1) or through a budding process) or more outgrowths with vascular differentiation. 3) A budding process seems to be involved also in strain P_1 Adriana where shoots develop from a green branched structure, vaguely similar to a deformed leaf, differentiated inside a friable white homogeneous mass of cells obviously different from the non-morphogenic brownish callus. These shoots resemble embryos since roots easily develop after their tranfer to a quarter strenght Ms medium.

Our preliminary study can be summarized as follows: i) The esterase activity is in the first place detected on the cotyledonary traces which appear to be determinant and influence the development of hypocotyl and root in this order. ii) If this polarity is not achieved an aberrant structure with one or more vascular bundles is formed and on which epidermis secondary embryos may arise. iii) Secondary embryos may develop from epidermal cells or by a budding process from a branched structure, cotyledons appearing first as shoots, then by a well developed ovoidal axis. Root is never present, being formed only after the tranfer of the embryos to fresh medium. iiii) Single normal roots may also differentiate from globular structures when esterase activity remains centrally located and a vascular root system is developed.

Such different developmental sequences are greatly influenced by environmental and cultural conditions which may easily change carrot and alpha-alpha embryos in structures intermediate between embryos and shoots. The knowledge of the relationships between environmental parameters and

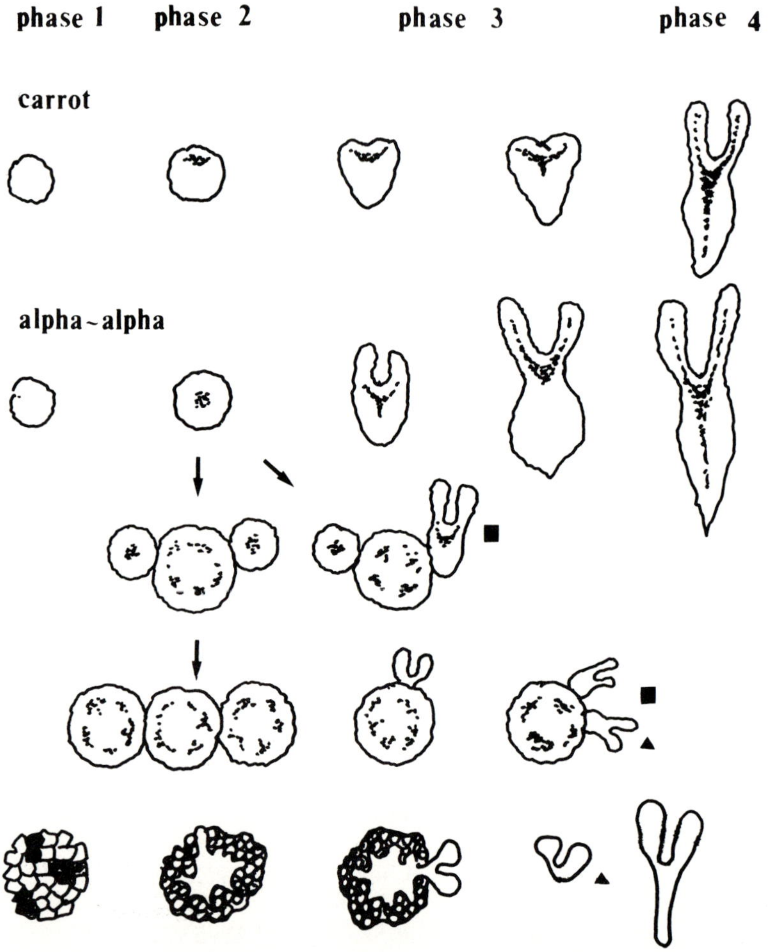

Developmental sequences of carrot and alpha-alpha somatic embryogenesis. ■ secondary embryogenesis ▲budding processess. In both cases roots develop only after transfer in diluted hormone-free medium.

forms is of great importance to understand the development and to define the conditions for differentiating normal em bryos also in recalcitrant species.Moreover, histological ana lysis should be carried on any time a clonal single cell origin of the embryo is required.

REFERENCES

Sung ZR, Fienberg A, Chorneau C, Borkird C, Furner I, Smith J, Terzi M, Lo Schiavo F, Giuliano G, Pitto L, Nuti Ronchi V (1984). Developmental Biology of embryogene sis from carrot. Plant Mol Biol Rep 2:3-14.

Gahan PB (1981). An early cytochemical marker of commitment to stelar differentiation in meristems from dicotyledono- us plants. Ann Bot 48:769-775.

Caligo MA, Nozzolini M, Nuti Ronchi V (1985). Proline and serine affect polarity and development of carrot somatic embryos. Cell Differentiation, in press.

Nuti Ronchi V, Giuliano G (1984). Temporal dissection of carrot somatic embryogenesis. In Novàk, Havel, Dolezel (eds): Proceeding of the International Symposium "Plant Tissue and Cell Culture. Application to Crop Improvement" Olomouc (CSSR) pp.77-87.

Leguay JJ, Guern J (1977). Quantitative effects of 2,4-Di- chlorophenoxyacetic Acid on growth of suspension-cultured Acer pseudoplatanus cells. Plant physiol 60:265-270.

Nuti Ronchi V, Caligo MA, Marazziti D, Terzi M (1982). A morphologic study of carrot somatic embryogenesis and its variously induced blocks. In Akio Fujiwara (ed): Proc 5th Intern Congress on Plant Tissue and Cell Culture, pp.117-118.

Terzi M, Giuliano G, Lo Schiavo F, Nuti Ronchi v (1982). Studies on plant cell lines showing temperature-sensitive embryogenesis. Embryonic Development, Part B: Cellular Aspects, Alan Liss, Inc; New York pp.521-534.

Progress in Developmental Biology, Part A, pages 195–198

FORM AND PATTERN: LINKAGE THROUGH CELLULOSE REINFORCEMENT AS SEEN IN FLOWERING

Paul B. Green

Department of Biological Sciences
Stanford University, Stanford, CA 94305

A direct biophysical relation between form and pattern can be sought on the surface of growing shoot apices. Tissue stresses concentrate at the surface as large tangential tensions. The directional yielding of the epidermis to these tensions will strongly influence growth and form. Tissues tend to yield at right angles to their reinforcement which is set by the orientation of cellulose in the primary walls (see Green and Brooks, 1978). While most cells in a cylindrical organ are transversely reinforced, analysis is simplified if emphasis is placed on the surface cells. Thus, as a useful first approximation, a cylindrical organ acquires its form, and elongates, due to the over-all hoop reinforcement of its surface (Fig. 1c).

The hoop reinforced condition of a new organ is readily derived from an earlier concentric ring pattern on the surface of the parent organ (Fig. 1a, b). The aim of this article is to explain the appearance of stamens, by way of such concentric patterns, in the flower of Crassula arborescens. Five stamens appear between, and interior to, five petals. The mechanism for stamens has two parts.

a. The enlarging base of a newly formed organ stretches the nearby dome to generate a new field of reinforcement tangential to the base (Fig. 1b, c). The new alignment, passing over many cells, is set up by cell responses involving oriented division and cellulose synthesis (Green, 1985).

b. Regions which have partial or rudimentary hoop

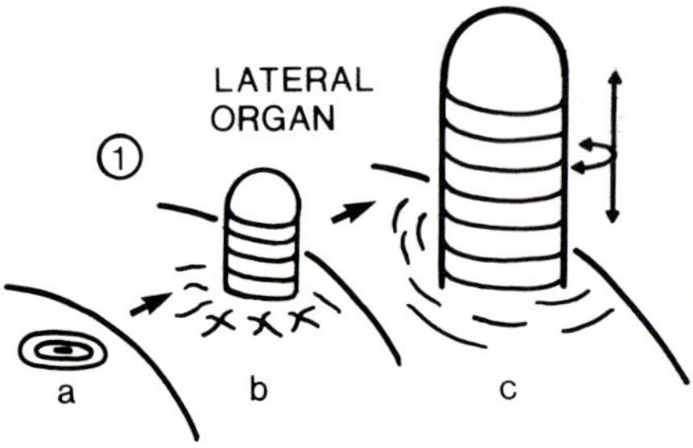

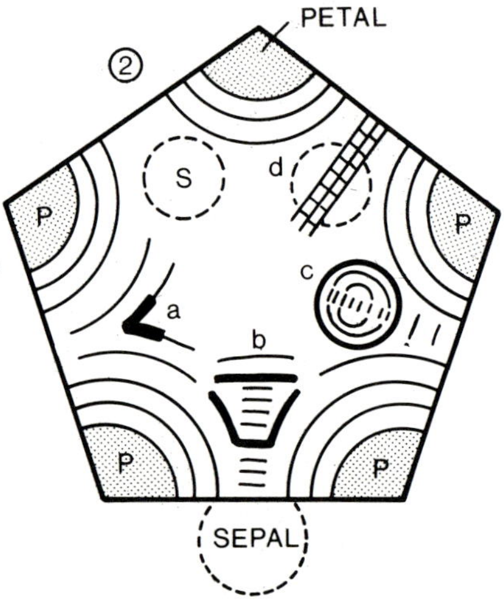

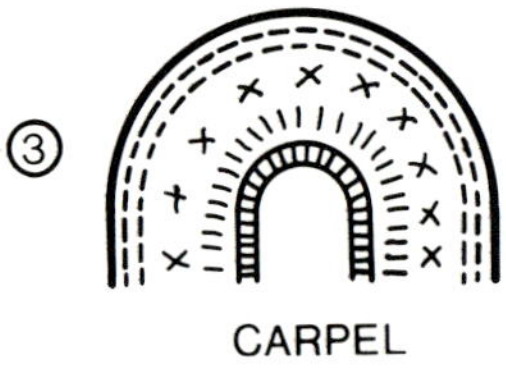

Figure 1. Organ initiation on a dome. A hoop reinforced area (a) emerges as a young primordium (b). Growth of its base stretches adjacent tissues; cells there form a new hoop field of reinforcement (c).

Figure 2. A floral dome. Five petals (p) are in a whorl. Each petal has made a circumferential field of reinforcement near it. If these merely contact (a), no new hoop pattern results. If, however, these fields combine with a field from a sepal, a new rough hoop field results (b). The region smooths and initiates a stamen field (c). This field has reinforcement over its center, shown by the radial files in (d). Diagram corresponds in part to Fig. 5.

Figure 3. The reinforcement lines on a young carpel of Graptopetalum paraguayense. A hairpin shaped file is reinforced transverse to the file axis. The file is like one on a leaf margin. The straight edges of the hairpin will later fuse. The bulk of the carpel has hoop reinforcement as indicated near the outer margin. See Fig. 6 and 7.

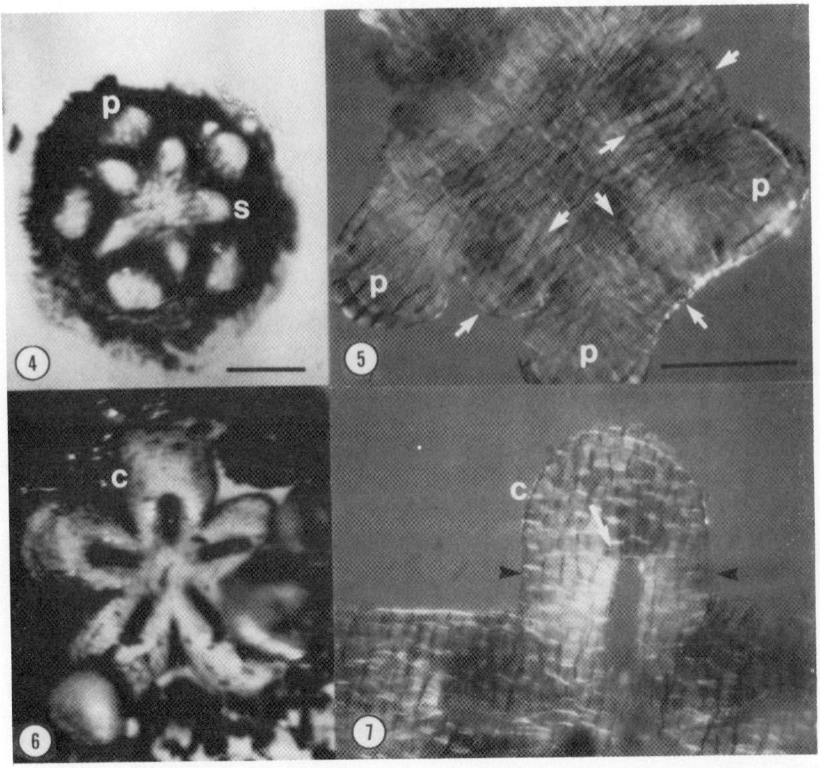

Figure 4. Negatively stained floral meristem of *C. arborescens*. Petals (p) are at the periphery; stamens (s) are at the vertices of the inner star. Bars show 0.1 mm.
Figure 5. Surface cells of the same meristem seen in polarized light. Regions darker than background have reinforcement running NE-SW on the page. Lighter regions run NW-SE. Opposing arrows show files, reinforced transversely to the file axis, for three stamen sites. These files pass through regions of opposite contrast, hence the site is hoop reinforced. See Fig. 2c, d.
Figure 6. Forming carpels (c) of *Graptopetalum paraguayense*. Central depression of each carpel is dark.
Figure 7. Surface cells of a carpel in polarized light. Dark regions have alignment N-S; light, E-W. White arrow shows a transversely reinforced file rimming the opening. Note the contrast reversal between the file and the periphery (dark arrows). This inidcates overall hoop reinforced structure as diagrammed in Fig. 3.

reinforcement of the right size will smooth in the gaps and emerge as a new organ.

The main proposal is that the new hoop reinforced region originates as a composite of two or more fields, each derived from a different older organ. In the simplest case the fields from two petals could set up only a V-shaped reinforcement site (Fig. 2a). If, however, a field from a sepal is also present, then four alignments are on hand and these approximate a hoop field rather well (Fig. 2b). When smoothed out, the structure is as shown in Fig. 2c, with the sepal derived files drawn in Fig. 2d.

The apical dome surface (Fig. 4), when removed and viewed in polarized light (Fig. 5) has alignments of cellulose compatible with the above propositions. Three stamens sites, between paired arrows in fig. 5, show alignment as diagrammed in Fig. 2c. The prominent cell files, see between the arrows, appear to establish bilateral symmetry for the anthers. At this stage there is no structure for carpels.

While special radial cell files, transversely reinforced, characterize forming stamens, the young carpel has distinctive files, transversely reinforced, running in a waved circumferential fashion (Fig. 6, 7). This structure, for a related species, is diagrammed in Fig. 3. The origin of this pattern is under study.

To summarize, the typical centripetal development of floral parts may have its basis in a cascade of biophysical inputs and responses that set up appropriate surface reinforcement patterns for the successive organs.

REFERENCES

Green, PB (1985). Surface of the shoot apex: a reinforcement field theory for phyllotaxis. J Cell Sci Suppl 1: in press.

Green PB, Brooks KE (1978). Stem formation from a succulent leaf: its bearing on theories of axiation. Amer J Bot 65:13-26.

Progress in Developmental Biology, Part A, pages 199–202

METABOLIC NETWORKS IN DEVELOPING PISTACHIO EMBRYOS (PISTACHIA VERA CV. KERMAN)

Don J. Durzan and K. Uriu

Department of Pomology, University of California, Davis, California 95616

INTRODUCTION

A novel set of 3-dimensional contour maps describing stages of embryonic development and metabolic networks of the pistachio has been constructed to characterize the effects of zinc deficiency. Our approach allows us to characterize the physiological states of each developmental stage and leads to more holistic interpretations of the control over growth and development than has been the case in tree biology.

We have identified 12 fruit-bearing trees (20-year-old) having a wide range of zinc deficiency (Uriu, 1982). The lack of zinc leads to empty or blank nuts. Embryo development was studied over 158 days in trees not having any signs of zinc deficiency (Durzan, unpublished). In normal trees, stages of development ranged from fertilization to complete filling of the nut by cotyledons. The composition of the developing nut was determined for free, naturally occurring nitrogenous compounds. In the pistachio, these totalled nearly 100 different compounds, i.e., free amino acids (AA), polyamines (PA), and monosubstituted guanidines (MSG). Composition is based on a computer-assisted analytical system with an automated amino acid analyzer employing specific post-column reactions (Dilley and Rocek, 1982; Natelson, 1984).

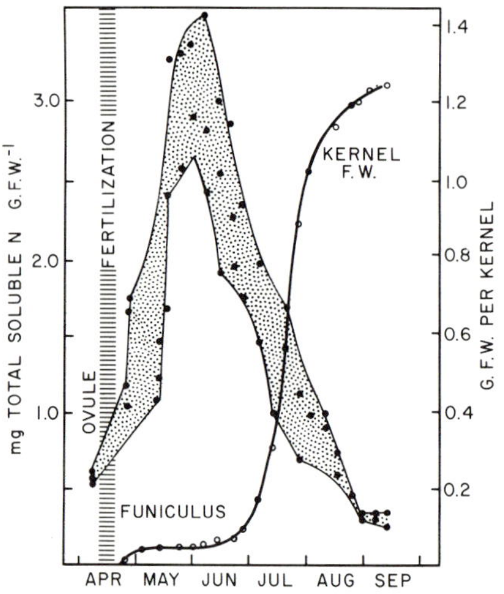

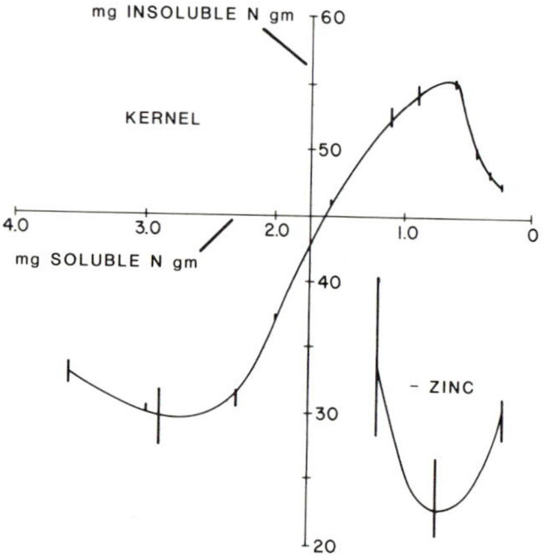

Figure 1. Growth in fresh weight and total soluble N in normally developing pistachio kernels.

Figure 2. Relationship between total soluble N and bound N (protein) during kernel development and effect of zinc deficiency on reducing the levels of N.

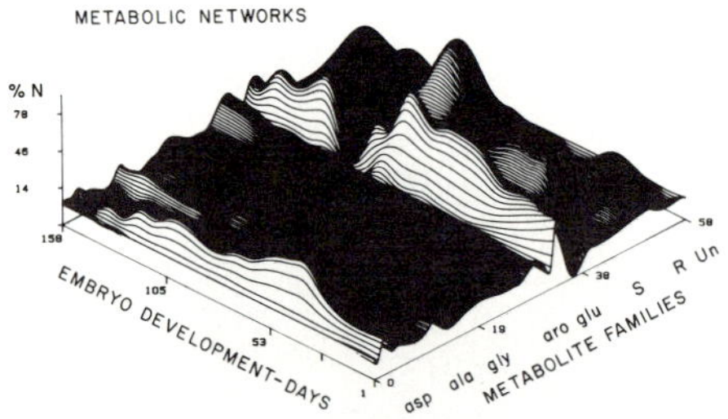

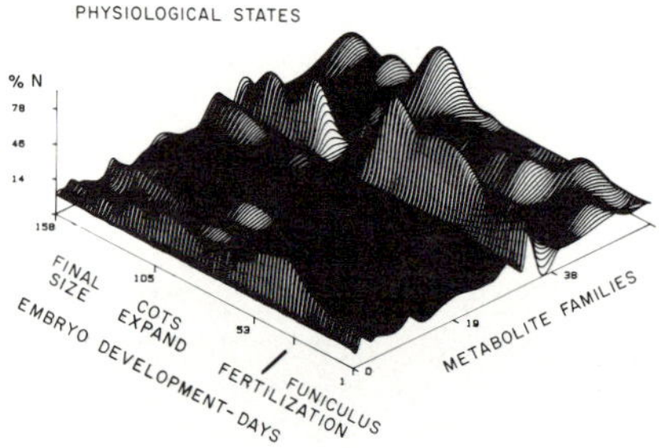

Figure 3. Metabolic networks arranged as amino acid families during normal kernel development (cf. Durzan and Steward 1983). Families: aspartic, alanine, glycine, aromatic, glutamic, sulfur (S), remaining (R) and unidentified (Un) compounds.

Figure 4. Stages or physiological states of embryonic development (Lin et al. 1984) may be characterized by integrations of all metabolite levels over networks.

RESULTS

The total alcohol soluble N accumulated in the funiculus and ovule (days 1 to 90 after fertilization) and declined as the embryo began rapid growth and development (days 90 to 158) (Fig. 1). The decline in total soluble N was related nonlinearly to an increase in bound (protein) N (Fig. 2). By contrast, in zinc-deficient trees, both total soluble N and protein N remained low and characterized the abortion of embryos or blank nut production. The correlation coefficient between blank production and zinc in leaves at harvest date (Sept. 9) was -0.939.

The percentage composition of free amino acids found in the total soluble N is described in two ways: 1) as families of metabolic networks based on the first amino acid formed by the incorporation of N into organic combination (Fig. 3) and 2) as physiological states of each stage of kernel development integrated over the metabolic networks (Fig. 4). Figures 3 and 4 are combined to produce a contour map characterizing quantitatively changes in developmental stages and composition of metabolic networks (Fig. 5).

The N metabolism of the normally developing ovule, funiculus and embryo (kernel) was dominated by the glutamate family and especially by arginine N. The soluble N in the funiculus is mainly asparagine. During embryo development, arginine levels give way to glutamate N even as the total soluble N declines. By contrast, with increasing zinc deficiency, the blank production increases and arginine N accumulates in the aborting embryo and especially in leaves (Fig. 6). Fully developed embryos are characterized with a high correlation coefficient between zinc and percent glutamate N (0.819).

CONCLUSION

Through computer-assisted analytical systems, the generation of 3-D contour maps have considerable diagnostic value for developmental systems. This approach has the potential of accommodating earlier concepts introduced by Waddington (1977), Thom (1975), and Zeleny (1977) to characterize morphogenetic systems.

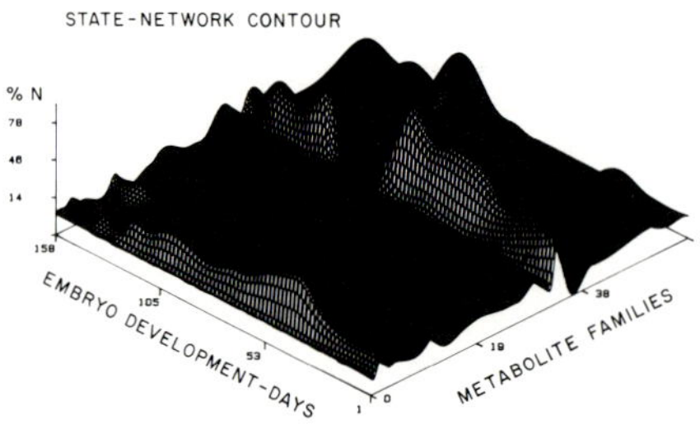

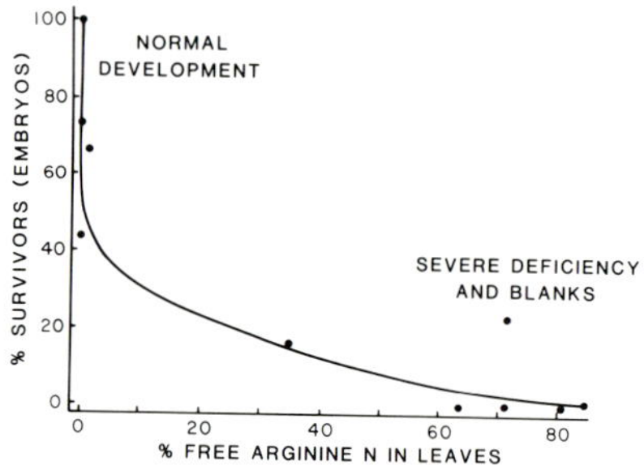

Figure 5. Contour map combining developmental stages and metabolic networks.

Figure 6. Threshold levels of arginine N in leaves and blank production (% surviving embryos) in trees having a range of zinc deficiency.

REFERENCES

Dilley KJ, Rocek PH (1982). Protein chemistry notes PCN-11 LKB Biochrome Ltd, Cambridge, England.

Durzan DJ, Steward FC (1983). Nitrogen metabolism. In Steward FC (ed): "Plant Physiology, A Treatise," Chap 2, Vol 8, New York: Academic Press, p 55-265.

Lin T-S, Polito VS, Crane JC (1984). Embryo development in "Kerman" pistachio. HortScience 19:105-106.

Natelson S (1984). Metabolic relationship between urea and guanidino compounds as studied by automated fluorimetry of guanidino compounds in urine. Clin Chem 30:252-258.

Thom R (1975). Structural stability and morphogenesis: An outline of a general theory of models. WA Benjamin, 348 pp.

Uriu K (1982). Diagnosis and correction of nutritional problems including the "crinkle-leaf" disorder. Ann Rept Calif Pistachio Industry, Fresno, CA (1982-1983), p 47.

Waddington CH (1977). "Tools for Thought." Basic Books, NY.

Zeleny BJ (1977). Self-organization in living systems. A formal model of autopoiesis. Int J Gen Syst 4:13-28.

Special Tenth Congress Lecture

Progress in Developmental Biology, Part A, pages 205–218

GEOFFROY AND THE HOMEOBOX

Stephen Jay Gould

Museum of Comparative Zoology, Harvard University, Cambridge, MA 02138

As Dickens said about another moment in French history, July, 1830 was both the best and the worst of times. During July 27-29, known ever since to Frenchmen as _Les Trois Glorieuses_, citizens manned the barricades in Paris and finally forced the abdication of despotic Charles X, younger brother of Louis XVIII. On July 31, on the balcony of the _Hotel de Ville_, the aging Marquis de Lafayette, America's friend in previous times of trouble, draped himself in the tricolor banner, embraced the Duc d'Orleans and proclaimed him Louis-Philippe, the new _constitutional_ monarch of France.

In the midst of this upheaval, a noted German observer wrote with brimming enthusiasm about a great Parisian battle: "We find in conflict here two such different ways of thinking...It has often happened before in the history of this antagonism, and the phenomenon must occur again and again...When such different elements are forced into contact, where they touch, every time an explosion must result."

The observer, we assume, must be writing about the age-old strife of monarchists and republicans, then playing its latest version on the streets of Paris. He was not. He described, in these dispatches, an argument then raging in l'Academie des Sciences--"of such great importance," he wrote, "that the history of science may never experience a second example."

The debate, running throughout 1830 in and around the

July revolution, featured France's two greatest zoologists as adversaries: Georges Cuvier and Etienne Geoffroy Saint-Hilaire. The observer was no mean journalist: Johann Wolfgang von Goethe. Geoffroy's son, writing later about the great debate, spoke of Goethe's involvement:

> At the very dawn of events that would overthrow a throne and change the face of France, he immersed himself in this peaceful revolution occurring in the pure realm of ideas.

Goethe was no incidental bystander, roped in on a lark. The great poet was also a fine scientist, a student of organic form and inventor, incidentally, of the word "morphology." He was also a partisan of Geoffroy's side in the great debate.

This famous struggle often portrayed quite wrongly as a debate about evolution (which Cuvier opposed to the core and Geoffroy supported in a fitful way), began as an argument about anatomy and ended up as an unresolvable conflict between the two most general and irreconcilable attitudes about the nature of organic form.

Geoffroy, a romantic and a dreamer, had a vision--perhaps the boldest, the most noble, the most comprehensive idea ever promoted in biology. As a <u>pure</u> morphologist, he sought nature's order in a single abstract plan for all organic design. He maintained, virtually as a motto, "there is, philosophically speaking, only a single animal." (Cubber, as a <u>functional</u> morphologist, denied the utility or reality of such abstractions and argued that an irreducible diversity of form matches the particular functions that organisms perform).

Goethe hoped against hope that Geoffroy could beat the silver-tongued Cuvier, one of France's greatest orators. For Goethe shared Geoffroy's vision and had, himself, advanced for plants the same general argument that Geoffroy now advocated for animals.

Goethe's idealistic morphology of plants, proposed late in the 18th century, tried to understand the varied parts, particularly the diverse structures of flowers, by conceiving them as so many expressions of one underlying, abstract theme that could generate all existing particulars through a set of geometrical transformations. For Goethe, this ideal source was the leaf itself.

Geoffroy applied the same vision to animals, arguing that vertebrae could perform for moving things the same role that Goethe ascribed to leaves as the foundation of form in plants. The dream that an apparently overwhelming diversity might be reduced to one underlying generality--a basis sought not in the mechanical terms of physics and chemistry, but in the ideal realm of pure form--motivated both men.

But scientific visions, no matter how noble, are only so good as their evidence--and Geoffroy proposed an elaborate set of hypotheses to affirm his claim that animals are built as repeated and modified vertebrae. In particular, he sought to establish a homology--that is, a true similarity based on common origin, rather than a mere analogy, or likeness developed independently by two groups in response to shared functional needs--between the complete segmentation of insects and the partial repetition of vertebrates (seen primarily in the vertebral column and paired ribs).

Such a proposed homology between the segments of an insect's *external* skeleton and a vertebrate's *internal* spine lead to some mighty peculiar comparisons, if we pledge to follow the basic model in a forthright and rigorous way--as Geoffroy certainly did. He wrote in 1822, with a firm assurance that others will interpret as bold or foolhardy according to their prejudices:

> I have never ceased to follow nature's great thought: the unity of organic composition; and, in the confidence that this fundamental truth has given me, I must grasp my problem in its greatest generality, approaching each part not with hesitation, but moving on the contrary with sureness.

Two comparisons, in particular, struck most biologists as especially hard to swallow, and helped to seal Geoffroy's fate. First, if the bones of a vertebrate's spine are homologous with external segments of insects, then insects must--quite literally--live *within* their own vertebrae. Goeffroy wrote in 1822: "every animal lives outside or inside its vertebral column" (*tout animal habite en dehors ou en dedans de sa colonne vertebrale*). Second, if the paired attachments to all "vertebrae" are also homologous, then the legs of insects are the same structure as the ribs of vertebrates--and insects walk on their ribs.

The comparison of insect and vertebrate caused difficulty enough for Geoffroy's vision of organic unity., But he wanted more and went further, provoking Cuvier's ire and his own downfall. Early in 1830, two of Geoffroy's young disciples tried to extend the grand homology by arguing that cephalopods (squids and their allies) are built as vertebrates doubled back on themselves--thus bringing the great phylumn of mollusks (including such overtly unsegmented creatures as clams and snails) into line with arthropods and vertebrates.

This was too much for Cuvier, the preeminent functionalist who understood that the strong similarities between squid and fish were analogies independently built as good hydrodynamic solutions to the problems of swimming. The great debate had begun.

Goethe, from the sidelines, rooted for Geoffroy, but most observers scored the match as a decided victory for Cuvier (although debates of such abstraction are not simply "won" or "lost" in conventional terms). And so history has judged this curious event. In our current classifications, complex animals are divided into two major branches, the protostomes (including insects) and the deuterostomes (including vertebrates)--and never the twain shall meet. As a graduate student in systematics 20 years ago, I learned no truth with greater assurance than the proposition, proved by Cuvier against Geoffroy, that the full and complete segmentation of arthropods has no homology with the limited and imperfect segmentation of vertebrates, confined primarily to muscles of the trunk and their associated skeletal and nervous systems. Goeffroy's vision of unity in organic design, died 150 years ago.

But big ideas developed by great thinkers rarely disappear completely. Few men of vision are so profoundly wrong that the motivating dream of an intellectual lifetime contains nothing worthy of later attention, often in an intriguingly altered form. Sure enough, last year's biggest news in developmental genetics has exhumed Geoffroy's old vision of homology between the segmentation of arthropods and vertebrates (although I have been astonished that no one, in scores of primary articles and secondary reports devoted to this subject of the so-called homeobox, has drawn the obvious connection with Geoffroy's ancient vision). The story, as we shall see, is particularly interesting because a superficial interpretation of this "homeobox" revives Geoffroy's dream in its overt form. We shall see, however, that this interpretation cannot stand--but that victory may yet be snatched from these jaws of Geoffroy's second defeat to affirm something deeper about his curious, yet profound, vision.

In the fruit fly Drosophila, best known of all complex animals in genetic terms, differentiation of segments (why some develop mouth parts, other wings and legs, still others abdominal structures) falls under the control of two gene complexes located on the right arm of the third chromosome. Each complex contains several genes. The antennapedia complex (called ANT-C for short) influences anterior segments of the head and thorax, while the bithorax complex (BX-C) works upon the fly's rear end, from the back part of the thorax through the numerous abdominal segments. Well-known mutations in several of these genes have the peculiar property of placing normal structures in the wrong places. (In the namesakes of the two complexes, flies with the antennapedia mutant grow legs where their antennae should be, while bithorax individuals develop a second pair of wings in opposition to their own name--flies are Diptera, or two-winged). Such mutations, replacing expected organs with homologous parts that should develop in other places are called homeotic--a term coined early in the century by William Bateson, who also gave genetics its name. (See my column of Oct., 1980 for more information on homeotic mutations).

Early in 1984, molecular geneticists announced a remarkable discovery about the fine structure of three genes in these homeotic complexes. We are living in the midst of a revolution in technique within genetics (spilling into public attention via debates about genetic engineering). The methods of research in so-called recombinant DNA have permitted us to learn, in a routine way (albeit with a vast amount of hard work), the sequence of base pairs that builds the genetic material of DNA. We can, in other words, now read the detailed structure of genes.

A very short region, only 180 base-pairs long and

located within the *antennapedia* gene, showed such remarkable similarity to a unit of identical length in another gene of the antennapedia complex (called *fushi tarazu*, or "futz" for short among the cognoscenti) that both must share a common evolutionary origin--the similarities are too complex and detailed for independent evolution under similar constraints of function (see my column of April, 1985 for an explicit discussion of when and why detailed molecular similarity must record homology, or unity of evolutionary origin). Soon afterward, the same short sequence was detected in a gene of the other homeotic complex, ultrabithorax of BX-C.

This short, highly conserved sequence received the name "homeobox" because it was first found only in genes of the homeotic complexes and because its trim, tight composition of but 180 base pairs suggested a discrete unit (a "cassette" as one discoverer wrote) that might be copied precisely and inserted into various parts of an organism's genetic material.

This discovery provoked a search for other copies of the homeobox within *Drosophila* genes. The work continues, but a recent report (a review by W.J. Gehring in *Cell*, January, 1985, pages 3-5) mentions more than 10 additional genes contining homeoboxes. The homeotic complexes have emerged as preferred "homes" for this conserved sequence--five genes of ANT-C, and three within BX-C, contain homeoboxes. The copies found outside the homeotic complexes lie in genes that also seem to code for some aspects of spatial organization in the embryonic segments.

The homeobox first inspired excitement because it seemed to confirm a longstanding hypothesis about the evolutionary origin of the homeotic complexes. Ed Lewis of CalTech, the world's expert on the homeotic genes of *Drosophila*, had proposed that the many genes of BX-C (at least eight), arrayed in linear sequence along the right arm of the third chromosome, arose as repetitions of a single ancestral gene (see my column of November, 1981 on the ability of some genes to make copies of themselves and move to new positions). The presence of such a highly conserved, shared sequence within so many homeotic genes provides strong support for Lewis' theory (an alternative, or complementary, idea might hold that only the homeobox "cassette" duplicates and inserts itself in new places).

The simple existence of such a highly conserved sequence in so many places provided enough excitement, but the story goes deeper--thus forging the link to Geoffroy's old vision. (I should say at the outset of this linear account of events that I am not describing discoveries in their chronological order, but in a logical sequence of my own construction that allows me to make sense of much fine work, proceeding in several laboratories and cascading simultaneously in many directions. All these discoveries occurred (or were published) during 1984--a good sign of how quickly an exciting field can move.

The presence of so many copies throughout the Drosophila genome does not prove that the homeobox does anything important for the organism (some repeated elements may act as "selfish DNA," able to replicate, move and insert, but performing no function for the larger organism containing them--again, see my column of November, 1981). An argument for function must first determine whether or not the homeobox gets transcribed into RNA and then translated to protein or enzyme--and the link to Geoffroy's vision requires a hypothesis of function. Much DNA is "silent," or untranscribed, and much may be "junk" (so-called) with reference to its utility for organisms.

The homeobox is, in fact, transcribed and translated--so this basic argument for function holds. Since DNA is a triplet code, the 180 base-pairs of the homeobox produce a so-called "homeodomain" of 60 amino acids. The strongest argument for important function rests upon the discovery that this homeodomain (the 60 amino acids) is even more highly conserved than the parent homeobox (the 180 base pairs of DNA). The foregoing statement may seem puzzling--that higher amino acid than code similarity indicates functional meaning for a stretch of DNA--but the justification arises from the nature of the DNA code.

The genetic code is, in technical parlance, "redundant" at the third position in most cases--that is, the first two base pairs of any triplet determine the amino acid that will be translated, while the third position can be occupied by any of four possible base pairs and still yield the same amino acid in translation. Thus, any mutation at the first or second position will cause the translation of an altered amino acid, but third position mutations are usually "silent" because they induce no change in the product--the amino acid --that the organism sees and uses.

Now, the homeodomain is substantially more similar among its several copies than the homeobox that codes for it (for the first copies discovered in Drosophila, homeobox similarities ranged between 75 and 79 percent of base pairs, while homeodomain conservation rose to 87 percent of amino acids). The basis of this difference must be--and indeed is, as we now know from direct sequencing of the base pairs and amino acids--that many silent mutations accumulate among the different copies of the homeobox to drive their overall similarity down, but that something is tightly conserving the amino-acid sequences by rejecting first and second position mutations that change the structure and function of proteins, but permitting the "invisible" silent mutations to accumulate in third positions. That "something" must be natural selection, operating to conserve the protein structure of the homeodomain. If natural selection is working so assiduously to conserve the homeodomain, then its protein must be doing something right for the organism--it must, in other words, perform an important function. (The comparison of silent versus translated mutations is a standard test for function in molecular genetics, not a new argument developed for the homeobox story.)

Thus, we know that the homeobox occurs in several Drosophila genes, mostly in the homeotic complexes, and that it codes for a protein with important functions. But what does this protein do? At first glance, but please remember that "first glances" can often be dangerously simplistic, we might be tempted to conjecture that the homeobox must be a key to the control and expression of segmentation. Why else should its copies lie in genes of the homeotic complexes and rarely elsewhere?

We might gain further clues about function of the homeobox by scanning other organisms to see where it does (and does not) occur. Initial results on other invertebrates might be viewed as enhancing the suspicion that homeoboxes are regulators of segmentation. The homeobox sequence has been found in other species of Drosophila, in Tenebrio molitor (a beetle) and in an earthworm of the genus Lumbricius--all segmented animals. It has not been detected in a nematode, a sea urchin and the bacterium E. coli, all unsegmented creatures (see W. McGinnis and others, Cell, volume 37, 1984, pages 403-408).

But the biggest surprise, and the source of popular appeal for the homeobox, arrived with its discovery in yet another group of organisms--our own. If *Drosophila* is the standard invertebrate of laboratories throughout the world, the South American toad *Xenopus* has been granted this dubious distinction among cold-blooded vertebrates (mice play the part for mammals). Inevitably, geneticists turned their search for homeoboxes upon *Xenopus*--with success. Homeoboxes have been found, so far, in two *Xenopus* genes. (Our knowledge of vertebrate development is so poor that we do not know, for certain, how any single gene operates to regulate the position or differentiation of structures in embryology. But we have some pitifully preliminary hints that the homeobox-containing genes of *Xenopus* may play some role in development. One gene, for example, produces RNA transcripts in egg cells and again at the crucial stage of gastrulation in early embryology--where the transcripts might be translated to proteins involved in the determination of embryonic cell types, see Muller and others, *Cell*, volume 39, 1984, pages 157-162).

The similarity between these *Xenopus* homeoboxes and those discovered earlier in *Drosophila* is startling. The corresponding homeodomain of one *Xenopus* gene is identical to the homeodomain of *antennapedia* in *Drosophila* at 59 of 60 amino acid positions! This frog-fly concordance is far stronger than the similarity between the two frog homeodomains (54 of 60 amino acids, with 23 silent third-position mutations in addition), or between any pair of homeodomains *within* Drosophila. Something is working awfully hard to conserve this similarity across one of the most profound phylogenetic gaps in the history of life (the common ancestry of insect and vertebrate predates the establishment of our adequate fossil record in the Cambrian explosion nearly 600 million years ago).

Further scrutiny then revealed homeoboxes in other vertebrate genes. A gene on mouse chromosome 6 contains a homeobox with 66-70 percent base pair similarity to *Drosophila* and 66 percent to *Xenopus* (W. McGinnis and others, *Cell*, volume 38, 1984, pages 75-80). Two human genes also sport homeoboxes, yielding an average of 90 percent amino acid similarity with the three homeodomains first discovered in *Drosophila* (see Levine and others, *Cell*, volume 38, 1984, pages 667-673).

As discovery cascaded upon discovery, many scientists and commentators asked what common theme lay behind these obviously interrelated observations. An intriguing possibility inevitably bubbled forth: homeoboxes are primarily found in genes that regulate the differentiation of segments in *Drosophila*; they are also present in other segmented invertebrates, and absent from unsegmented creatures. Vertebrates also have homeoboxes and at least some form of imperfect segmentation. Of course, we had all learned that segmentation in insects and vertebrates can only be analogous--a product of independent evolution in lineages long separate. Or must segmentation be only analogous? Could it possibly be homologous after all. The homeoboxes of insects and vertebrates are clearly homologous. If they regulate segmentation...well, by God, could Geoffroy have been right all along and after all (following 150 years of calumny and the kind of dismissal that so erased his name from the history of science that not a single geneticist remembered this link with the past while so many discussed the potential homology of segmentation)?

Press reports poured forth. Some began to speak, in overblown terms, about the Rosetta stone of development, biology's greatest mystery. Scientists tended to be more cautious, but several floated the heretical idea of homology for the segmentation of insects and vertebrates--Geoffroy's common ground plan of animal development. W. McGinnis and colleagues wrote, for example (*Cell*, volume 38, 1984, p. 679):

> If the conserved homeodomain in fruit flies, frogs, mice and humans is involved in the control of segmental development, then it is possible that the segmentally organized animals in both the protostome and deuterostome classes had a common ancestor, and that the metameric body plan has evolved only once in metazoa. (Metamerism refers to the complete segmentation of arthropods; metazoa are the multicellular animals.)

This explicit vindication of Geoffroy remains a possible interpretation of the homeobox, but proper caution has begun to replace overextended enthusiasm, and other less flashy (but in many ways more intriguing)--and (I suspect) more probable--explanations have now been offered.

The superficial interpretation--if it's in homeotic genes and homeotic genes regulate segments--is logically flawed, in fact one of the classically false syllogisms of basic courses. We might as well argue that sneakers produce a high field goal percentage because all great basketball players wear them.

Sure, the homeobox performs an important function, and (from the genes it chooses as bedfellows) it probably has something to do with the embryology of spatial organization. But why should that "something" be the actual control of segmentation itself (the only argument that would affirm Geoffroy's vision in its explicit form)? Think of the more basic activities that genes must superintend before we can even talk about such overt morphological phenomena as segmentation (just as basketball players have to cover their feet with something before they can display any fancy artistry with drives and jump shots)--particularly the production of proteins in the right places, at the right times and in the right amounts. Perhaps the homeobox works for one of these deeply basic functions acting as a vital prerequisite for anything so much later (in a causal chain) and so much more specific as segmentation.

My friends Elizabeth C. and Rudolf A. Raff (Nature, volume 313, 1985, page 185) have suggested such a more general and more "mundane" (their words) function for the homeobox. Translation of RNA to proteins occurs in the cytoplasm. But proteins that will act as basic regulators of development by turning other genes on and off must get back into the nucleus (where the genes reside on their chromosomes). Perhaps, they suggest, the homeodomain simply acts as a "nuclear tag" attaching to other regulating proteins that must enter the nucleus and, perhaps, facilitating their entry. I do not know whether the Raffs' hypothesis has any validity, but it represents the *kind* of proposal more likely to prove useful than the superficially attractive idea that homeoboxes build segmentation directly.

In fact, we have some hints of evidence for more basic function. Homeoboxes have also been found in two of the so-called mating-type genes of yeast--genes that seem to control the stable determination of cell types in these organisms. Now yeast, obviously, are not segmented. (This argument, however, does not preclude the notion that homeoboxes regulate segmentation in insects and vertebrates. After all, the homeobox didn't arise mysteriously from nothing. It must have developed from some precursor in an ancestral unsegmented organism. This ancestral proto-homeobox may have worked in one way for ancestors and then been co-opted as a regulator of segmentation in descendants. I discussed the important evolutionary principle of such quirky shifts in function, known to biologists by the awful and senseless name "preadaptation," in last month's column.)

We may, however, read the discovery of homeoboxes in yeast in a more positive light--not as a problem to finesse in preserving the hypothesis that homeoboxes regulate segmenntation, but as overt support for the conjecture that homeoboxes perform a more basic function common to yeast and segmented organisms--some primary control in the differentiation of cell types, for example.

An intriguing hint in this positive direction comes from the apparent homology between proteins coded by the yeast mating genes and proteins in bacteria that may perform the the most basic regulatory activity of all--binding to DNA. (If (If some genes work by regulating the turning on and off of others, then their products--proteins--must be able to bind with the DNA of other genes). Could the homeobox code for such a central protein function as DNA binding? Such a utility would represent function at the most basic level of organic construction--a role far more general than the production of segments. By the time segments form in insects and (imperfectly) in vertebrates, most of what really matters in the regulation of development has already occurred.

So where does this claim for function more general than segmentation leave Geoffroy's dream of a common ground plan for all organisms? Just when his most radical notion of homology between segmentation of insects and vertebrates seemed on the verge of revival in molecular form, along come all the nay-sayers and doubting Thomases of biological caution to point out that presence of a homeobox in a homeotic gene doesn't mean function for segmentation. These modern Cuviers, it seems, have again crushed the dream of unity after such a short and fitful flicker. Or have they?

What was Geoffroy really after? He had a vision of unity and he expressed it in the deepest and most general terms that his century knew--the generating rules of overt morphology. But Geoffroy dreamed of something deeper than external form--some underlying basis that would produce similarities of form as a necessary and superficial expression of its work. Indeed, though he invoked form faute de mieux, Geoffroy always sought a deeper unity that would generate form as a product. After comparing the segments of a cow and a crab, he once wrote in 1807: "I have always insisted upon an identity which has extended to the least important relation of all, that of form." And, towards the end of his life, following his rout by Cuvier, he wrote in 1835:

> I have, moreover, regenerated this principle (unity of plan and composition), and obtained for it universality of application, by showing that it is not always the organs as a whole, but merely the materials composing each organ that can be reduced to unity.

In other words, Geoffroy dreamed of a unity deeper than the homology of overt segmentation (which he viewed only as the best empirical evidence he could find--the closest approach that techniques of his time would permit to the underlying basis and mystery of development). Somewhere, wherever good biologists go, I think that Geoffroy's spirit is now rejoicing that the homeobox may unite insects and vertebrates in a developmental unity more basic than their final forms--and that the most general rules regulating the conversion of DNA code to adult organism may be common to all creatures.

Surely, in some sense, Geoffroy was correct. Segmentation in insects and vertebrates probably is only analogous, but all organisms have a common evolutionary origin on this planet, and all must therefore share some basic rules of construction inherited as consequences of this history. There must be "unity of plan and composition" at some higher level that transcends the simple and obvious identity of chemical building blocks.

And perhaps Goethe was right after all to ignore the tumultous events in the streets of Paris and to argue instead that an abstract debate in the French Academy of Sciences had far more ultimate importance in human history.

The art of finding timeless essences in apparent trifles, and ignoring contemporary, engulfing tumults as ephemeral, marks the kind of perception that we call genius. Umberto Eco has captured this vital theme as the centerpiece of his wonderful mystery novel, *The Name of the Rose*. William of Baskerville, that medieval Sherlock, is surrounded by the portentious events of his century, particularly the dispute between papacies of Rome and Avignon. But he solves the Abbey murders because only he (and his antagonist) understnd that these contemporary doings, however enveloping, will pass and be forgotten in time--and that the only important object in the long run, however invisible to all others amidst such tumult, is the library's copy, the only in existence, of a potentially anti-Christian work of Aristotle. Goethe had his priorities straight as well--and the plea of his dying worlds rings through the ages: *Mehr Licht* (more light).

Variations in RNA Processing During Development and Translational Controls

Progress in Developmental Biology, Part A, pages 221-234

HORMONAL ASPECTS OF THE REGULATION OF DOPA DECARBOXYLASE IN *DROSOPHILA MELANOGASTER*

Ross B. Hodgetts, William C. Clark, David D. Eveleth Jr.*, R. Daniel Gietz, Charlotte A. Spencer and J. Lawrence Marsh*
Department of Genetics, University of Alberta, Edmonton, Alberta T6G 2E9
*Developmental Biology Center, University of California, Irvine, California 92717

INTRODUCTION

Complexity of form in multicellular organisms is achieved via chemical signalling between spatially separated cells. The steroid hormones constitute one of the most intensively studied classes of chemical messenger and it is generally believed that their mode of action occurs primarily at the level of transcription of specific genes.

The currently accepted model of transcriptional control via steroid hormones (for review, see Anderson, 1984) requires that the nuclear steroid-receptor complex becomes bound to a specific DNA sequence within the control region of a gene or a battery of genes (Scheidereit et al., 1983; Mulvihill et al., 1982; Chandler et al., 1983; Compton et al., 1983). The accessibility of this particular DNA sequence to the hormone-receptor complex is presumed to result from the altered chromatin structure which is known to surround responsive genes (Shermoen and Beckendorf, 1982; Burch and Weintraub, 1983). Binding of the hormone-receptor complex in turn may then facilitate a loosening of the nucleosome structure, allowing entry of RNA polymerase and specific gene expression.

Although elegant in its simplicity, this model of transcriptional control cannot adequately explain all the effects elicited by steroid hormones in many target tissues. The fruit-fly, *Drosophila melanogaster*, provides a particularly good model system for addressing some of these unanswered questions and the regulation of a number of genes by the steroid molting hormone, 20-OH-ecdysone, is currently under investigation. The analysis, by Ashburner and his co-workers, of the effects of this hormone on salivary gland chromosome puffing patterns (Ashburner et al., 1974; Richards, 1980), revealed a variety of responses within one target tissue. These workers were able to treat explanted glands *in vitro* with 20-OH-ecdysone and define several classes of affected genes. The *intermolt* genes were active at the time of gland explantation from mature larvae, and were turned off by exposure to 20-OH-ecdysone. The *early* genes were initially inactive, but within minutes of hormone treatment, puffing was observed, and since it was independent of protein synthesis, a direct effect of the hormone-receptor complex on gene transcription was inferred. The *late* genes were activated only after a lag of 3-4 hours following exposure to 20-OH-ecdysone, and activation required protein synthesis during the period of early puff activity. Finally, the *mid prepupal* genes, unlike the three classes above, required a pulse of hormone for their expression and were actually inhibited in the continuous presence of hormone.

These studies were significant in illustrating the different effects of a single hormone in a given target tissue. Unfortunately, with the exception of the intermolt puffs, which code for the Sgs or "glue" proteins (Akam et al., 1978; Crowley et al., 1983), gene products are unknown. We have chosen to study the dopa decarboxylase *(Ddc)* gene-enzyme system in *Drosophila*. The classic experiments of Karlson amd Sekeris have shown that the appearance of DDC in the epidermal cells of the mature dipteran larva is dependent on 20-OH-ecdysone (Karlson and Sekeris, 1966; Fragoulis and Sekeris, 1975a). Also, DDC has been subjected to intensive genetic, developmental and molecular analysis in *Drosophila*. In this paper we will show that in the mature larval epidermis, the *Ddc* gene responds to the hormone in a manner similar, but not identical, to that described for the early salivary gland ecdysteroid genes. In contrast, in imaginal disc cells and perhaps also in the epidermal cells of embryos, *Ddc* gene expression is akin to a typical mid prepupal salivary gland gene. Thus, one structural gene *(Ddc)* exhibits the properties of at least two of the classes of hormonally controlled salivary gland genes. We will conclude with an analysis of a naturally-occurring genetic variant, Ddc^{+4}, which alters the pattern of DDC gene expression in a stage and tissue specific manner. It is hoped that a combination of DNA sequencing and DNA-mediated genetic transformation using *Ddc*$^{+4}$ will reveal the various mechanisms of ecdysteroid action at this locus and will contribute to more refined molecular models for steroid hormone action.

SPATIAL AND TEMPORAL PATTERNS OF DDC ACTIVITY

High levels of DDC activity can be detected in the epidermis at five stages in the life cycle when the sclerotization (hardening) of cuticle occurs (Kraminsky et al., 1980; Marsh and Wright, 1980). Five major peaks of 20-OH-ecdysone also occur during the life cycle (Kraminsky et al., 1980). A comparison of hormone and enzyme profiles indicates only one stage (pupariation) when 20-OH-ecdysone and DDC activity coincide; at all other stages, there is a significant lag (10-48 hr) between hormone and enzyme maxima. The dopamine produced by DDC is rapidly converted to N-acetyl dopamine (Hodgetts and Konopka, 1973) which is the presumed sclerotizing agent in Diptera (Sekeris and Karlson, 1962). DDC is also present in neural tissue of *Drosophila* (Dewhurst et al., 1972; Livingstone and Tempel, 1983) where dopamine presumably functions as a neurotransmitter. Finally, both ovaries (Wright et al., 1981) and the larval proventriculus (Marsh and Wright, 1979) have detectable levels of DDC. The function(s) of DDC in ovaries is unknown, but in proventriculus the enzyme is probably involved in the sclerotization of the peritrophic membrane (Rizki, 1956).

THE GENETIC SYSTEM SPECIFYING DDC

Despite the complex developmental profile of epidermal DDC activity and the occurrence of the enzyme in at least three other tissues, genetic evidence demonstrates that a single structural gene specifies the enzyme in *Drosophila*. Wright and his co-workers (Wright et al., 1976; Wright et al., 1982) have isolated recessive lethal and temperature-sensitive mutations in this gene which have been shown to eliminate DDC activity in the epidermis of mature embryos, mature larvae and young adults (Gietz and Hodgetts, 1985). Activity in brain tissue (Wright, 1977; Livingstone and Tempel, 1983), and in the proventriculus (Marsh and Wright, 1979) is likewise affected in these mutants, suggesting that both neural and epidermal DDC molecules are the product of the same structural gene. The genetic data are supported by immunological evidence. Antigenically indistinguishable molecules, as revealed by Oucterlony double diffusion, are produced in the epidermis of embryonic, larval and adult stages (Hodgetts et al., 1985). Furthermore, all the DDC activity in adult brains can be immunoprecipitated by the antibody raised against the larval enzyme (Spencer, unpublished). Finally, genomic Southern analysis and *in situ* hybridization results (Hirsh and Davidson, 1981) likewise point to the existence of just one structural gene, located in chromomere $37C_{1,2}$ on the left arm of chromosome 2.

To date, very little is known about regulatory loci that may influence the temporal or spatial patterns of DDC expression. Sherald and Wright (1974) isolated two strains, R^E and R^S, which were shown to have roughly 50% more DDC activity than a wild type (Marsh and Wright, 1979). The mutations map just proximal to the bristle marker *hk*, on chromosome 2L, which places them very close to the DDC structural gene. Another variant strain, designated *Ddc*$^{+4}$, has been more thoroughly characterized (Estelle and Hodgetts, 1984a,b). *Ddc*$^{+4}$ is tissue specific, affecting epidermal, but not neural, DDC expression, and maps adjacent to the structural gene for DDC. This strain is also developmentally specific, exhibiting a 40-50% overproduction of DDC mRNA at all stages in the life cycle except at pupariation, when DDC mRNA levels are about 50% of normal. The variations in mRNA levels detected in this strain are presumably responsible for the analogous changes in the amount of DDC activity and protein levels observed at these stages. Further analysis of this variant is presented later in this paper.

EVENTS IN THE EPIDERMIS AT PUPARIATION

The first evidence of a role for the molting hormone in the control of DDC expression was reported by Karlson and Sekeris (1962) who demonstrated the induction of DDC activity by 20-OH-ecdysone in isolated abdomens of *Calliphora*. Later, Chen and Hodgetts (1974) demonstrated the premature appearance of DDC in last instar larvae of *Sarcophaga* in response to the injection of 20-OH-ecdysone. Finally, Fragoulis and Sekeris (1975b) showed that injection of 20-OH-ecdysone into ligated abdomens of mature *Calliphora* larvae caused an elevation in the level of translatable DDC mRNA.

We have further established the role of 20-OH-ecdysone in DDC induction at pupariation using the temperature-sensitive *Drosophila* mutant, *ecd*^{1}, isolated by Garen et al. (1977). This mutant, when maintained at the restrictive temperature of 29°C is unable to produce the high levels of 20-OH-ecdysone normally seen at pupariation. At this temperature, the mutant larvae are unable to pupariate, and are likewise unable to produce DDC activity (Kraminsky et al., 1980). However, feeding 20-OH-ecdysone to *ecd*^{1} larvae maintained at 29°C effects the elevated ecdysteroid titer normally seen at this time, and pupariation ensues. Furthermore, DDC translatable mRNA is detectable as early as 2 hr after feeding and DDC activity begins to increase in the epidermis 4 hr after feeding (Kraminsky et al., 1980).

Since this rapid response to the hormone is reminiscent of the early puff class of genes in the salivary glands, we asked whether the epidermal DDC response shared the other property of the classic early salivary puff gene: namely, insensitivity to inhibitors of protein synthesis. We have developed larval feeding conditions which inhibit incorporation of ^{35}S-methionine into total protein by more than 98%. Likewise, hormonally-induced levels of DDC enzyme activity at pupariation are inhibited by more than 95% (Clark et al., 1985). In these experiments, where 20-OH-ecdysone was fed to mature *ecd*^{1} larvae maintained at 29°C, DDC transcript levels, monitored by hybridization to a cloned genomic fragment of the *Ddc* gene, were observed to increase 25-fold over a 6-7 hr period. Inhibition of protein synthesis in these experiments, reduced the increase in DDC transcripts to 6-fold over control organisms maintained at 29°C in the absence of 20-OH-ecdysone. Although inhibition of protein synthesis reduced DDC mRNA levels, the considerable transcript accumulation in the absence of protein synthesis clearly distinguishes epidermal DDC in puparia from the late genes of the salivary glands, whose expression is totally inhibited by cycloheximide.

For comparison, the transcript levels of a typical early salivary gland gene were monitored in these experiments, in collaboration with Ken Burtis at Stanford University, who has cloned DNA from the 74EF early puff region (Burtis et al., 1984). While induced levels of the 74EF transcript increased 3.6-fold over control values in a 6-7 hr period following feeding of hormone in the absence of inhibitors of protein synthesis, transcript levels actually increased 4.4-fold when protein synthesis was inhibited. The results demonstrated that *Ddc* is not behaving exactly like the early gene at 74EF, since the inhibition of protein synthesis markedly decreased the induced level of DDC mRNA.

Therefore, we conclude that two mechanisms, operating together, account for the significant increase in DDC mRNA in the mature larval epidermis following a rise in the ecdysteroid titer. An immediate effect, independent of protein synthesis, results from the direct action of the steroid-receptor complex, perhaps at the level of transcription. Secondarily, DDC mRNA continues to accumulate under the action of a protein or proteins synthesized during the period of elevated 20-OH-ecdysone. As judged from our data, these two effects contribute roughly equally (on a logarithmic scale) to the total accumulation of DDC seen during the first 6-8 hr following exposure of the epidermis to 20-OH-ecdysone.

EVENTS DURING EMBRYOGENESIS

The relationship between 20-OH-ecdysone and *Ddc* gene expression just described for the mature larval epidermis presents an interesting contrast with events at embryogenesis. Figure 1 in Kraminsky et al. (1980) shows the developmental profiles of both 20-OH-ecdysone and DDC activity *in vivo* at these two stages. Unlike pupariation, embryogenesis is characterized by a significant lag between the maximum ecdysteroid titer (10 hr) and the DDC activity maximum (22 hr). Several possible explanations for the lag seen in embryos have been examined. Excluded was the possibility that embryonic DDC is encoded by a different gene from that producing the enzyme at pupariation (Gietz and Hodgetts, 1985; Hodgetts et al., 1985). Likewise, a second possibility, that the ecdysteroid peak in embryos was not 20-OH-ecdysone, the hormone identified at pupariation (Borst et al., 1974) was also excluded (Hodgetts et al., 1985). In an attempt to provide a unifying explanation for the effects of 20-OH-ecdysone on DDC gene expression in embryos and mature larvae, we sought evidence for the production in embryos of DDC mRNA at the time of maximum hormone titer, rationalizing the lag in the appearance of enzyme activity by postulating a block in the translation of the mRNA. Northern analyses of embryonic poly-A^+ RNA have ruled out this explanation of the lag, as significant DDC transcript accumulation occurs only in late stage embryos (Beall and Hirsh, 1984); Gietz

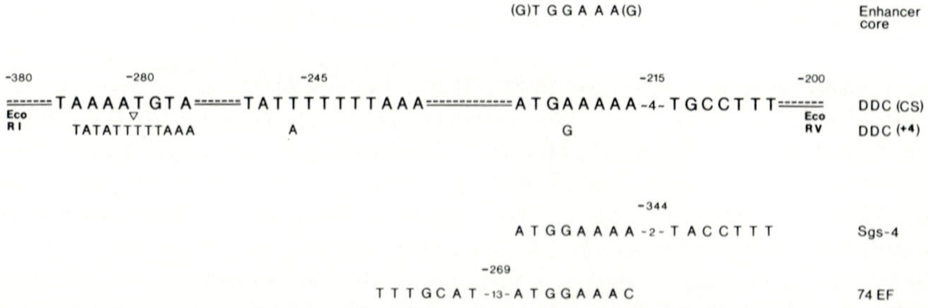

Figure 1. DNA sequence comparisons between Canton-S and *Ddc*$^{+4}$ upstream of the *Ddc* gene. DNA sequences upstream of *Ddc* were determined by the dideoxy method of Sanger et al. (1977). The numbers refer to distances from the transcript start sites (O). For the *Ddc* gene, this was taken from Scholnick et al. (1983). The Sgs-4 sequence was taken from Muskavitch and Hogness (1982) and the 74EF sequence from Möritz et al. (1984). The consensus sequence for the enhancer core is given by Weiher et al. (1983).

and Hodgetts, 1985). Our conclusions from this study were that two different stage specific mechansims must control the expression of the *Ddc* gene in the epidermis of late stage embryos and mature larvae. We hypothesize that in embryos expression of the *Ddc* gene actually may require a falling ecdysteroid titer, a notion that we pursue further in the following section.

It should be noted that the significant hybridization response ascribed to DDC mRNA in 0-4 hr embryonic RNA (Gietz and Hodgetts, 1985) is accounted for by hybridization to a transcript of similar size, encoded from the opposite strand of a transcription unit immediately downstream of *Ddc* (Spencer et al., in preparation). Our conclusion, therefore, that DDC mRNA is contributed maternally to the developing embryo (Gietz and Hodgetts, 1985) has to be reassessed in view of the low levels of DDC transcript we detect now using strand-specific RNA probes.

EVENTS IN IMAGINAL DISCS

Because the embryonic stages are inaccessible to experimental manipulation, we wanted to examine another stage at which a lag occurred between high levels of 20-OH-ecdysone and subsequent DDC gene expression. Earlier we had shown that imaginal discs of the flesh-fly, *Sarcophaga bullata*, produced DDC after a lag of 1-2 days when cultured *in vitro* in the presence of 20-OH-ecdysone (Chen and Hodgetts, 1974). We therefore undertook a study of DDC expression in mass-isolated imaginal discs of *Drosophila* in collaboration with Dr. J. Fristrom, at Berkeley.

Imaginal discs isolated from wandering third instar larvae of *Drosophila* and cultured in the continuous presence of 20-OH-ecdysone failed to produce DDC (Clark et al., 1985). This stands in sharp contrast to the results just described from *Sarcophaga* and we have no explanation for the difference at present. On the other hand, *Drosophila* discs exposed to a 6 hr pulse of 20-OH-ecdysone did show DDC activity about 6 hr following removal of the hormone. Quantitation of DDC transcripts by dot hybridization in RNA samples extracted from cultured discs revealed that DDC mRNA appears concurrently with enzyme activity - at about 6 hr following hormone withdrawal. The results parallel those from embryos, and demonstrate that the lag is not caused by translational blocks on transcripts produced during exposure to 20-OH-ecdysone.

The similarity between the expression of the DDC gene in embryos and imaginal discs goes further. Both Beall and Hirsh (1984) and Gietz and Hodgetts (1985) detected a transcript in late embryos that was not seen in mature larvae. Designated species 5 by Gietz and Hodgetts (1985), we have also observed this transcript in Northern analysis of RNA extracted from imaginal discs expressing DDC (Clark et al., 1985).

We postulate that in embryos and in newly-everted imaginal discs, DDC activity appears in response to a falling ecdysteroid titer, reminiscent of the mid prepupal salivary gland puff genes. We should note that similar conclusions have been reached by Hiruma and Riddiford (1985) in their analysis of larval-larval molts in *Manduca sexta*.

DDC^{+4}, A PUTATIVE REGULATORY VARIANT

The phenotype of this strain may reflect the two different mechanisms controlling DDC expression that we have outlined above. At pupariation, where a high titer of 20-OH-ecdysone has an immediate and direct effect on DDC gene expression, *Ddc*$^{+4}$ is an underproducer of DDC mRNA, CRM, and enzyme activity. At the other stages in the life cycle (hatch, larval molts and adult eclosion), when DDC activity appears only after 20-OH-ecdysone levels have decreased to near zero, *Ddc*$^{+4}$ is an overproducer. Conventional genetic mapping located the sites of both overproduction and underproduction of DDC activity near the structural gene (Estelle and Hodgetts, 1984a). Cloning of the variant DDC region and subsequent restriction mapping revealed a number of differences between the variant and the Canton-S laboratory strain (Estelle and Hodgetts, 1984b). Six of the seven differences noted in the variant consisted of small (<100 base pairs) insertion-deletion polymorphisms in intronic or flanking non-coding regions. Of particular interest were two small insertion polymorphisms lying within or near the 5' untranslated region of the variant *Ddc* gene (polymorphisms 4 and 5).

We have sequenced the variant DNA in this region (Spencer, unpublished) and have compared the sequence to that of the wild-type Canton-S strain (Eveleth and Marsh, unpublished). The complete sequences will be published elsewhere but several interesting observations are summarized below and in Figure 1. One major polymorphism in *Ddc*$^{+4}$ is in the duplication of a 12 base pair AT-rich region roughly 280 base pairs upstream of the larval transcript start site as positioned by Scholnick et al. (1983). The duplicated region is found inserted about 30 base pairs upstream of its normal location. This duplication probably accounts for most of the size difference ascribed to polymorphism 4 by Estelle and Hodgetts (1984b). (The size difference described as polymorphism 5 in Estelle and Hodgetts (1984b) is not due to a single event but rather to an accumulation of one and two base pair changes). Although the biological significance of such an AT-rich duplication is not known at present, we note that the progesterone receptor preferentially binds to an 18 base pair AT-rich sequence which lies between -135 and -247 base pairs upstream of the ovalbumin gene (Compton et al., 1983).

A second feature of interest in *Ddc*$^{+4}$ concerns the sequence 5'-ATGAAAA-3', which bears considerable homology to the canonical enhancer sequence of mammalian viruses (Weiher et al., 1983), and is located about 15 base pairs upstream of the EcoRV site shown in Figure 1. The variant contains an A to G transition in this sequence converting it to 5'-ATGGAAA-3' which resembles the enhancer consensus sequence even more than does the Canton-S sequence.

It is interesting to speculate that this single base pair transition, creating a more precise enhancer-like sequence, may contribute to the up-promotion of transcription which is noted in this variant at all but one stage in development. If so, this would help define the control region operating on the *Ddc* gene at these four stages - but not at pupariation, which appears to be subject to more direct hormonal control.

We have noted (Figure 1) that similar enhancer-like sequences occur upstream of the intermolt puff gene, *Sgs-4* (McGinnis et al., 1983; Muskavitch and Hogness, 1982), and the early puff gene, 74EF (Möritz et al., 1984), both responsive to regulation by 20-OH-ecdysone. A single base pair substitution at position -344 in a naturally-occurring variant of the *Sgs-4* gene is correlated with a 50% reduction in *Sgs-4* RNA, which suggests that these DNA sequences are important for regulation (McGinnis et al., 1983).

Clearly, the functional significance of these and other DNA sequences in *Ddc* regulation must be determined. DNA sequence transformation in whole organisms or in cultured cells should permit such functional tests for regulatory sequences governing the *Ddc* gene.

CONCLUDING REMARKS

We have discussed in this paper the complex developmental regulation of the *Ddc* gene in *Drosophila* in the context of current models of steroid hormone action. At pupariation, an increasing titer of 20-OH-ecdysone induces transcript accumulation within several hours. At other stages, high levels of the hormone have no such effect and indeed, *Ddc* gene activation may require a falling hormone titer or other regulatory factors. Our pursuit of the mechanisms of steroid hormone action will be through an analysis of the multifunctional control region operating on the *Ddc* gene using both genetic and molecular techniques. We feel that a clearer picture of hormone action will be forthcoming from these studies on stage-specific genetic controls effected by the steroid molting hormone, 20-OH-ecdysone.

REFERENCES

Akam ME, Roberts DB, Richards GP, Ashburner M (1978). *Drosophila:* The genetics of two major larval proteins. Cell 13: 215-225.

Anderson JN (1984). The effect of steroid hormones on gene transcription. In Goldberger RF, Yamamoto K (eds.): "Biological Regulation and Development", Vol. 3 B. New York: Plenum Press, pp 169-212.

Ashburner M, Chihara C, Meltzer P, Richards G (1974). Temporal control of puffing activity in polytene chromosomes. Cold Spring Harbor Symp Quant Biol 38: 655-662.

Beall CJ, Hirsh J (1984). High levels of intron containing RNA's are associated with expression of the *Drosophila* dopa decarboxylase gene. Mol Cell Biol 4: 1669-1674.

Borst DW, Bollenbacher WE, O'Connor JD, King DS, Fristrom JW (1974). Ecdysone levels during metamorphosis of *Drosophila melanogaster*. Dev Biol 39: 308-316.

Burch JBE, Weintraub H (1983). Temporal order of chromatin structural changes associated with activation of the major chicken vitellogenin gene. Cell 33: 65-76.

Burtis KC, Thummel CS, Hogness DS (1984). Genetic and molecular analysis of an ecdysone-inducible gene from *Drosophila melanogaster*. Genetics 107: S15.

Chandler VL, Maler BA, Yamamoto K (1983). DNA sequences bound specifically by glucocorticoid receptor *in vitro* render a heterologous promoter hormone responsive *in vivo*. Cell 33: 489-499.

Chen TC, Hodgetts RB (1974). The appearance of dopa decarboxylase activity in imaginal discs of *Sarcophaga bullata,* undergoing development *in vitro*. Dev Biol 38: 271-284.

Clark WC, Doctor J, Fristrom JW, Hodgetts RB (1985). Early and delayed responses of the dopa decarboxylase gene to 20-OH-ecdysone in the epidermis of *Drosophila melanogaster*. Dev Biol (submitted).

Compton JG, Schrader WT, O'Malley B (1983). DNA sequence preference of the progesterone receptor. Proc Natl Acad Sci USA 80: 16-20.

Crowley TE, Bond MW and Meyerowitz EM (1983). The structural genes for three *Drosophila* glue proteins reside at a single polytene chromosome puff site locus. Mol Cell Biol 3: 623-634.

Dewhurst SA, Croker SG, Ikeda K, McCaman RE (1972). Metabolism of biogenic amines in *Drosophila* nervous tissue Comp Biochem Physiol 43 B: 975-981.

Estelle MA, Hodgetts RB (1984a). Genetic elements near the structural gene modulate the level of dopa decarboxylase during *Drosophila* development. Mol Gen Genet 195: 434-441.

Estelle MA, Hodgetts RB (1984b). Insertion polymorphisms may cause stage specific variation in mRNA levels for dopa decarboxylase in *Drosophila*. Mol Gen Genet 195: 442-451.

Fragoulis EG, Sekeris CE (1975a). Induction of dopa (3,4-dihydroxy-phenyl-alanine) decarboxylase in the blowfly integument by ecdysone. Biochem J 146: 121-126.

Fragoulis EG, Sekeris CE (1975b). Translation of mRNA for 3,4-dihydroxyphenylalanine decarboxylase isolated from epidermis tissue of *Calliphora vicina* R.-D. in an heterologous system. Eur J Biochem 51: 305-316.

Garen A, Kauvar L, Lepessant J-A (1977). Roles of ecdysone in *Drosophila* development. Proc Natl Acad Sci USA 74: 5099-5103.

Gietz RD, Hodgetts RB (1985). An analysis of dopa decarboxylase expression during embryogenesis in *Drosophila melanogaster*. Dev Biol 107: 142-155.

Hirsh J, Davidson N (1981). Isolation and characterization of the dopa decarboxylase gene of *Drosophila melanogaster*. Mol Cell Biol 1: 475-485.

Hiruma K, Riddiford LM (1985). Hormonal regulation of dopa decarboxylase during a larval molt. Dev Biol 110: (in press).

Hodgetts RB, Konopka RJ (1973). Tyrosine and catecholanime metabolism in wild-type *Drosophila melanogaster* and a mutant, ebony. J Insect Physiol 19: 1211-1220.

Hodgetts RB, Clark WC, Gietz RD, Sage BA, O'Connor JD. Stage specific mechanisms regulate the expression of the dopa decarboxylase gene during *Drosophila* development. Arch Insect Biochem Physiol: (in press).

Karlson P, Sekeris CE (1962). Zum Tyrosinstoffwechsel der Insekten IX. Kontrolle des Tyrosinstoffwechsels durch Ecdyson. Biochem Biophys Acta 63: 489-495.

Karlson P, Sekeris CE (1966). Ecdysone, an insect steroid hormone and its mode of action. Recent Progr Hormone Res 22: 473-502.

Kraminsky GP, Clark WC, Estelle MA, Gietz RD, Sage BA, O'Connor JD, Hodgetts RB (1980). Induction of translatable mRNA for dopa decarboxylase in *Drosophila:* an early response to ecdysterone. Proc Natl Acad Sci USA 77: 4175-4179.

Livingstone MS, Tempel BL (1983). Genetic dissection of monoamine neurotransmitter synthesis in *Drosophila*. Nature 303: 67-70.

Marsh JL, Wright TRF (1979). Control of dopa decarboxylase expression during development in *Drosophila*. In Axel R, Maniatis T, Fox CF (eds.): "Eucaryotic Gene Regulation". ICN-UCLA Symposium on Molecular and Cell Biology, Vol. XIV. New York: Academic Press, pp 183-194.

Marsh JL, Wright TRF (1980). Developmental relationship between dopa decarboxylase, dopamine acetyltransferase, and ecdysone in *Drosophila*. Dev Biol 80: 379-387.

McGinnis W, Shermoen AW, Heemskerk J, Beckendorf SK (1983). DNA sequence changes in an upstream DNAase 1-hypersensitive region are correlated with reduced gene expression. Proc Natl Acad Sci USA 80: 1063-1067.

Möritz TH, Edström JE, Pongs O (1984). Cloning of a gene localized and expressed at the ecdysteroid regulated puff 74EF in salivary glands of *Drosophila* larvae. EMBO J 3(2) 289-295.

Mulvihill ER, LePennec J-P, Chambon P (1982). Chicken oviduct progesterone receptor: location of specific regions of high-affinity binding in cloned DNA fragments of hormone-responsive genes. Cell 24: 621-632.

Muskavitch MAT, Hogness DS (1982). An exandable gene that encodes a *Drosophila* glue protein is not expressed in variants lacking remote upstream sequences. Cell 29: 1041-1051.

Richards G (1980). Ecdysteroids and puffing in *Drosophila melanogaster*. In Hoffmann J (ed.): "Progress in Ecdysone Research". Amsterdam: Elsevier/North Holland, pp 363-378.

Rizki MTM (1956). The secretory activity of the proventriculus of *Drosophila melanogaster*. J Exp Zool 131: 203-222.

Sanger F, Nicklen S, Coulsen AR (1977). DNA sequencing with chain-terminating inhibitors. Proc Natl Acad Sci USA 74: 5463-5467.

Scheidereit C, Geisse S, Westphal HM, Beato M (1983). The glucocorticoid receptor binds to defined nucleotide sequences near the promoter of mouse mammary tumor virus. Nature 304: 749-752.

Scholnick SB, Morgan BA, Hirsh J (1983). The cloned dopa decarboxylase gene is developmentally regulated when reintegrated into the *Drosophila* genome. Cell 34: 37-45.

Sekeris CE, Karlson P (1962). Zum Tyrosinstoffwechsel der Insekten VII. Der Katabolische Abbau des Tyrosins und die Biogenese der Sklerotisierungssubstanz, N-acetyl-dopamin. Biochem Biophys Acta 62: 103-113.

Sherald AF, Wright TRF (1974). The analog inhibitor, α-methyl dopa, as a screening agent for mutants elevating levels of dopa decarboxylase in *Drosophila melanogaster*. Mol Gen Genet 133: 25-36.

Shermoen AW, Beckendorf SK (1982). A complex of interacting DNAase 1-hypersensitive sites near the *Drosophila* glue protein gene, Sgs-4. Cell 29: 601-607.

Weiher H, König M, Gruss P (1983). Multiple point mutations affecting simian virus 40 enhancer. Science 219: 626-631.

Wright TRF (1977). The genetics of dopa decarboxylase and α-methyl dopa sensitivity in *Drosophila melanogaster*. Amer Zool 17: 707-721.

Wright TRF, Bewley GC, Sherald AF (1976). The genetics of dopa decarboxylase in *Drosophila melanogaster* II. Isolation and characterization of dopa-decarboxylase-deficient mutants and their relationship to the α-methyl-dopa-hypersensitive mutants. Genetics 84: 287-310.

Wright TRF, Steward R, Bentley W, Adler PN (1981). The genetics of dopa decarboxylase in *Drosophila melanogaster* III. Effects of a temperature sensitive dopa decarboxylase deficient mutation on female fertility. Dev Genet 2: 223-235.

Wright TRF, Black BC, Bishop CP, Marsh JL, Pentz ES, Steward R, Wright EY (1982). The genetics of dopa decarboxylase in *Drosophila melanogaster* V. *Ddc* and *l(2)amd* alleles: Isolation, characterization and intragenic complementation. Mol Gen Genet 188: 18-26.

Progress in Developmental Biology, Part A, pages 235–238

DIFFERENTIAL INDUCTION OF DOPA DECARBOXYLASE ACTIVITY BY 20-OH ECDYSONE IN SUBLINES OF THE DROSOPHILA Kc CELL LINE

Ruth E. Swiderski and John D. O'Connor

Department of Biology, University of California, Los Angeles, California 90024

INTRODUCTION

Dopa decarboxylase (DDC; EC 4.1.1.28) is an enzyme of critical importance in insect development. It is responsible for the production of compounds which cross-link and pigment the insect cuticle as well as for neurotransmitters. The appearance of DDC activity during Drosophila development is regulated in a stage- and tissue-specific manner by the insect steroid hormone 20-OH ecdysone (20-OHE; Marsh and Wright, 1980, Kraminsky et. al. 1980). The induction of DDC activity by 20-OHE was investigated using the Drosophila Kc cell line and several Kc-derived sublines. Kc cells, derived from 6-12 hr Drosophila embryos, are hormone-responsive cells in which 20-OHE is neither synthesized nor metabolized and are thereby well-suited for the manipulation of hormone levels. The cells respond to physiological concentrations of 20-OHE (4×10^{-7} M) and exhibit such well-characterized responses as changes in cell morphology, mitotic arrest in the G_2 phase of the cell cycle, and the induction of acetylcholinesterase and β-galactosidase (reviewed by Cherbas et. al. 1984). Given the ease of handling and availability of large numbers of cells, the Kc cell line was chosen as a model system in which to study the induction of DDC by 20-OHE.

RESULTS

The Kc cell line and three Kc-derived sublines (1G9, 7E10(0), and 7E10(4)EC were categorized according to their capacity to induce DDC activity when cultured in the con-

Table 1

The response of Kc cells and Kc sublines to 20-OHE and the induction of DDC activity

Cell Type	Time After Addition of 20-OHE (hr)	DDC Specific Activity[1] (units/mg protein)	% Cells in G_2 Arrest[2]	Presence of 20-OHE Receptor	Capacity to Induce DDC Activity
Kc	0	0.003	63	+	–
	24	0.003	87		
	48	0.003	91		
	72	0.003	94		
	96	0.002	94		
	120	0.004	94		
1G9	0	0.001	61	+	–
	24	0.004	90		
	48	0.002	**		
	72	0.001	**		
	96	0.001	**		
	120	0.003	**		
7E10(0)	0	0.047	73	+	–
	24	0.036	79		
	48	0.027	81		
	72	0.034	83		
	96	0.021	80		
	120	0.010	81		
Br^-	0	0.019	66	–	–
	24	0.011	70		
	48	0.007	56		
	72	0.018	81		
	96	0.014	83		
	120	0.012	82		
7E10(4)EC	0	0.17	59	+	+
	24	0.4	92		
	48	0.24	94		
	72	0.84	91		
	96	4.4	88		
	120	4.32	88		

[1] determined by liquid cation exchange DDC assay (7E10(4)EC) and by $^{14}CO_2$ microdiffusion DDC assay (Kc, 1G9, 7E10(0), Br^-). One unit of activity corresponds to 1 nanomole L-Dopa decarboxylated/hr at 30°C.

[2] standard deviation: +/- 5-7%.

** cell clumping induced by 20-OHE made G_2 arrest assay impossible.

tinuous presence of a physiological concentration (4×10^{-7}M) of 20-OHE (Table 1). One subline, 7E10(4)EC, exhibited a 26-fold increase in DDC activity at 96 hr. In contrast, Kc, 1G9, and 7E10(0) cells were classified as DDC non-inducible based on their failure to increase DDC activity following the administration of 20-OHE. In order to determine whether the failure of Kc, 1G9, and 7E10(0) cells to induce DDC activity was associated with the loss of other well-characterized Kc cell responses to hormone, several parameters of 20-OHE responsiveness were tested. The 20-OHE receptor was detected in all Kc cells used in this study with the exception of the receptor-deficient Br- subline. Kc, 1G9, 7E10(0) and 7E10(4) EC cells were categorized as hormone-responsive as demonstrated by their hormone-induced mitotic arrest in the G2 phase of the cell cycle and the morphological changes observed as the consequence of hormone treatment. Furthermore, the failure of Kc, 1G9, and 7E10(0) cells to induce DDC activity was not due to deletion of the Ddc structural gene. All Kc cells used in this study contained essentially equivalent Ddc gene copy number as quantified by genomic DNA blots. In addition, studies using restriction endonuclease digestion of genomic DNA revealed no gross rearrangement of the Ddc gene in the cells and no restriction endonuclease polymorphism unique to either the DDC inducible or DDC non-inducible cells was observed.

All hormonally naive cells contained a basal amount of Ddc transcript, a fraction of which was localized on polysomes sedimenting $>$80 S. The increased sensitivity of immunoblots revealed that the DDC polypeptide was indeed present though enzyme activity was undetectable. The relative amount of steady state Ddc transcript present in total cellular poly(A) RNA populations isolated from cells exposed to 20-OHE was quantified by densitometric tracing of autoradiograms derived from RNA blots. There was no increase in Ddc mRNA due to 20-OHE in the DDC non-inducible cells. In contrast, 7E10(4)EC cells treated with 20-OHE for 96 hr exhibited a 6-fold increase in the amount of Ddc transcript which was correlated temporally with a 26-fold increase in DDC enzyme activity. The 20-OHE-induced 7E10(4) EC Ddc transcript was able to direct the *in vitro* synthesis of a polypeptide that was immunoprecipitated with antiserum directed against *Drosophila* epidermal DDC. This result, in conjunction with DDC depletion experiments using newly eclosed adult *Drosophila* and 20-OHE-treated 7E10(4)EC cells suggests that the Kc subline induces the activity of *bona fide* DDC under the influence of 20-OHE.

The use of a *Drosophila* cell line has been instrumental in studying the response of hormonally naive cells to a physiological concentration of 20-OHE. Preliminary evidence suggests that the hormone may have a pleiotropic effect on 7E10(4)EC cells. In addition to stimulating transcription at the Ddc locus, 20-OHE may also act at the post-transcriptional level, such as mRNA stability or translational efficiency. Given the nature of the experiments described herein, more quantitative investigation will resolve these possibilities.

REFERENCES

Cherbas L, Fristrom JW, and O'Connor JD (1984). The action of ecdysone in imaginal discs and Kc cells of Drosophila melanogaster. In Hoffmann J, Porchet M (eds): "Biosynthesis, Metabolism and Mode of Action of Invertebrate Hormones", Berlin: Springer-Verlag, p. 305.

Kraminsky GP, Clark WC, Estelle MA, Gietz RD, Sage BA, O'Connor JD, and Hodgetts RB (1980). Induction of translatable mRNA for dopa decarboxylase in Drosophila: an early response to ecdysterone. Proc Nat Acad Sci 77: 4175.

Marsh JL, and Wright TRF (1980). Developmental relationship between dopa decarboxylase, dopa acetyl transferase, and ecdysone in Drosophila. Dev Biol 80: 379.

Progress in Developmental Biology, Part A, pages 239–242

DEVELOPMENTAL REGULATION OF TRANSCRIPT SPLICING IN DROSOPHILA MELANOGASTER

Scott Falkenthal

Department of Genetics and the Molecular, Cellular, and Developmental Biology Program, The Ohio State University, Columbus, Ohio 43210

INTRODUCTION

Developmental alterations in the pattern of RNA splicing are the results of unique temporal and spatial programs of gene expression during development which in a number of cases are used to generate protein diversity. The alterations in splicing patterns mechanistically are the result of 1) alternative use of promotors, and 2) alternative use of polyadenylylation sites (reviewed by Mahler, 1983). In contrast to this I wish to summarize recent results on the developmentally controlled splicing of the myosin alkali light chain (MLC-ALK) primary transcript of Drosophila and to present an alternative model for the control of this process.

RESULTS

The structure of the MLC-ALK gene is shown in Figure 1 (Falkenthal et al., 1984; 1985). Transcription begins at a single initiation site at the 5' end of exon 1 and terminates at either of two polyadenylylation sites in exon 6. The first polyadenylylation site is the preferred site, however both are used at all times during development that the gene is expressed. Note: the same primary transcript is generated by this gene at all developmental stages. During the late pupal stage two populations of MLC-ALK mRNA exist with different sequence arrangements; E1, E2, E3, E4, E5, E6; and E1, E2, E3, E4, E6 with E5 not a part of the mature mRNA. The former mRNA is expressed during the late embryonic, larval, and late pupal stages; whereas the latter mRNA is found exclusively during the late pupal stage. This pupal specific mRNA also has tissue specificity; it is localized in the indirect flight muscle but is absent in the musculature of the head. Conversely the exon 5 containing MLC-ALK mRNA is not detected in the indirect flight muscle and is the only MLC-ALK mRNA that accumulates in the musculature of the head. It should be noted that these two mRNAs encode slightly different proteins. Translation termination is at the penultimate codon in exon 5, whereas termination is 14 codons into exon 6 for the mRNA lacking exon 5 sequences. This results in the synthesis of the two polypeptides that differ in sequence only at their carboxyl-terminal ends.

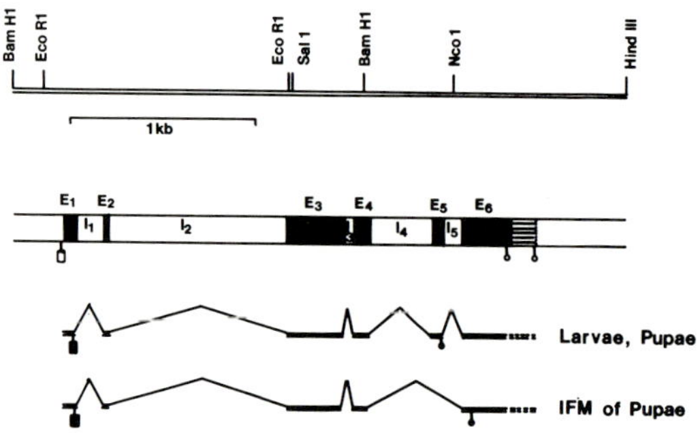

Figure 1. Structure of the MLC-ALK Gene. Transcription initiation is denoted by the open square; polyadenylylation sites are denoted by open circles. Translation initiation is denoted by the closed square and translation termination is denoted by the closed circle.

DISCUSSION

To begin to understand the regulation of alternative processing of primary transcripts it is requisite to know if all the primary transcripts of a particular gene are colinear. Non-colinearity can be due to trans-acting factors which recognize alternate promotors or polyadenylylation sites. The manner in which the transcript is then spliced is not dependent upon trans-acting factors that act upon the splice site, but rather upon cis-regulatory sequences which govern the folding of the primary transcript. The MLC-ALK transcripts deviate from the above scenario. The primary transcripts of this gene are colinear at all developmental times, but are spliced differently depending upon temporal and spatial parameters. The molecular basis for regulating which splice site is utilized must directly involve negative or positive trans-acting factors that ultimately interact with sequences within the intron.

To identify possible regulatory sequences involved in splice site choice the sequences of the MLC-ALK introns were compared (Falkenthal et al., 1985). All the 5' splice sites are identical and conform to the concensus eukaryotic sequence (Mount, 1982). The introns all contain three conserved sequence domains in the 3' region of the intron; 1) the branch point sequence, 2) the ultimate AG dinucleotide at the intron:exon boundary, and 3) a stretch of polypyrimidine residues adjacent to the AG dinucleotide. The only exception to this structural motif is intron 4, the intron which preceeds the alternatively spliced exon 5. It does not contain a polypyrimidine tract; in its place is a tract of purine bases. This deviation from the consensus structural motif is significant because in vitro experiments have shown that the polypyrimidine tract is required for the cut made at the 5' exon/intron boundary and the formation of the branched intermediate (Reed and Maniatis, 1985).

On the basis of the importance of the polypyrimidine tract for the process of RNA splicing and its replacement in intron 4 by a polypurine tract, we hypothesize that a positive trans-acting regulatory factor exists which converts the 3' end of intron 4 to an active splicing substrate (Figure 2). In the absence of the transacting factor, such as in the indirect flight muscle, the 3' end of intron 4 is not an active splicing substrate, whereas the 3' end of intron 5 is, leading to the ligation of exon 4 to exon 6. Why does exon 6 not splice to exon 5 in the absence of the trans-acting factor? A simple explanation is that the assembly of the putative positive trans-acting factor with the MLC-ALK primary transcript alters its secondary structure so that exon 5 and exon 6 are in close proximity favoring the splicing event. In the absence of the factor the 5' splice site of intron 5 would be buried in the secondary structure making it in accessible for splicing, whereas the 5' splice site of exon 4 would be in a favorable conformation for exon 4-exon 6 splicing to occur.

Two recent technological breakthroughs, the ability to synthesize ug quantities of specific RNA transcripts in vitro and the ability to make nuclear extracts with splicing activity, have made it possible to test the validity of our model (Reed and Maniatis, 1985).

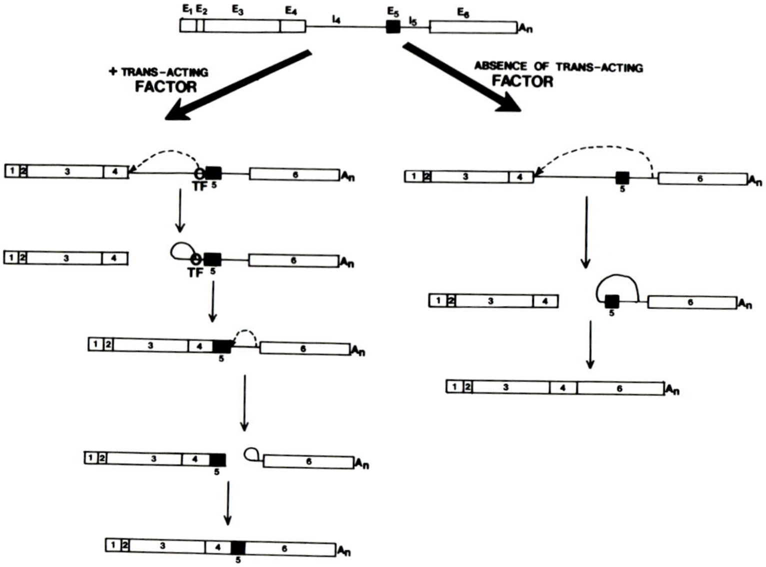

Figure 2. Proposed pathway for alternate splicing in which TF is the transacting-factor.

REFERENCES

Falkenthal S, Parker VP, Mattox WM, Davidson, N (1984). Drosophila melanogaster has only one myosin alkali-light chain gene which encodes a protein with considerable amino acid sequence homology to chicken myosin alkali light chains. Molec Cell Biol 4:956.

Falkenthal S, Parker VP, Davidson N (1985). Developmental variations in the splicing pattern of transcripts from the Drosophila gene encoding myosin alkali light chain result in different carboxyl-terminal amino acid sequences. Proc Natl Acad Sci USA 82:449.

Mahler HR (1983). The exon:intron structure of some mitochondrial genes and its relation to mitochondrial evolution. Int Rev Cyto 82:1.

Mount S (1982). A catalogue of splice junction sequences. Nucleic Acids Res 10:459.

Reed R, Maniatis T (1985). Intron sequences involved in lariat formation during pre-mRNA splicing. Cell 41:95.

Progress in Developmental Biology, Part A, pages 243–247

THE USE OF RNA PROBES FOR DETECTION OF ISOFORM TRANSITIONS DURING SKELETAL MUSCLE DEVELOPMENT

John R. Coleman and Charles P. Ordahl

Division of Biology and Medicine, Brown University, Providence, RI 02912 (JRC) and Department of Anatomy, University of California, San Francisco, CA 94143 (CPO)

INTRODUCTION

Cell and tissue differentiation during development is accompanied by changes in gene expression culminating in synthesis of an array of protein molecules characteristic of a given cell type and responsible for its functional specializations. As shown dramatically in construction of the contractile apparatus during skeletal muscle development, many tissue specific proteins are turning out to be isoforms of proteins distributed ubiquitously among many cell types. Further, tissue specific proteins may exist in families of isoforms which exhibit tightly regulated transitions in expression during differentiation and functional maturation (see review by Caplan et al, 1983). Such protein isoform families may be encoded in multimember gene families and regulated at the transcriptional level (e.g. actin, Fyrberg et al, 1983), or they may be encoded by single genes and the transitions regulated by alternative RNA processing (e.g. myosin heavy chain, Rozek and Davidson, 1983; skeletal troponin T, Medford et al, 1984; cardiac troponin T, Cooper and Ordahl, 1985).

We have begun to investigate isoform transitions in avian skeletal muscle development by cloning developmentally regulated chicken cDNA sequences into the SP6 vectors designed by Melton and co-workers (1984). These are transcribed to form single stranded RNA probes which show high sensitivity in RNA blot and in situ hybridization analyses. We report here results on three isoform families exhibiting distinct regulatory patterns during avian skeletal and cardiac muscle development: the skeletal muscle isoforms of creatine kinase (MCK) and α-actin, and a cardiac isoform of troponin T (TnT).

RESULTS AND DISCUSSION

Chicken cDNA sequences were inserted into SP6 vectors to yield isoform specific probes. Inserts were derived from the 3' untranslated ends of skeletal α-actin (Fornwald et al, 1982) and cardiac TnT (Cooper and Ordahl, 1984), and a portion of the coding sequence of MCK (Ordahl et al, 1984). The skeletal α-actin insert contains the terminal 261 base pairs from the 3' untranslated end plus 60 bp from pBR322, excised from the Bcl I to the Eco RI site. The cardiac TnT insert contains 132 bp from the 3' untranslated end, 10 bp of coding sequence and 30 bp from pBR322, excised at Eco RI and Hind III sites. The MCK insert includes 250 bp of coding sequence, from the Hpa II to the Pst I site, plus a flanking 20 bp from a pUC polylinker. SP6 transcripts of the inserts include part of the vector polylinker, thus the actin, MCK and TnT probes are approximately 350, 180 and 275 bases in length, respectively (Figure 1). Probes transcribed in 12μM^{32}P-GTP are predominantly run-off transcripts of uniform length with a specific activity of approximately 7 x 10^8dpm/μg.

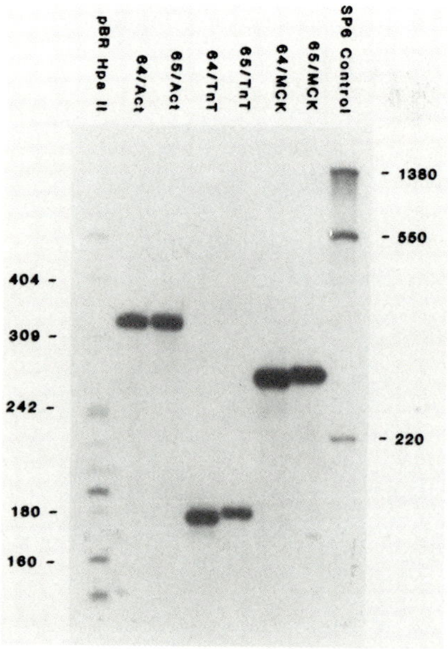

Figure 1. Transcripts of sequences from chicken skeletal muscle cDNA's cloned into pSP64 and pSP65. Each lane contained 10^4 dpm of transcript electrophoresed in a 6% acrylamide-urea sequencing gel and autoradiographed for 3 hours.
Act: skeletal α-actin;
TnT: cardiac troponin T;
MCK: M-creatine kinase.

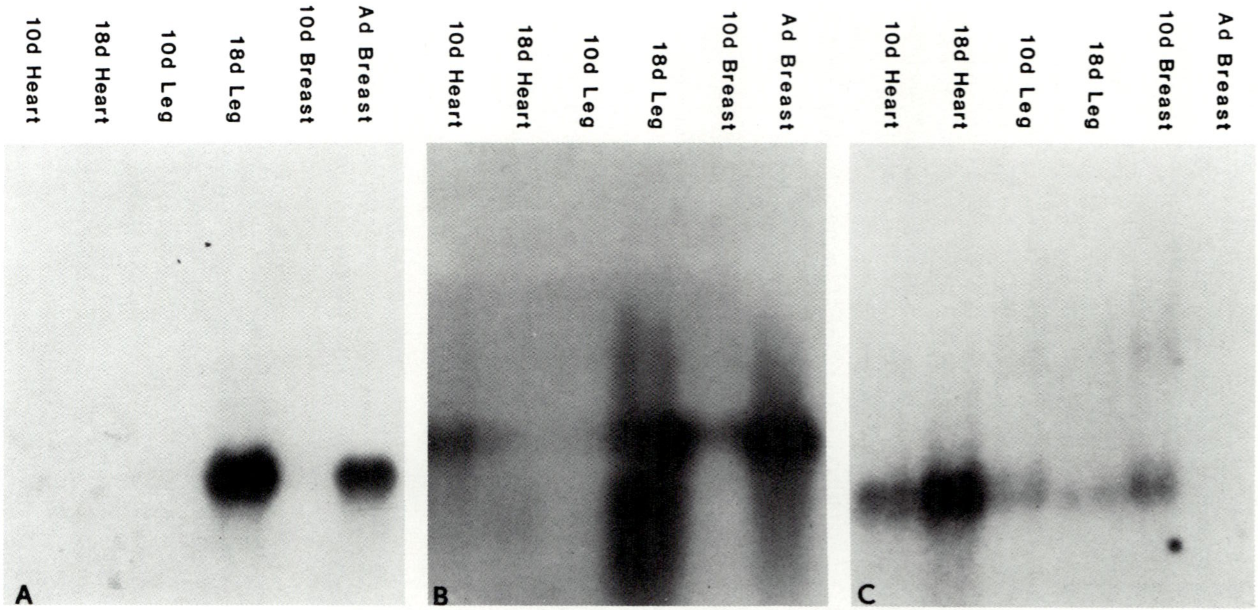

Figure 2. RNA blots probed with SP6 transcripts. Each lane contained 5 μg poly(A^+) RNA from the embryonic or adult chicken tissue indicated, electrophoresed in 1.25% agarose-formaldehyde and blotted onto Gene Screen. Blots were hybridized overnight at 52°C and washed at 65°C in 50mM NaCl, 20mM $NaPO_4$ (pH 6.5), 1mM EDTA, 0.1% SDS. A: M-creatine kinase, probed with 10^7dpm pSP64/MCK, 2.5 hr exposure; B: Skeletal α-actin, probed with 10^6dpm pSP65/Act, 7 hr exposure; C: Cardiac troponin T, probed with 10^5dpm pSP64/TnT, 44 hr exposure. The same blots hybridized with transcripts from pSP65/MCK, pSP64/Act and pSP65/TnT showed no bands, thus these transcripts consist of non-complementary mRNA sequences.

Figure 2 illustrates blots of RNA prepared from embryonic cardiac and leg skeletal muscle, as well as embryonic and adult breast muscle, probed with the SP6 transcripts described above. MCK mRNA increases dramatically in skeletal muscle during late embryonic development. Longer exposures of this blot demonstrated that MCK mRNA is present in 10 day embryonic leg and breast muscle as well, but in relatively low abundance. MCK mRNA was not detectable in cardiac RNA, even at longer exposures. Skeletal α-actin mRNA shows a similar pattern in skeletal muscle, but differs from the MCK pattern in that it is present in the embryonic heart but decreases in abundance during heart development. This pattern of regulation is similar to that reported for mouse muscle by Minty et al (1982). It is likely that the major isoform present early in skeletal muscle development is cardiac α-actin (Eldridge and Paterson, 1984; Ordahl, 1985), with skeletal α-actin becoming predominant later.

The TnT sequence investigated here was originally isolated as a cDNA clone complementary to mRNA that is transiently expressed during skeletal muscle development, and was later found to code for TnT present in adult heart but absent from adult skeletal muscle (Cooper and Ordahl, 1984). These findings are confirmed here, with cardiac TnT mRNA increasing sharply in abundance in embryonic heart, but decreasing in abundance in skeletal muscle. Thus its pattern of regulation is the converse of that shown by skeletal α-actin mRNA.

It is likely that the regulatory transitions illustrated here involve changes in transcriptional activity of individual genes. Interestingly, Cooper and Ordahl (1985) have demonstrated that cardiac TnT undergoes a transition between two isoforms in chicken cardiac muscle and that this is governed at the level of RNA processing, as is the case for rat skeletal muscle TnT (Medford et al 1984). The evolution of such disparate mechanisms for regulation of isoform transitions remains an enigma and probably reflects both the need for tissue-specific (i.e., developmental) regulation and the specific functional properties of distinct protein isoforms.

(Supported by grants from the Muscular Dystrophy Association to JRC and CPO, NSF Grant Number DCB 82-15972 to JRC and NIH Grant Number GM-32018 to CPO).

REFERENCES

Caplan AI, Fiszman MY, Eppenberger HM (1983). Molecular and cell isoforms during development. Science 221:921-927.

Cooper TA, Ordahl CP (1984). A single troponin T gene regulated by different programs in cardiac and skeletal muscle development. Science 226:979-982.

Cooper TA, Ordahl CP (1985). A single cardiac troponin T gene generates embryonic and adult isoforms via developmentally regulated alternate splicing. J Biol Chem, in press.

Fornwald JA, Kuncio G, Peng I, Ordahl CP (1982). The complete nucleotide sequence of the chick α-actin gene and its evolutionary relationship to the actin gene family. Nucleic Acids Res 10:3861-3876.

Fyrberg EA, Mahaffey JW, Bond BJ, Davidson N (1983). Transcripts of the six *Drosophila* actin genes accumulate in a stage- and tissue-specific manner. Cell 33:115-123.

Medford RM, Nguyen HT, Destree AT, Summers E, Nadal-Ginard B (1984). A novel mechanism of alternative RNA splicing for the developmentally regulated generation of troponin T isoforms from a single gene. Cell 38:409-421.

Melton DA, Krieg PA, Rebagliati MR, Maniatis T, Zinn K, Green MR (1984). Efficient *in vitro* synthesis of biologically active RNA and RNA hybridization probes from plasmids containing a bacteriophage SP6 promoter. Nucleic Acids Res 12:7035-7056.

Minty AJ, Alonso S, Caravatti M, Buckingham ME (1982). A fetal skeletal muscle actin mRNA in the mouse and its identity with cardiac actin mRNA. Cell 30:185-192.

Ordahl CP (1985). The skeletal and cardiac α-actin genes are co-expressed in early embryonic striated muscle. Submitted for publication.

Ordahl CP, Evans GL, Cooper TA, Kunz G, Perriard J-C (1984). Complete cDNA-derived sequence of chick muscle creatine kinase. J Biol Chem 259:15224-15227.

Paterson BM, Eldridge JD (1984). Alpha-cardiac actin is the major sarcomeric isoform expressed in embryonic avian skeletal muscle. Science 224:1436-1438.

Rozek CE, Davidson N (1984). *Drosophila* has one myosin heavy chain gene with three developmentally regulated transcripts. Cell 32:23-34.

Progress in Developmental Biology, Part A, pages 249–252

C-ABL IS DIFFERENTIALLY EXPRESSED DURING MOUSE SPERMATOGENESIS

Debra J. Wolgemuth and Carola Ponzetto

Dept. of Human Genetics and Development and Center for Reproductive Sciences, Columbia University College of P & S, New York, N.Y. 10032

INTRODUCTION

c-*Abl*, the cellular homolog of the transforming gene of the Abelson murine leukemia virus, is a member of the class of proto-oncogenes coding for membrane-associated proteins with a highly conserved tyrosine kinase domain. c-*Abl* is apparently ubiquitously transcribed at low levels in mouse tissues, giving rise to two mRNAs of 6.2 and 8.0 kb in size (Muller et al.,1982; Wang and Baltimore,1983). Mature testis has the highest level of c-*abl* expression of any tissue examined, whether of adult or embryonic origin (Muller et al., 1982). Furthermore, an additional transcript of 4.7 kb represents the most abundant c-*abl* mRNA species in this tissue (Muller et al.,1982; Ponzetto and Wolgemuth,1985). This transcript is missing from RNA isolated from testis of immature animals, which lack germ cells in meiosis, while it appears at high abundance in post-meiotic germ cells of the mature testis (Ponzetto and Wolgemuth, 1985).

As a part of our ongoing investigations on the role of c-*abl* expression in the male germ line, we sought to 1) confirm our earlier observation on the developmental specificity of appearance of the novel 4.7 kb c-*abl* transcript; 2) examine embryonic mouse tissues for the presence of the 4.7 kb c-*abl* transcript; and 3) extend this analysis to germinal and other androgen-responsive tissues from a different rodent species, the rat.

MATERIALS AND METHODS

Mature Swiss Swebster mice were used as source of testis and liver. Purified populations of male germ cells at different stages of spermatogenesis were isolated as described by Wolgemuth et al.(1985). Day 6 embryos (day of vaginal plug was designated day 0 of gestation) were dissected free of uterine tissue but included extra-embryonic tissues. Day 11 embryos were dissected free of maternal and extra-embryonic tissues. Mature Copenhagen x Fisher F1 male rats were used as source of testis and prostatic tumor. The tumor is an androgen-dependent highly anaplastic variant (White et al., 1983) of the original Dunning R 3327 tumor (Dunning, 1963).

The probe used for this analysis,p *abl*-sub3, a v-*abl* clone (Wang and Baltimore, 1983), was generously provided by S.Goff. RNA was isolated from cells and tissues using the LiCl precipitation method of Cathala et al.(1983), and poly(A)+ selected according to Aviv and Leder (1972). RNA samples were electrophoresed on denaturing 1% agarose-2.2 M formaldehyde gels. Gels were blotted onto nitrocellulose paper and hybridized according to Maniatis et al.(1982). The probe was labeled by nick-translation (Weinstock et al., 1978) using ^{32}P nucleoside triphosphates. Hybridized filters were exposed to X-ray film for autoradiography.

RESULTS AND DISCUSSION

Two transcripts of 6.2 and 8.0 kb in size are visualized upon Northern blot analysis of poly(A)+RNA from all tissues examined. This is illustrated in Fig. 1a in which RNAs from day 6 embryos, day 11 embryos, liver, and adult testis were analyzed. The higher molecular weight bands visualized in liver (Fig.1a, lane 3), probably represent incompletely spliced precursors of the mature c-*abl* mRNAs. Only mature testis contains an additional c-*abl* mRNA of 4.7 kb (Fig.1a, lane 4).

The pattern of expression of the c-*abl* gene during spermatogenesis was then considered, analyzing total RNA obtained from populations of purified germ cells. The 4.7 kb transcript was highly abundant in RNA obtained from early spermatids (post-meiotic) but was absent in RNA from enriched populations of pachytene spermatocytes (early meiotic) (Fig. 1b). Altogether these observations show that while both the higher molecular weight c-*abl* mRNAs are expressed in embryonic and adult tissues (e.g. liver and testis), the

4.7 kb transcript is confined to post-meiotic cells in the male germ line.

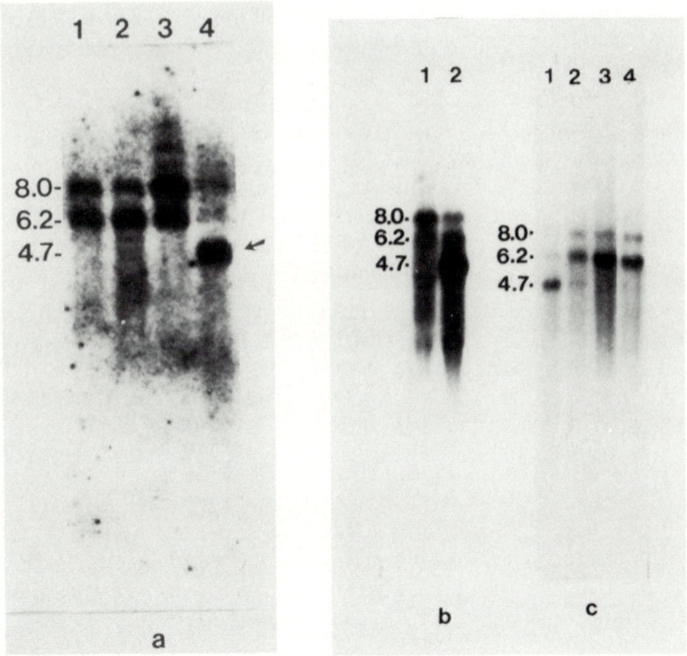

Figure 1. Northern blots of a) Poly(A)$^+$RNA from day 6 embryo (lane 1, 10ug), day 11 embryo (lane 2, 10 ug), liver (lane 3, 10 ug), and testis (lane 4, 5ug). b) Total RNA from pachytene spermatocytes (lane 1, 20ug), and early spermatids (lane 2, 20ug). c) Poly(A)$^+$RNA from mouse testis (lane 1), rat testis (lane 2), rat prostatic tumor (lane 3) and rat prostate (lane 4). All sample 5 ug.

Fig. 1c shows the same type of analysis performed on rat tissues. The rat testis also contains a 4.7 kb c-abl transcript, albeit at apparently lower levels. This may be due to a differential homology of the murine probe for the three c-abl mRNAs, the lower molecular weight transcripts possibly being the most divergent between species.

Since androgens are involved in the development of the male germ line, we sought to examine rat prostate and rat prostatic tumor as additional examples of androgen-responsive tissues of the male reproductive system. The absence of the 4.7 kb c-abl transcript from these sources confirms the male germ line specificity of this mRNA and serves to underscore our efforts to understand the function of c-abl in male germ cells.

Supported by grants HD 18122 and HD 05077 from the NIH and a grant from the I.T. Hirschl Trust.

REFERENCES

Aviv H, and Leder P (1972). Purification of biologically active globin mRNA by chromatography on oligothymidilic acid-cellulose. Proc Natl Acad Sci USA 69: 1408.

Cathala G, Savouret J-F, Mendez B, West B L, Karin M, Martial J A, and Baxter J D (1983). A method for isolation of intact translationally active ribonucleic acid. DNA 2: 329.

Dunning W F (1963). Prostate cancer in rats. Natl Canc Inst Monograph 12: 351.

Maniatis T, Fritsch E F, and Sambrook J (1982) Molecular Cloning, Cold Spring Harbor Laboratory, CSH, NY.

Müller R, Slamon D J, Tremblay J M, Cline M J, and Verma I (1982).Differential expression of cellular oncogenes during pre- and post-natal development in the mouse. Nature 299:640.

Ponzetto C, and Wolgemuth D J (1985). Haploid expression of a unique c-abl transcript in the mouse male germ line. Mol Cell Biol 5: 1791.

Wang J Y J, and Baltimore D (1983). Cellular RNA homologous to the Abelson murine leukemia virus transforming gene: expression and relationship to the viral sequence. Mol Cell Biol 3:7

Weinstock R, Sweet R, Weiss M, Cedar H, Axel R (1978). Intragenic DNA spacers interrupt the ovalbumin gene. Proc Natl Acad Sci USA 75: 1299.

White R deV, Deitch A D, Tomashewsky P, and Olsson C A (1983). Flow cytometry in prostatic adenocarcinoma. World J Urol 1:11

Wolgemuth D J, Gizang Ginsberg E, Engelmyer E, Gavin B J, and Ponzetto C (1985). Separation of mouse testis cells on a CelsepTM apparatus and their usefulness as a source of high molecular weight DNA or RNA.

Progress in Developmental Biology, Part A, pages 253–258

TWO-STEP CONTROL OF δ CRYSTALLIN EXPRESSION DURING RETINAL TRANSDIFFERENTIATION IN CULTURE

David I. de Pomerai, Andrew Carr and David Ellis

Department of Zoology, University of Nottingham, University Park, Nottingham NG7 2RD, U.K.

INTRODUCTION

Chick embryo neuroretinal cells transdifferentiate extensively into lens and pigment cells when cultured *in vitro* in Eagle's Minimal Essential Medium supplemented with 5% Foetal Calf serum and 5% Horse serum (FH medium). The production of lens cells (monitored as accumulation of δ crystallin) is strongly inhibited when FH medium is supplemented with high levels of glucose (18mM final; FHG medium; de Pomerai and Gali, 1982), or when using medium 199 plus 10% foetal calf serum (F199 medium ; Gali and de Pomerai, 1984). In both of these "blocking" media, δ crystallin accumulation is reduced by approximately 100 fold as compared to cultures in FH medium. In this report, we show that δ crystallin transcripts do not accumulate detectably in FHG cultures, while they are present but remain confined to the nuclei in F199 cultures. F199 cultures maintained under continuous heat-shock conditions (43°C) show extensive transdifferentiation into lens and concomitant δ crystallin accumulation. Transient heat-shock treatment of 30-day F199 cultures can apparently coinduce δ crystallin along with heat-shock protein synthesis.

MATERIALS AND METHODS

Materials

Medium components were from GIBCO-Europe Ltd and plastic

disposables from Nunc Ltd. (^{35}S)-methionine was from Amersham, as were the (^{32}P)-nucleotides used for nick translation. The pδCR2 clone was a generous gift from J. Piatigorsky, NIH.

Cell Culture

Cultures were prepared from 9-day chick embryo neural retina (NR) cells, and kept in media FH, FHG or F199 (see above), as described previously (de Pomerai and Gali, 1982; Gali and de Pomerai, 1984). Cultures were maintained in a humid atmosphere of 95% air:5% CO_2 at 37°C (or at 43°C in heat shock experiments), the medium being changed every 2-3 days.

Culture Labelling

Cultures were washed in saline, then incubated in culture medium containing 10μCi/ml of (^{35}S)-methionine for 5 hours at 37° or 43°C. After washing, labelled cultures were homogenised in saline and the soluble proteins subjected to SDS-polyacrylamide gel electrophoresis and fluorography, as described in detail elsewhere (Carr and de Pomerai (1985) Devl Biol, in press).

Northern Blotting

RNA was extracted from whole cultured cells or from nuclear or cytoplasmic fractions, and subjected to agarose gel electrophoresis (using ribosomal RNAs as size markers). Such gels were Northern blotted onto nitrocellulose sheets and probed with a δ crystallin cDNA probe (pδCR2), (^{32}P)-labelled by nick translation. These methods are also described in detail by Carr and de Pomerai (1985, Devl Biol, in press).

RESULTS AND DISCUSSION

As shown in fig. 1, δ crystallin mRNA is undetectable in the cytoplasmic RNA of 14-day cultures maintained in all three media. By 40 days, however, δ crystallin mRNA is apparent in the cytoplasm of transdifferentiating FH

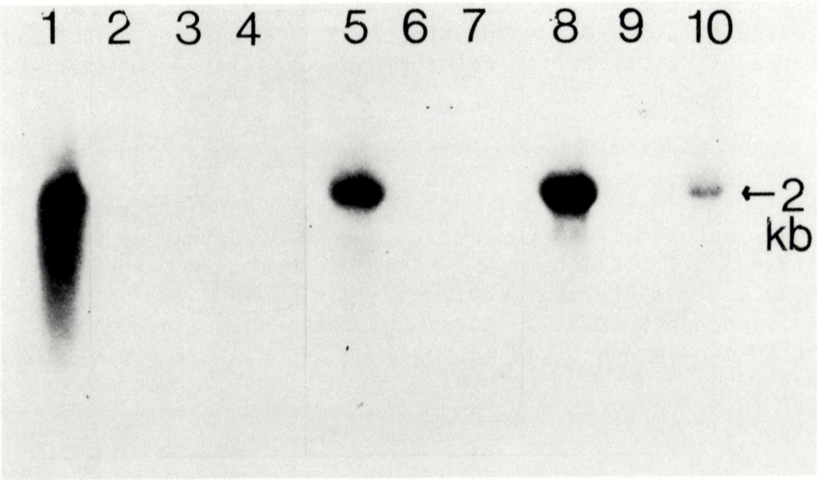

Figure 1. Northern blot of RNA from lens and NR cultures, probed with (^{32}P)-labelled nick-translated pδCR2. 10μg RNA per lane except for lane 1 (0.1μg). Lane 1, newly-hatched chick lens poly (A)$^{+}$ RNA. Lanes 2 to 4, cytoplasmic RNA from 14-day cultures in FH (2), FHG (3) and F199 (4) media. Lanes 5 to 7, cytoplasmic RNA from 40-day cultures in FH (5), FHG (6) and F199 (7) media. Lanes 8 to 10, total cellular RNA from 40-day cultures in FH (8), FHG (9) and F199 (10) media.

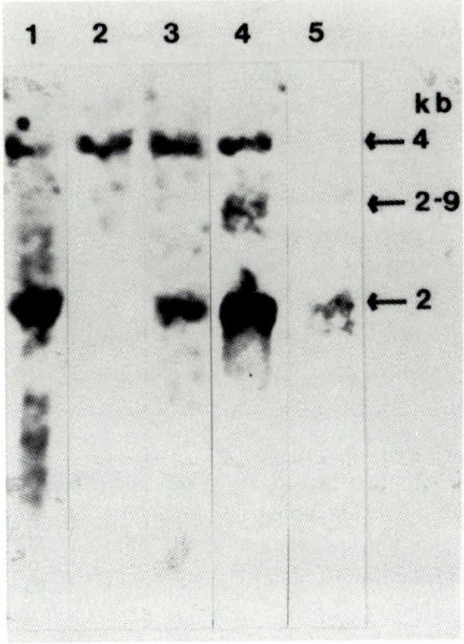

Figure 2. Northern blot of nuclear RNAs extracted from ribonucleoprotein particles from isolated nuclei, probed with (^{32}P)-labelled pδCR2. 10μg RNA per lane. Lane 1, 6-day embryo total lens. Lane 2, 21-day NR culture in FH medium. Lanes 3 to 5, 40-day cultures in FH (3), F199 (4) and FHG (5) media.

cultures, but not in "blocked" FHG or F199 cultures. If total cellular RNA is used instead of cytoplasmic RNA, low levels of δ crystallin transcripts can also be detected in F199 cultures, but not in FHG cultures. In order to investigate this point further, we extracted RNA from ribonucleoprotein particles present in isolated nuclei, using vanadyl ribonucleoside complexes (BRL) to inhibit ribonuclease activity. Both 2kb (message-sized) and 4kb (processing intermediate) δRNA species are detected in the nuclei of 6-day embryonic lenses and of 40-day cultures maintained in F199 or FH media (fig. 2). In addition, a 2.9kb intermediate is detectable in F199 culture nuclei. Only trace amounts of the 2kb species can be found in FHG culture nuclei. Of particular interest is the finding that 4kb processing intermediates, but not 2kb processed messengers, are detectable in the nuclei of 21-day cultures in FH medium, i.e. at a stage before the onset of overt transdifferentiation (circa 30 days). This accords with previous suggestions (Bower et al., 1983) that the onset of δ gene transcription may precede the development of efficient processing mechanisms (for δ transcripts) during NR transdifferentiation *in vitro*. The absence of larger δ RNA precursors (the two identified δ crystallin genes are about 10kbp long; Piatigorsky, 1984) may be attributable to rapid processing down to the 4kb intermediate.

Despite a report of limited sequence homology between four small *Drosophila* heat-shock (stress) proteins and mammalian α crystallins (Ingolia and Craig, 1982), we have been unable to substantiate any general relationship between stress-protein induction and transdifferentiation into lens in cultures of NR cells maintained in FH medium (Carr and de Pomerai (1985) Dev Growth and Differ, in press). Specifically, crystallins are not co-induced along with stress proteins following heat-shock (43°C) or sodium arsenite treatment of such cultures at any stage. Nevertheless, we find that cultures maintained throughout at the heat-shock temperature of 43°C in medium F199 are able to transdifferentiate extensively into lens cells (see fig. 3), and actively synthesise δ crystallin (fig. 4A), whereas parallel F199 cultures at 37°C do not. Furthermore, protein-labelling experiments following transient heat shock (8 hr at 43°C) of 30-day cultures in F199 medium, suggest that proteins comigrating with lens crystallins (specifically δ) may be co-induced along with the 70 and 85kd stress proteins (fig. 4B). Future work will attempt: (i) to confirm that

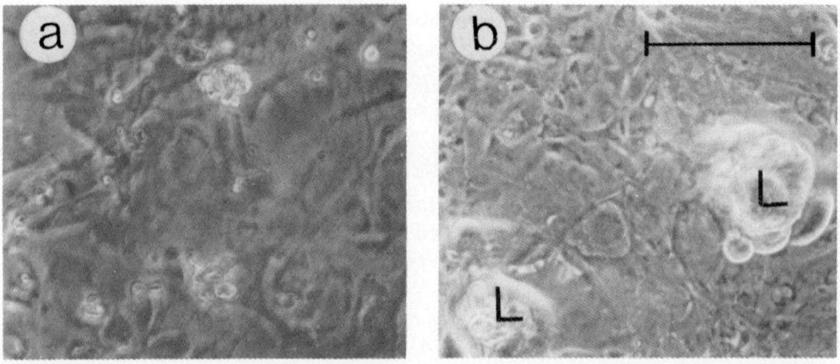

Figure 3. Phase contrast photographs of cultures maintained for 35 days in medium 199: (a) at 37°C throughout; (b) at 43°C throughout. Lentoid bodies (marked L) are absent from cultures at 37°C, but develop widely in cultures at 43°C. Bar represents 100μm.

Figure 4. (^{35}S)-methionine labelled proteins (synthesised over a 5 hr period) were run on SDS-PAGE and fluorographed (20,000cpm/lane).

A. Lanes 1 and 2, NR cultures maintained for 30 days at 43°C in FH (1) or F199 (2) media. Lane 3, chick lens.

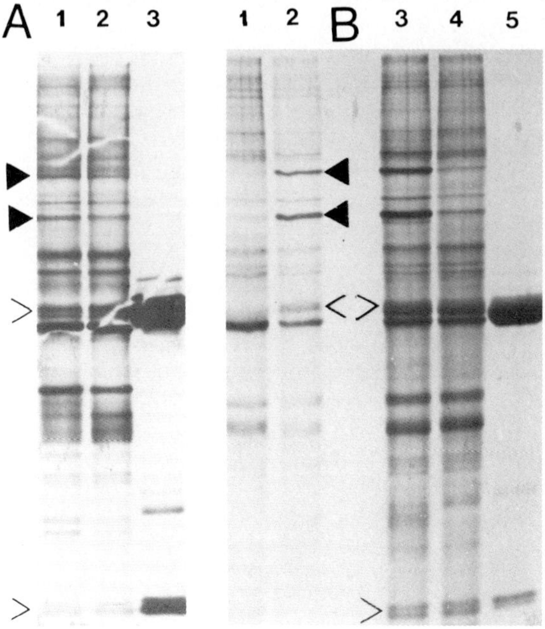

B. Lane 1, 30-day culture in F199 labelled at 37°C. Lanes 2 & 3, 30-day cultures in F199 (2) or FH (3) heat-shocked from 37°C to 43°C for 8 hr total. Lane 4, 30-day culture in FH labelled at 37°C. Lane 5, chick lens. Solid arrowheads, 85 and 70kd stress proteins. Open arrowheads, 50kd δ and 19-20kd α crystallins.

the 50kd protein co-induced by heat shock is indeed δ crystallin (e.g. by immunoprecipitation); (ii) to determine the relative time-course of induction for stress proteins and putative δ crystallin in this system; (iii) to determine whether this is a genuine stress response or merely a temperature-dependent effect (e.g. by stressing cultures with sodium arsenite at 37°C); and (iv) to determine whether the apparent co-induction of δ crystallin following heat-shock is due to the export and translation of δmRNAs which normally remain confined to the nuclei in F199 cultures at 37°C.

ACKNOWLEDGEMENTS

This work was supported by the Cancer Research Campaign, and by SERC studentships awarded to D.E. and A.C. We are grateful to Wendy Lister for typing the manuscript.

REFERENCES

Bower DJ, Errington LH, Pollock BJ, Morris S, Clayton RM (1983). The pattern of expression of chick δ-crystallin genes in lens differentiation and in trans-differentiating cultured tissues. EMBO J 2: 333.

de Pomerai DI, Gali MAH (1982). A switch for transdifferentiation in culture: effects of glucose on cell determination in chick embryo neuroretinal cultures. Devl Biol 93: 534.

Gali MAH, de Pomerai DI (1984). Differential effects of culture media on normal and foreign differentiation pathways followed by chick embryo neuroretinal cells in vitro. Differentiation 25: 238.

Ingolia TD, Craig EA (1982). Four small Drosophila heat shock proteins are related to each other and to mammalian δ crystallin. Proc Natl Acad Sci (USA) 79: 2360.

Piatigorsky J (1984). Lens crystallins and their gene families. Cell 39: 620.

Progress in Developmental Biology, Part A, pages 259–270

VARIATIONS IN mRNA PATTERNS DURING DEVELOPMENT OF CYTOLYTIC T-CELLS DURING HEPATITIS VIRUS INFECTION

Anwar A. Hakim

Department of Histology. School of Dentistry
Loyola University Medical Center. Maywood
Illinois 60153

INTRODUCTION

Infection with hepatitis B virus (HBV) produces a broad spectrum of clinical manifestations ranging from absence of overt disease and minimal aberration of liver function tests to acute fulminant disease which is usually fatal. Symptoms and signs of overt disease commonly include malaise,anorexia weight loss, hepatomegaly and jaundice. In the United States alone, it is estimated that the life time risk of contracting an HBV infection is 5% with an estimated 200,000 persons infected annually. There is a pool of 400,000 to 800,000 infectious carriers in this country of which 25% develop chronic active hepatitis B (CAH-B), many of whom will progress to cirrhosis. An estimated 4,000 die each year in US from HBV-related cirrhosis and another 800 die from HBV-related liver cancer(Center for Disease Control 1982). Chronic HBV-infections frequently follow mild or subclinical acute illness. Acquired resistance to hepatitis B virus (HBV) infection appears to be intimately related to the immune response directed against the hepatitis B virus surface antigen (HBsAg) (Melnick et al.1975; McAuliffe et al.1980; Krugman et al. 1979; Vyas 1981; Szmuness et al.1980). In vitro response to HBsAg of peripheral blood lymphocytes from recipients of hepatitis B vaccine have been described in several reports (Celis and Chang 1984a,b; Chang et al.1984; Cupps et al.1984 and Tong et al. 1975). The present studies, examined the changes in mRNA synthesis, its correlation with in vitro mRNA-encoded pro-protease synthesis and cytotoxicity of the cloned lymphocytes before and after incubation with the viral surface antigen (HBsAg). The data indicate that development of cytotoxic cells correlates with the synthesis of mRNA.

Materials and Methods

Effector Cells- Peripheral blood lymphocytes (PBL) from donord without(HBsAb-sero negative) and with (HBsAb-sero positive) circulating antibodies to the hepatitis B virus surface antigen(HBsAg) were prepared by the standard Ficoll-Hypaque gradient centrifugation, followed by separation on Percoll columns, then cloned in presence of Interleukin-2 (IL-2) in presence and absence of HBsAg(Hakim 1985a,b).The cells were carried on in serum-free medium.

Serum-Free Medium: he culture medium consisted of RPMI-1640 enriched with 1% Solco Seryl(Basle, Switzerland), 4mM glutamine, 3.024 g/l $NaHCO_3$, antibiotics, Insulin (5 ug/ml), transferrin (35 ug/ml) ethanolamine ($2x10^{-5}$M), bovine serum albumin (2.5 ug/ml), fatty acids (linoleic, oleic and palmitic acids 1 ug/ml of each) and Interleukin-2 (IL-2) 5 units/ml.

Target Cells- Two cell lines were used, a hepatocellular carcinoma cell line LPC/PRF/5 previously established and described by Alexander et al.(1976) and a leukemia-derived K-562 specific for Natural Killer cells. These cells were labeled with ^{51}Cr-sodium chromate and used as described by Brunner et al.(1968) and Hakim (1975).

Isolation of Functionally Active mRNA- mRNA preparations were obtained using the guanidium isothiocyanate/CsCl gradient followed by fractionation on Oligo(dT) cellulose column (Hakim 1979). The poly(A).mRNA fractions were translated in a cell-free protein synthesis system (Hakim 1983).

Beginning with 15 ug of HBsAg-enriched poly(A).RNA as template, double stranded cDNA was synthesized. The DNA was made blunt-ended with S1 nuclease and fractionated by electrophoresis in 1% agarose using a buffer system consisting of 40 mM tris.HCl(pH 8.0).20 mM sodium acetate.18 mM NaCl and 1 mM EDTA. DNA of size greater than 1000 base pair was electroeluted from the agarose gel and purified by DEAE-cellulose chromatography. The double stranded cDNA was annealed into Pstl site of PBR 322 using a homopolymeric polyCpolyG tailing technique, and the resulting recombinant plasmids were used to transform E. Coli K-12 strain RPI. Tetracycline selected transformants were screened for clones carrying HBsAg.cDNA by in situ hybridization. Electrophoresis in the presence of SDS was generally carried out using the discontinuous Tris/glycine buffer system of Laemli (1970).

RESULTS

Peripheral blood lymphocytes (PBL) from HBsAb-sero positive HBsAg vaccine recipients and/or in vitro incubation of the PBL from the HBsAb-sero positive with HBsAg increased the PBL cytotoxicity against K-256 cells and the mRNA encoding for a pro-protease. HBsAg had no effect on PBL from HBsAb-sero negative healthy donors. Figure 1 is an autoradiogram of sodium dodecyl sulfate polyacrylamide gel electro phoresis (SDS-PAGE) of the cell-free translation products It shows a time-dependent protein synthesis which varied by the action of mRNA source.

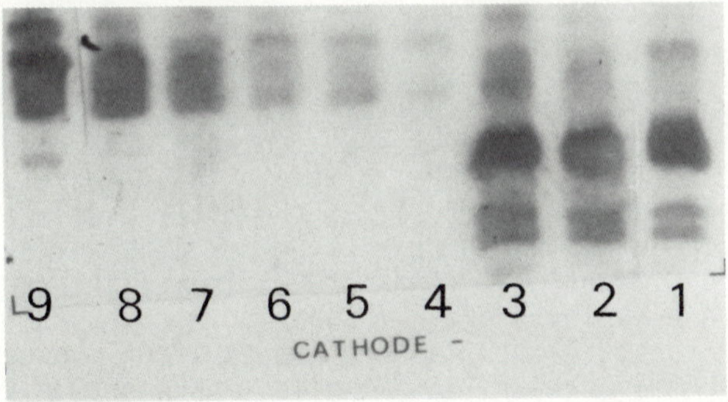

Fig. 1. Auto-radiogram of Sosium Dodecyl Sulfate Polyacrylamide gel electrophoresis (SDS-PAGE) of the cell-free translation products encoded by mRNA from PBL of HBsAb-sero positive HBsAg-vaccine recipients prior(Lane 1) and after(lanes 2,3) incubation with 10 ug of HBsAg for 24 hours at 37°C in 5% CO_2 humidified air environment. The Immunoprecipitated proteins encoded by mRNA from PBL of HBsAb-sero positive vaccine recipients prior (lane 4)), 12 hours (lanes 5,6) and 24 hours (lanes 7, 8, 9) incubation with HBsAg.

Figure 2 shows the electrophoretic patterns of the translation products that are affinity labelled with N -benzyloxycarbonyl-L-lysine thiobenzyl ester, i.e. ^{35}S-methionine labeles with a protease active center.

PBL from HBsAb-sero positive healthy vaccine recipients before and after incubation with HBsAg showed an increased mRNA synthesis encoding for two major glycoproteins of Molecular weights at 23,500 and 27,000 daltons, similar to those encoded by mRNA from the hepatitis B virus, i.e. hepatitis B surface antigens (HBsAg).These results indicate that the viral surface antigen (HBsAg), whether in vivo by vaccination or incubated in vitro alters the synthetic activities of the peripheral blood lymphocytes, i.e. cytotoxic lymphocytes.

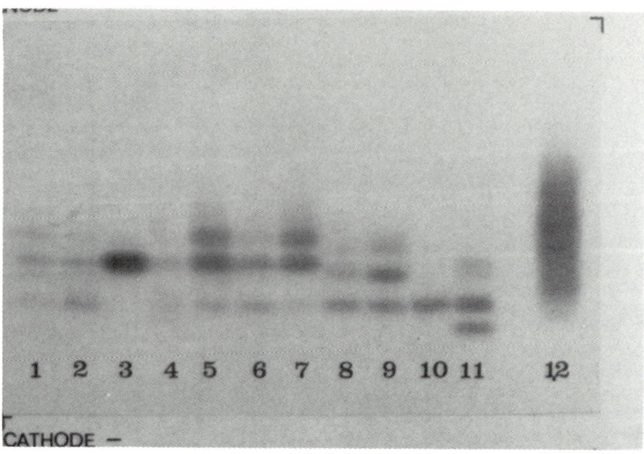

Fig.2. Affinity labeling of the protease with diisopropylfluorophosphoridate (^{3}H-DFP). Affinity labelling of PBL from HBsAb-sero negative (lanes 1,2), and after vaccination(lane 3). PBL from HBsAb-sero positive before (lane 4) and after vaccination (lanes 5,6). mRNA of PBL from HBsAb-sero negative vaccine recipients after incubation with HBsAg (lane 7), HBsAb-sero positive vaccine recipients after incubation with HBsAg (lane 9) or with PHA (lane 10). mRNA of PBL from patients with chronic active hepatitis B (lane 11).

Figure 3 shows that poly(A) RNA from PBL of HBsAb-sero positive pre-incubated with HBsAg (lane C), and from HBsAb-sero positive HBsAg-vaccine recipients (lane H)contained, whereas similar preparations from PBL of HBsAb-sero negative (lane G) did not contain products hybridizable with DNA from HBV.

Figure 4 shows poly(A)RNA preparations from PBL of patients with various types of hepatitis hybridizable with ^{32}P-labeled DNA from viral particle from sera of patients with hepatitis Non A.Non B (lanes B and C); hepatitis A (lanes G and H) and hepatitis B (lanes K, L, M, N, and O).

When examined on SDS-PAGE, the immunoprecipitated products by mRNA from PBL of patients with CAH-B and CAH-A each had four major protein bands,with the strongest intensity at Molecular weights of 27,000 and 40,000; and 23,500 daltons respectively. The mRNA from PBL of patients with Non A.Non B encoded five different protein bands. These results indicate variations in mRNA and their translation products between Hepatitis A, B and NonA.Non B.

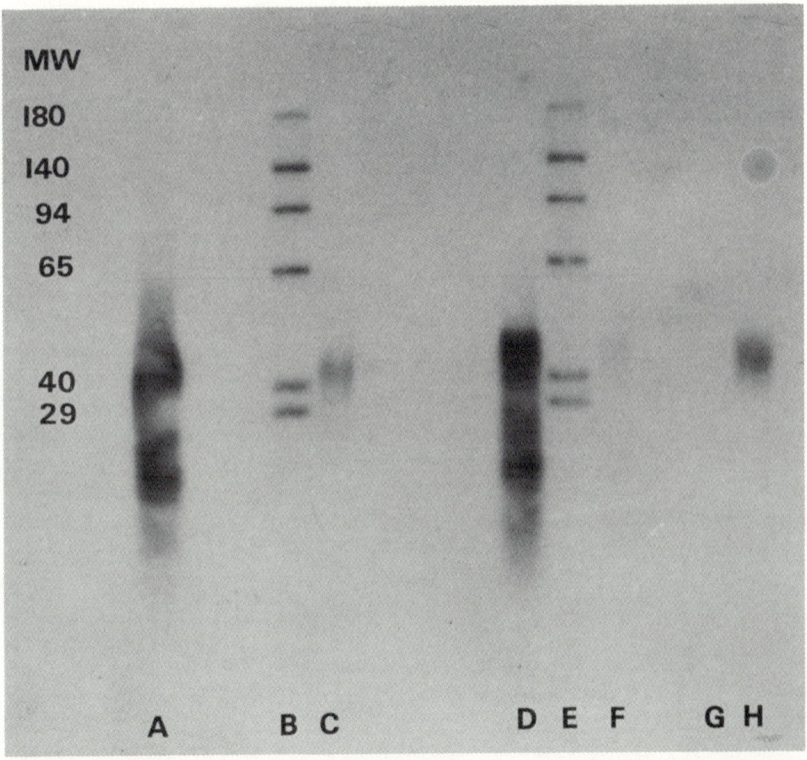

Fig. 3. RNA gel blot hybridization of lymphocyte Poly(A)RNA transferred to nitrocellulose paper. Poly(A)-enriched RNA was electrophoresed on 1.2% agarose gel after glyoxal denaturing treatment. The nitrocellulose paper was hybridized with the nick-translated cDNA probe, which contains sequence specific to HBsAg.mRNA. Nitrocellulose was exposed for 93hr using Dupont lightening plus screen at -70°C. Lanes B and E are molecular markers (MWx10^{-2}Kb) represent restriction fragments of $BamH_1$ and Sac 1 digest of PBR-322 plasmids. Lanes C, F, G, H show the hybridizable products of 10 ug respectively of Ploy(A).RNA from PBL of HBsAb-sero positive incubated with HBsAg; HBsAb-sero negative incubated with HBsAg; HBsAb-sero negative HBsAg vaccine recipients; and HBsAb-sero positive HBsAg vaccine recipients.

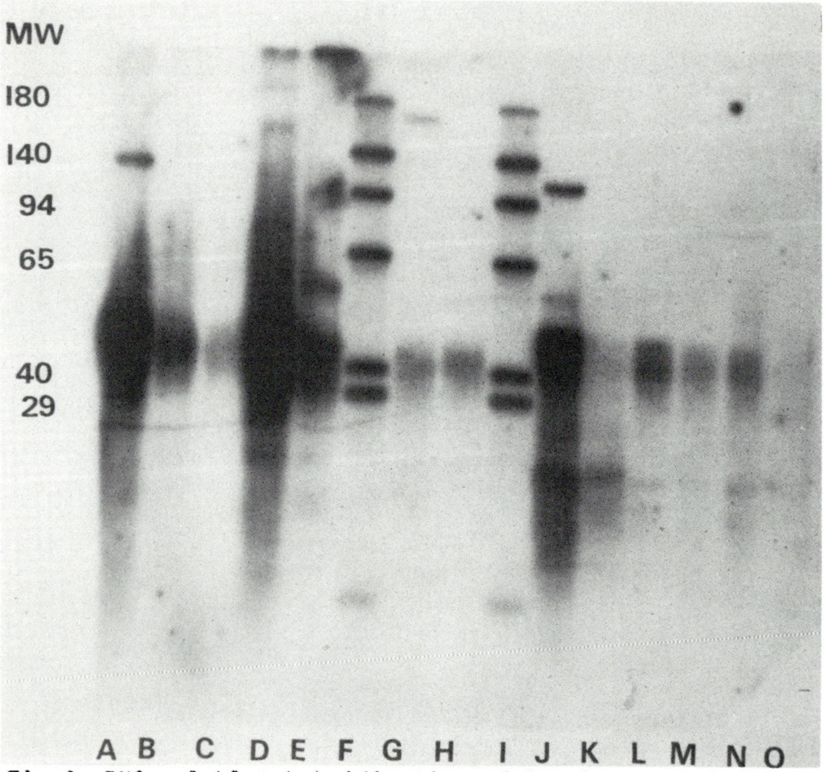

Fig.4. RNA gel blot hybridization of lymphocyte Poly(A).RNA transferred to nitrocellulose paper. Poly(A)enriched RNA was electrophoresed on 1.2% agarose gel after glyoxal denaturing treatment. The nitrocellulos paper was hybridized with the nick-translated cDNA probes from the viral particl es from patients with:a. NON A.Non B hepatitis (lanes B-convalescing and C-Ill); b. Hepatitis A (lanes G-Ill, and H-convalesscing); and c. Hepatitis B (K-hepatitis B carrier; L-PBL from HBsAb-sero positive vaccine recipients incubated with HBsAg; M-PBL from HBsAg-sero positive; N-HBsAb-sero positive vaccine recipients; and)-PBL from patients with Non A.NOn B. hepatitis pre-incubated with HBsAg. Lanes F and I molecular markers ($MWx10^{-2}Kb$) represent restriction fragments of BamHl and Sacl digests of PBR-322 plasmids.

Lane K in the RNA gel block (Fig.4), the RNA from HBsAg carriers has several RNA bands, whereas lane L, the RNA from PBL of HBsAb-sero positive-HBsAg vaccine recipients has one major strong band. The PBL from the HBsAg-carriers had approximately one fourth the cytotoxicity of the PBL from the vaccine recipients. The cause of this difference remains unknown. But, on Oligo(dT) cellulose columns, the RNA from PBL of the vaccine recipients had higher Peaks I (materials not retained) and Peak II (materials retained) by the column than the RNA from PBL of the HBsAg-carriers (Fig.5).

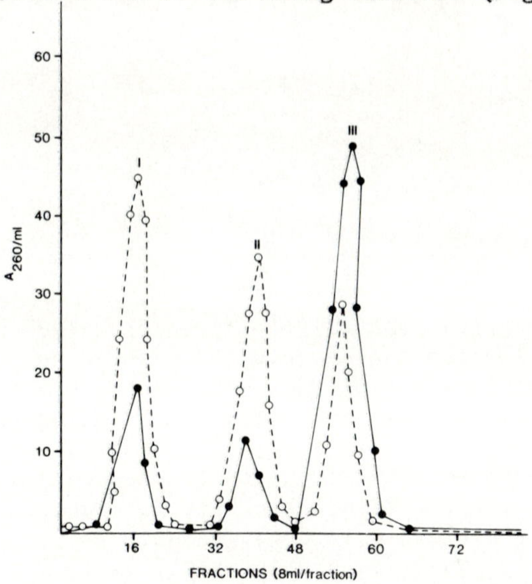

Figure 5. Oligo(dT)-cellulose column chromatography of RNA isolated from PBL of HBsAb-sero positive-Vaccine recipients (------------------) and the RNA from HBsAg-carriers(_____) Peak I materials that was not retained by the column. Peaks II and III are materials eluted with 0.01 Tris.Cl +0.1M KCl and 0.01 M Tris.Cl, respectively.

The retained peak II from both columns was examined in a cell-free-rabbit erythrocyte lysate system, using ^{3}H-leucine as the marker. The elution patterns in figure 6 show differences in the proteins encoded by the Poly(A)RNA in Peak II shown in Fig.5. The peak in RNA from PBL of the HBsAg-carriers encoded greater amounts of the 18S and 28S, whereas the Peak II in the RNA from the healthy vaccine recipients encoded an additional 12S peak (Fig.6). Indicating differences in the synthetic activities of the RNA present in Peak II.

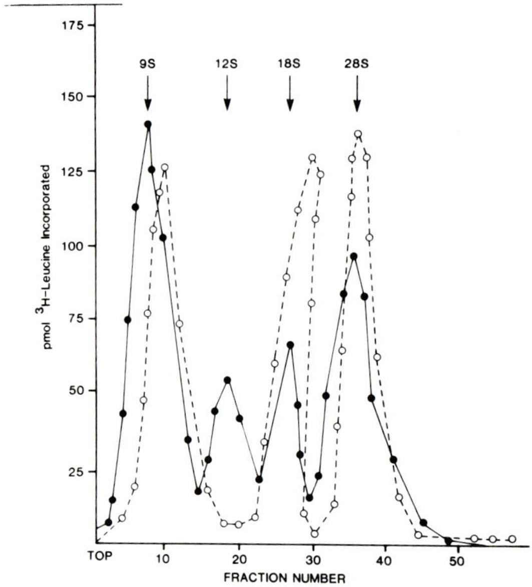

Figure 6. In vitro Peak II Poly(A)RNA directed protein synthesis. ^{3}H-leucine uptake into the protein encoded by Peak II.Poly(A)RNA from PBL of HBsAb-sero positive-HBsAg vaccine recipients(———) and from PBL of HBsAg-carriers (-------------).

The immunoprecipitated proteins encoded in vitro by peak II of the Poly(A)RNA from PBL of vaccine recipients and HBsAg carriers were applied onto Conacanavalin A (Con A)-Sepharose 4B affinity columns. The elution profiles in Fig.& indicate that Peak I-Poly(A)RNA from PBL of vaccine recipients differed from Peak II-Poly(A)RNA of the PBL from HBsAg carriers by encoding an extra (glyco)protein that binds to Con.A-Sepharose 4B .

DISCUSSION

Infection with hepatitis B virus (HBV) may result in either acute or chronic infection. Chronically infected patients may develop chronic hepatitis or become carriers i.e. having circulating viral surface antigen (HBsAg) with no inflammatory liver disease. Variations in host response may influence the duration of infection and the type of inflammatory liver disease that ensues. The described studies demonstrated for the first time variations in the biosynthesis and biofunction of mRNA from peripheral blood lymphocytes (PBL) of healthy adults with (sero positive) and without (sero negative) circulating antibodies to the hepatitis virus surface antigen before and after vaccination

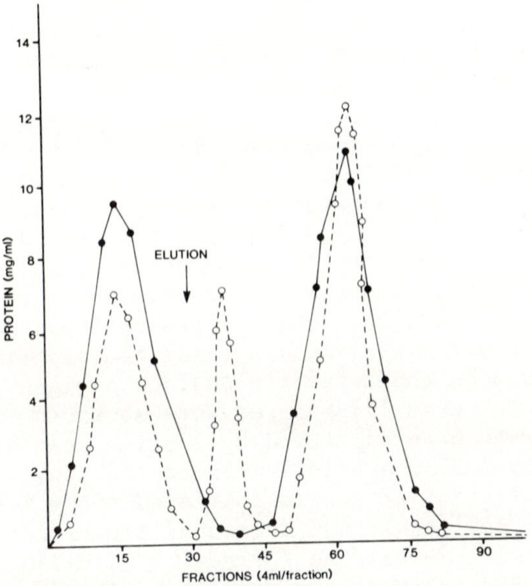

Figure 7. Affinity Chromatography of the immunoprecipitated in vitro synthesized glycoproteins on Concanavali A(Con.A)-Sepharose 4B Columns. The columns (1.6x12 cm) were equilibrated with 0.08 M sodium acetate buffer, pH 6.5. The samples were also dialyzed against the same buffer. The translation mixtures were applied to the columns and fractions of 4.0 ml were collected. After washing, the columns were eluted initially with 0.1 M -methylmannoside containing 0.05 M sodium acetate, pH 6.5, 0.1 M NaCl and 2 mM EDTA. The pattern (--------------) represents the proteins in the translation mixture mediated by Peak III Oligo(dT)-cellulose from PBL of HBsAb-sero positive HBsAg-vaccine recipients; and pattern (———————) glycoproteins in translation mixture encoded by Peak III Oligo(dT)-cellulose from PBL of HBsAg-carriers.

with the hepatitis surface antigen, and of patients with chronic active hepatitis(CAH), i.e. continuous carriers of the viral antigen (HBsAg).

PBL from healthy adults and from patients with CAH responsive to HBsAg were cloned and cultured for a brief period in a serum-free chemically-defined medium supplemented with the lymphokine-Interleukin-2 (IL-2). These lymphocytes maintained their initial proliferative and mitogenic responsiveness.

PBL of the same HBsAb-sero positive healthy adults, before and after vaccination differed in the amount of the Poly(A) RNA and its mediated synthesis, and in cytotoxicity. Vaccination with the viral surface antigen increased the three parameters. When incubated with PBL from HBsAb-sero positive vaccine recipients, HBsAg increased the amount of Poly(A) RNA and its mediated protein synthesis, and the PBL cytotoxicity. These results indicate that PBL activation by the viral surface antigen is accompanied by an increased mRNA and its mediated protein synthesis. Some of the proteins encoded by this mRNA are : One a pro-protease identified by an in-situ affinity labelling, and two proteins similar to those found in the hepatitis surface antigen in Molecular weight. Cultured lymphocytes were shown to transform and proliferate when activated by lectins such as Con A (Ling and Kay 1975). A brief exposure (20 min) to a sufficient concentration of lectin initiated entry into the G1 phase of the cell cycle and increased the radioactive labelling of RNA with ^{3}H-uridine (Milner 1977, Milner 1978). Changes in mRNA patterns occurred in cultured human lymphocytes in response to Con A (Forsdyke 1980). Resting and in-vitro growing lymphocytes also showed changes in protein synthesis (Lester et al 1981).

RNA gel blot hybridization showed that lymphocytes Poly (A) RNA hybridizes with ^{32}P-labeled DNA of the viral particles sedimenting from serum of the same patient as the PBL. It is suggested that in blood , the viral particles alter the PBL mRNA biosynthesis and its biofunction. Several types of virus including Measles virus are able to infect peripheral blood leukocytes in vitro (Hyypia et al 1985). In the present study, the gel blot hybridization analysis indicated that hepatitis virus are able to infect the PBL. Furhter investigations are in progress to confirm this possibility.

The lymphocyte Poly(A)-enriched RNA coding for a pro-protease increased in PBL from HBsAb-sero positive at 30 days post vaccination and/or after in vitro incubation with the viral surface antigen (HBsAg). This increase in the Poly(A) RNA correlated with a rise in the PBL cytotoxicity against K-562 a leukemia derived cell line specific for Natural Killer cells (Hakim- Cell. Mol. Biol. In Press). Therefore, development of cytotoxic lymphocytes induced by hepatitis B viral infection is modulated by induced variations in cellular Poly(A) enriched RNA biofunctions.

REFERENCES

Alexander, H., Bay, E.M., Geddes, E.W., Lacastus, G.(1976) Establishment of a continuously growing cell line from primary carcinoma of the liver. S. Afr. Med. J. 50,2124.

Brunner, K.T., March, J., Cerottini, J.C., Chapuis, B.(1968) Quantitative assay of the lytic action of immune lymphoid cells, or ^{51}Cr-labeled allogeneic target cells in vitro: Inhibition by Isoantibody and drug. Immunology, 14, 181.

Center for Disease Control (1982). The hepatitis B vaccine (Hepatovax-B). Morbifity, Mortality, Weekly Review 26June

Celis, E., Chang, T.W.(1984a) Antibodies to hepatitis B surface antigen potentiate the response of human T-lymphocyte clones to the same antigen. Science 224;227-229.

Celis, E., Chang, T.W.(1984) HBsAg-serum protein complexes stimulate immune T-lymphocytes more efficiently than do pure HBsAg. Hepatology 4: 1116-1123.

Chang, T.W., Celis, E., Miller, R.W.(1984) In vitro response to HBsAg of peripheral blood lymphocytes from recipients of hepatitis B vaccine. Hepatology 5: 824-829.

Cupps, T.R., Gerin, J.L., Purcell, R.H.(1984) In vitro Antigen-induced antibody reponses to hepatitis B surface antigen in man. J. Clin. Invest. 74: 1204-1213.

Hakim, A.A.(1985a) Cytotoxicity of human peripheral blood T-lymphocyte clones activated by hepatitis B virus surface antigen. Experientia-In Press.

Hakim, A.A.(1985b) Regulation of protein synthesis in hepatitis B virus surface antigen (HBsAg) activated peripheral blood lymphocytes. Expt. Cell Biology- In Press.

Hakim, A.A., Siraki, C. .(1985) Protein glycosylation of the cell membrane alters the biofunction of the mRNA translation products. UCLA Symposia, April 5.

Hakim, A.A., Siraki, C.M., Joseph, C.E.(1985) Terminal galactose residues are cytotoxic lymphocyte signals for target cell recognition. Fed. Amer. Soc. Expt. Biol.,Fed. Proc. 44/4, 927.

Hakim, A.A., Siraki, C.M., Joseph, C.E.(1984) Fibronectin protects circulating cytotoxic cells from exogeneous cytotoxic cells from exogeneous adenosine 5'-triphosphate. Cancer Detection Prevention 7:42:056.

Hakim, A.A.(1975) Mechanisms by which tumor avoids destruction by the immune system. BCG-catalyzed increase in IgA blocking activity of lymphocyte-mediated cytotoxicity. Klin. Wschr. 53: 881-885.

Hakim, A.A. (1979) Modulation of Immune RNA biological activities by the cell surface. Fed. Amer. Soc. Biol. Fed. Proc. 38: 1342.

Hakim, A.A. (1983) Cell-free transability of mRNA regulates cAMP-dependent protein kinase. Fed. Amer. Soc. Exp. Biol. Fed. Proc. **42**: 603.

Hyypia, T., Korkiamaki, P,,Vainionpaa, R.(1985). Replication of Measles virus in human lymphocytes. J. Exp. Med. **161**: 1261.

Krugman, S., Overby, L.R., Mushwar, I.K.(1979) Viral hepatitis, Type B : Studies on Natural history and prevention reexamined. N. Engl. J. Med. **300**: 101.

Laemli, U.K. (1970) Cleavage of structural proteins during the assembly of the head of bacteriophage T4. Nature **227**: 680.

Ling, N.R.,Kay, J.E.(1975) In lymphocyte stimulation. Elsevier Publishing Co. Amsterdam 147.

McAulife, V.J., Purcell, R.H., Gerin, J.L. (1980). Type B hepatitis : A review of current prospects for a safe and effective vaccine. Rev. Infect. Dis. **2** : 470.

Melnick, J.L., Dreeman, G.R., Hollinger, F.B.(1976). Approaching the control of viral hepatitis B. J. Infect. Dis. **176**: 210.

Szmuness, W., Stevens, C.E., Harley, E.J. (1980) Hepatitis B vaccine. Demonstration of efficiency in a controlled clinical trial in a high risk population in the United States. N. Engl. J. Med. **303** : 833.

Tong, M.J., Wallace, A.M., Peters, R.L., Reynolds, T.B.(1975) Lymphocyte stimulation in hepatitis B infection. N. Engl. J. Med. **293**: 318.

Vyas, G.N. (1981) Molecular Immunology of the hepatitis B surface antigen (HBsAg). In: Maupas P., Guesry, P. (eds) Hepatitis B vaccine. Amsterdam: Elsvier/North Holland 222.

Pattern Formation in Higher Organisms

Progress in Developmental Biology, Part A, pages 273-287

REGENERATION WITHIN THE INSECT THORACIC SEGMENT

Vernon French

Department of Zoology, University of Edinburgh, Kings Buildings, West Mains Road, Edinburgh EH9 3JT, Scotland, U.K.

INTRODUCTION

The development of most (if not all) animals involves extensive interactions which give cells information concerning their position and lead them to adopt appropriate developmental fates (Wolpert, 1969). The interactions mediating pattern formation can be studied by depriving cells of their normal neighbours or placing them against cells which are normally far away. These and other sorts of experiment show that interactions during the syncytial and early cellular stages of insect embryogenesis result in the determination of the segment pattern. Within the segment, however, further cellular interactions are involved in forming the detailed patterns of the segment and appendages (e.g. legs and wings), in controlling growth and, in insects which can replace missing parts, in directing regeneration. Particular attention has been paid recently to the layer of epidermal cells which secrete the overlying richly-patterned cuticle of the larval legs of various insects and the legs and wings of *Drosophila*.

In most insects the embryonic thoracic segments form limb buds which grow, become segmented, secrete the cuticle of the functional larval legs, and subsequently grow and secrete a new cuticle before each larval moult. The larva either develops directly into a similar adult (e.g. cockroaches, crickets) or undergoes metamorphosis with the larval leg developing into a rather different adult leg (e.g. beetles). In other holometabola (including *Drosophila*) some embryonic thoracic cells invaginate to form the leg and wing imaginal discs which grow but do not differentiate in the larva. In the puparium the discs evaginate, secrete cuticle and form the thoracic segment of the adult fly, while the larval cells die.

The interactions between cells from different positions can be studied in regeneration following amputation and grafting experiments on the larval legs, or comparable fragmentation and mixing experiments on imaginal discs. I shall briefly describe 2 recent models of limb regeneration and then discuss the relationship between the appendage and the surrounding body segment.

THE POLAR CO-ORDINATE MODEL

The Polar Co-ordinate Model (French et al, 1976; Bryant et al, 1981) proposes that the epidermal cells of the larval leg or the imaginal disc have positional values characteristic of (i) their proximal-distal level and (ii) their position around the circumference (Fig. 1). Positional values are stable unless the epidermis is disrupted, causing non-neighbouring cells to interact and producing intercalation and distalization.

1) The Shortest Intercalation Rule - Intercalary Regeneration

"When cells differing in proximal-distal or in circumferential positional value interact, they are stimulated to divide, producing new cells with the shortest set of intermediate values".

Intercalary regeneration can be directly demonstrated in both axes of the cockroach leg. When proximal and distal levels of a segment (A and E in Fig. 1) are grafted together, there is growth at the junction and the mid segment (B,C,D) is formed (Bohn, 1970; Bullière, 1971). Similarly, when different circumferential positions are confronted (French, 1978), local growth produces the shortest intervening arc of circumference (e.g. 5,6,7 forms between positions 4 and 8). Usually both edges contribute, although they may do so very unequally (e.g. distal cells form most of a proximal-distal regenerate) and in circumferential intercalation cells seem unable to cross between anterior and posterior compartments (French, 1980).

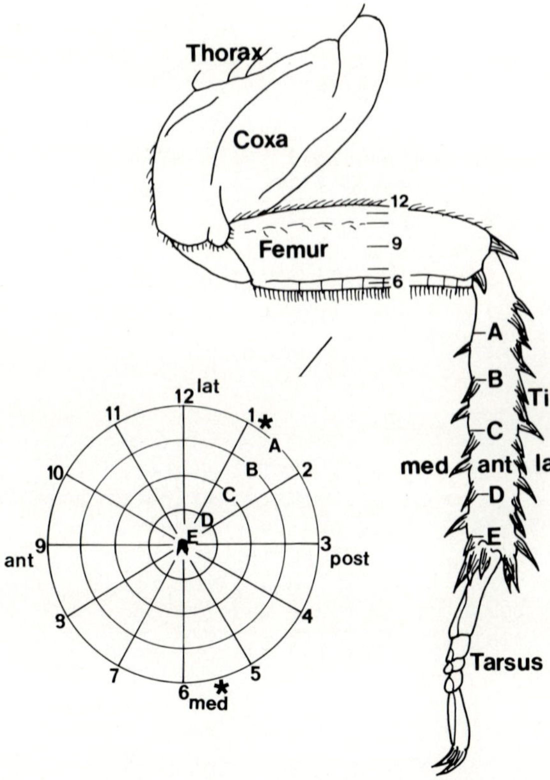

Figure 1. The larval cockroach leg shown in anterior view and in polar co-ordinate model representation. Epidermal cells have positional values corresponding to their proximal-distal level (A-E) and their circumferential position (1-12). Ant, post, med and lat are the anterior, posterior, medial and lateral faces of the leg, and asterisks mark the postulated compartment borders.

Imaginal discs cannot be grafted accurately, but intercalary regeneration can be demonstrated by mixing different fragments (Haynie and Bryant, 1976). The regeneration/duplication behaviour of different "pie-slice" sectors of discs is consistent with shortest-route intercalation between the 2 cut edges, provided that the distribution of values is fairly even around the wing disc but uneven around the leg disc (see French et al, 1976). In some leg disc fragments, one edge contributes all or most of the new tissue and compartment borders can be crossed early in intercalation (Abbott et al, 1981). The polar co-ordinate model proposes that intercalation involves only local cell division at a graft/host junction or site of healing, and recent histological studies on the cockroach leg (Anderson and French, 1985) and imaginal discs (Kiehle and Schubiger, 1985; O'Brochta and Bryant, 1986) are in general agreement with this prediction.

2) The Distalization Rule - Distal Regeneration

"If intercalation produces cells with values identical to those of adjacent cells, the new cells adopt a more distal value".

If the distal end of a larval leg or the central part of an imaginal disc is removed, distal regeneration occurs. A normal leg stump or a complete disc periphery regenerates all missing distal structures while a symmetrical double-half leg stump or a duplicating peripheral disc fragment regenerates to a variable extent and often remains distally incomplete. The polar co-ordinate model proposes that healing at the distal tip stimulates intercalation (except where healing occurs across a line of symmetry) and the new cells are forced to a more distal value (see Bryant et al, 1981). Distal regeneration should therefore involve only local cell division at the tip, but Truby (1983) showed that cell division spreads proximally after amputation of the cockroach leg, suggesting that this explanation is inadequate.

3) Supernumerary Regeneration

The intercalation and distalization aspects of the model readily explain the supernumerary regeneration which occurs when the distal part of the larval leg is grafted onto the opposite leg stump (Bohn, 1965; Bullière, 1970; Bart, 1971). After reversal of the anterior-posterior (A-P) axis, supernumeraries form at the junction in A and P positions, while after medial-lateral (M-L) axis reversal they form in M and L positions. These grafts will confront cells with different circumferential values and intercalation will occur to different extents at different points on the junction, forming circumferences (and hence distal regenerates) at the 2 sites where opposite values are confronted (Fig. 2). If compartments are respected

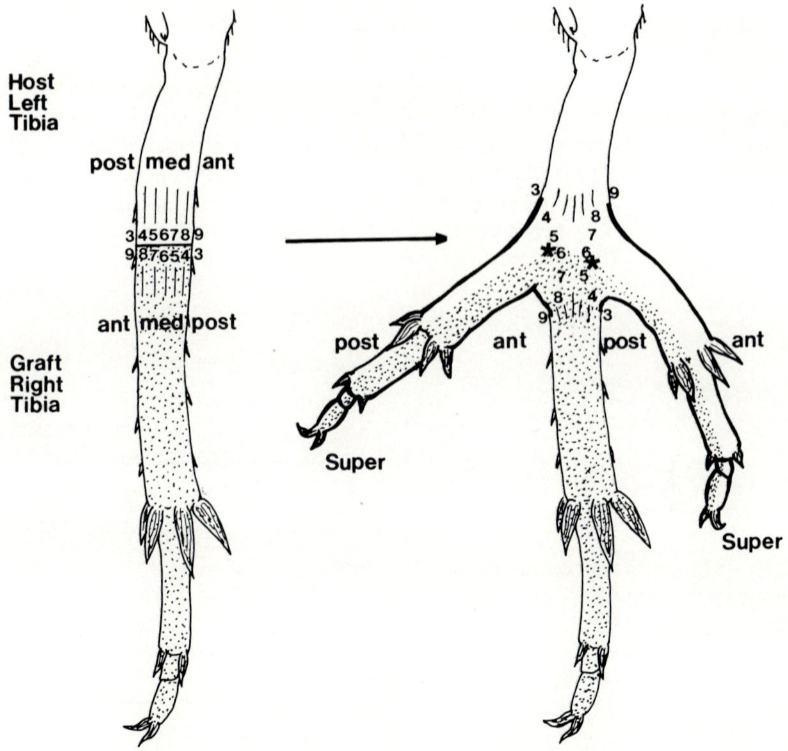

Figure 2. Graft at tibial level between left and right cricket legs, reversing the anterior-posterior (*ant*-*post*) axis. Shortest route intercalation will generate complete circumferences at *ant* and *post* positions because values *8,7,6,5,4* are formed medially and *10,11,12,1,2* are formed laterally. Supernumeraries are reliably half-host, half-graft in origin because intercalation is restricted at compartment borders (asterisk).

during regeneration, A-P supernumeraries will be reliably half-and-half in origin, while M-L supernumeraries will be variable: this result is found in grafts between cricket legs differing in structure or pigmentation (French, 1984; French and Toussaint, in prep.).

4. Normal Development

The polar co-ordinate model was derived from analysis of regeneration. It proposes that the epidermis bears a detailed map of stable positional values and responds to a discontinuity by cell division and change of positional value in accordance with 2 rules. The same cellular behaviour may govern growth and pattern formation in normal development if the appendage primordium is first formed with a few of the proximal circumferential values (French et al, 1976; Bryant and Simpson, 1985).

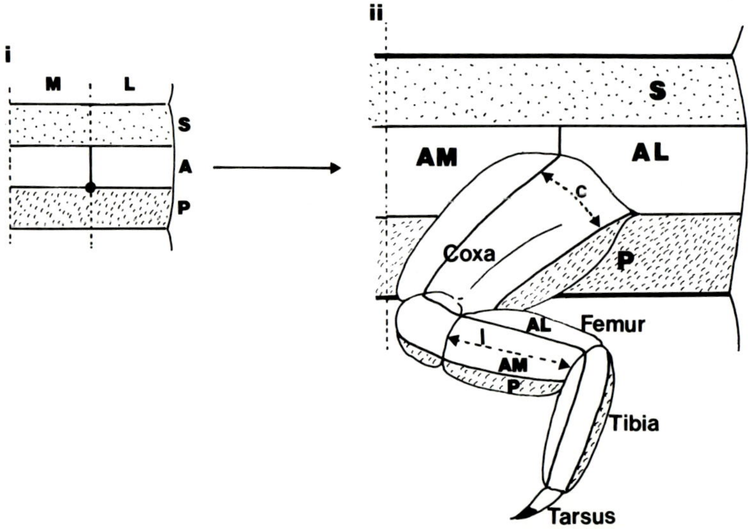

Figure 3. The Boundary Model of leg development. Segmentation forms a repeating pattern of 3 cell states (or 'compartments'), S, A and P, and segment borders form along the P/S boundary. Limbs are initiated in thoracic segments where the A/P boundary is crossed by that between AM and AL 'compartments'. This intersection is the source for a morphogen gradient whose decreasing levels determine concentric rings of tissue as leg tarsus, tibia, femur and coxa. Within a leg segment interactions between 'compartment' boundaries (e.g. AM/AL and AL/P) and between segment

boundaries generate circumferential (c) and proximal distal (l) positional values.

THE BOUNDARY MODEL

In the Boundary Model (Meinhardt, 1983, 1984) the segment consists of transverse and longitudinal 'compartments' (which may, but need not, correspond to lineage restrictions) and the appendage forms at the intersection of 'compartment' boundaries, where AM, AL and P 'compartments' meet (Fig. 3). A morphogen gradient specifies leg segments and subsequent interactions between 'compartment' and segment boundaries generate detailed positional information within the epidermis. The particular competition/activation mechanism suggested for the interactions (Meinhardt, 1983) explains many of the features of intercalary regeneration in legs and discs and, although it is not directly coupled to cell division, it is certainly compatible with the observations of local cell division. If the leg is amputated, the AM, AL and P 'compartments' will meet at the healing tip, forming a morphogen gradient and causing the end of the stump to be respecified as distal segments. This will also occur at 2 opposite positions on the junction between host and an A-P or M-L reversed graft, so the Boundary Model, like the Polar Co-ordinate Model, can explain the supernumeraries forming after these grafts (see Fig. 2). The models do, however, differ in their predictions for some other leg grafts and in their explanation of the relationship between the limb and the rest of the thoracic segment.

THE LEG AND THE THORACIC SEGMENT

Most of the work on insect leg regeneration has only been concerned with the spatial organization of the appendage, rather than the entire thoracic segment including the ventral sternites, lateral pleura and wing primordium, and dorsal tergites. Similarly, *Drosophila* imaginal discs are formed from a small part of the embryonic segment, so the organization of leg disc plus wing disc is unlikely to equal that of the entire thoracic segment in other insects.

We are currently examining the relationship between the leg and the rest of the thorax in the beetle, _Tenebrio_. If the larval leg is removed, it is regenerated from the surrounding tissue, and if the entire right leg is removed at the base and grafted to the site of the left leg, 2 supernumerary legs are formed (Fig. 4). Reversal of the graft A-P axis results in A and P supernumeraries which are consistently half-graft/half-host in origin, while M-L reversal produces M and L supernumeraries which are of variable origin (French, 1986). These results are similar to those found at a more distal level in the legs of _Tenebrio_ and other insects (Fig. 2) and suggest that the arrangement of positional values and A and P compartments extends back from the leg onto the surrounding thorax.

If a transverse strip of thorax is removed from the intersegmental region, thorax posterior to one leg can interact with that anterior to the next. This experiment, first done by Bohn (1974) on the cockroach, can result in the regeneration of a supernumerary leg of reversed A-P orientation (Fig. 5), and a series of such extirpations defines the extent of the regions which can interact to give this result (French and Rowlands, in prep.).

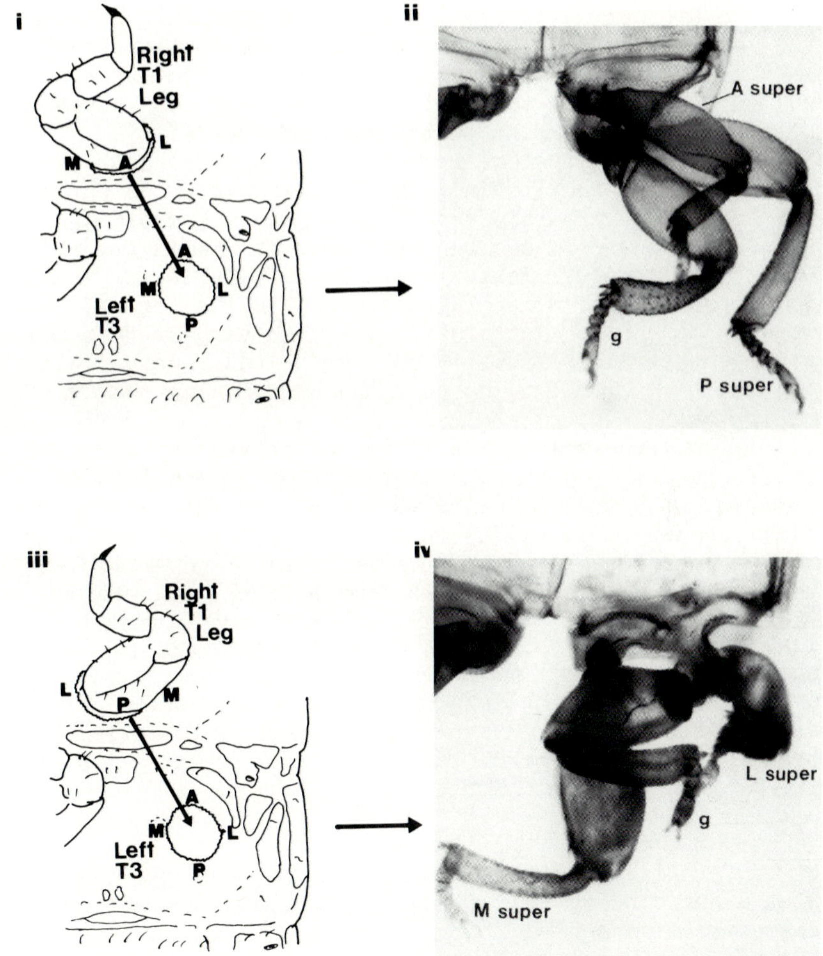

Figure 4. Grafts of the Tenebrio right larval leg onto the left side of the ventral thorax. Reversal of the graft anterior-posterior (A-P) axis (i) gives a resulting adult (ii) with the graft leg (g) plus anterior and posterior supernumeraries (A Super, P super). Reversal of the medial-lateral (M-L) axis (iii) gives medial and lateral supernumeraries (M super, L super).

The results of the strip extirpations and the A-P and M-L reversed leg grafts are all compatible with both the Polar Co-ordinate and Boundary Models of leg regeneration. The Polar Co-ordinate Model proposes that the map of

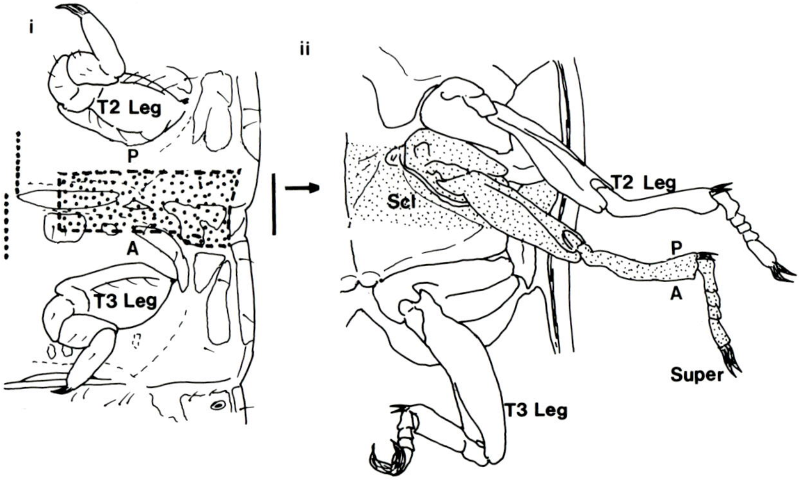

Figure 5. Excision of transverse strips of ventral thorax between mesothoracic (T2) and metathoracic (T3) legs of *Tenebrio*. Removal of a strip of tissue midway between the legs (solid bar in (i)) results in regeneration of a supernumerary leg (Super) and sclerites (Scl.) in reversed anterior-posterior (A-P) orientation (ii). Removal of a similar-sized strip from a more anterior or posterior position (dotted bars in (i)) results in regeneration of the excised parts, giving a normal pattern.

circumferential and radial values (Fig. 1) covers the leg plus a surrounding ring of thorax, with maps in adjacent segments being separated by an intersegmental region (French et al, 1976). Reversing one axis at the level of the base of the leg (Fig. 4) or more distally (Fig. 2) will give equivalent results, and any excision which completely removes the intersegmental region will result in intercalation between the posterior of one map and the anterior of the next, forming a complete circumference and hence a supernumerary leg (Fig. 5). In the Boundary Model the arrangement of AM, AL and P 'compartments' on the leg is an extension of that on the ventral thorax (Fig. 3) so, again, operations at the leg base will give the same results as those done more distally. A strip extirpation which completely removes the S 'compartment' will result in an ectopic confrontation of AM, AL and P, and hence in the regeneration of an extra leg (Meinhardt, 1984).

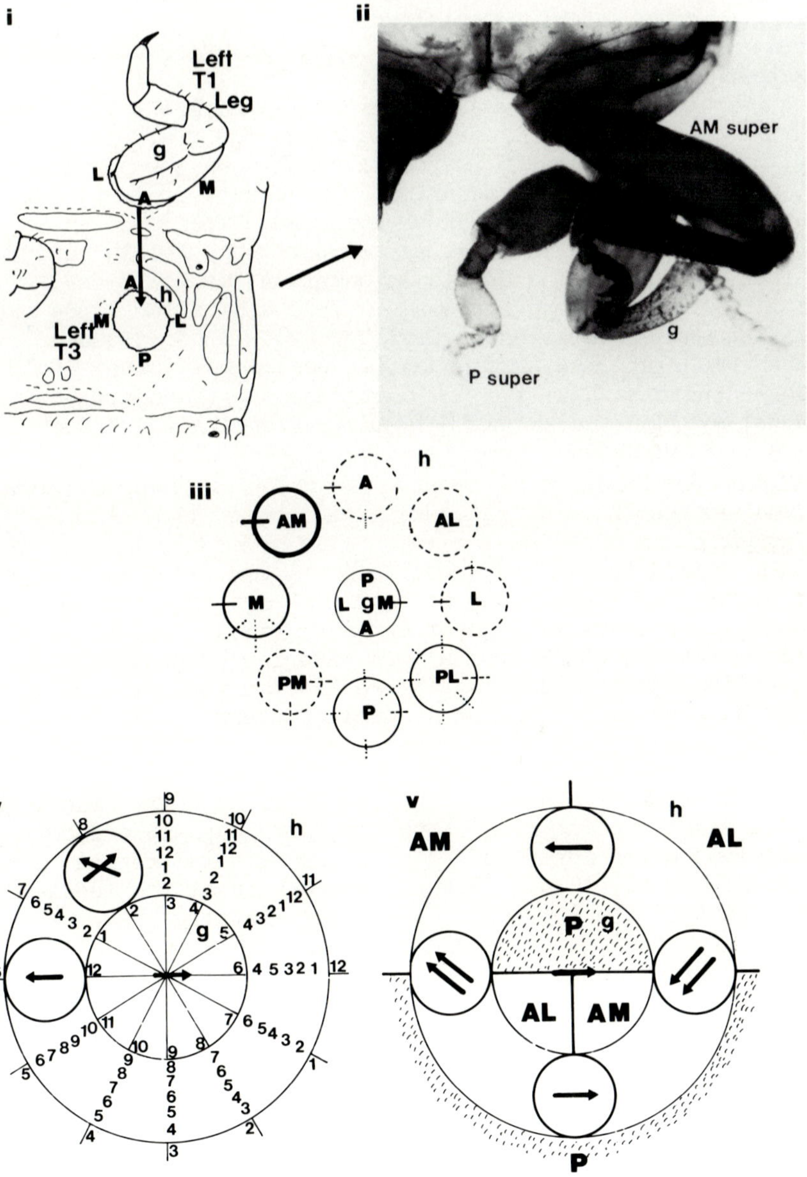

Figure 6. 180° rotation of the Tenebrio larval leg, reversing both anterior-posterior (A-P) and medial-lateral (M-L) axes. (ii) - a resulting adult with rotated graft leg (g) plus anterior-medial supernumerary (AM Super) orientated with claws pointing medially, and a posterior supernumerary (P Super) orientated laterally. (iii) - summary of the positions and orientations of the 162 supernumerary legs. The inner circle shows the grafted leg with reversed axes and the outer circles on the surrounding thorax (h) are supernumeraries. The heavy circle (AM) is a frequent position (> 33% of supernumeraries) while the light dashed circles (e.g. L) are positions rarely occupied. Lines on the circles denote orientation (medial side); solid lines are very frequent (> 75%) and dotted lines are infrequent orientations. (iv) and (v) - predictions of the 2 models. The arrangement of positional values (Polar Co-ordinate Model - iv) or 'compartments' (Boundary Model - v) on the thorax (h) and grafted leg (g) are shown and circles between them denote predicted sites and orientations (arrow pointing medially) of supernumeraries (see text).

The 2 models make different predictions in the case of a 180° leg rotation, which reverses both the A-P and M-L axes. When the rotation is done at the base of the *Tenebrio* leg (Fig. 6), 2 supernumeraries (occasionally 1 or 3) are formed, in a wide variety of positions (French, 1986). The most frequent position is AM, and the most frequent combination of 2 supernumeraries are AM and M, and AM and PL. AM legs are always orientated with claws pointing medially, but those in other positions have variable orientation (Fig. 6iii).

In the Polar Co-ordinate Model supernumeraries result from patterns of intercalation at the graft/host junction. If circumferential values are evenly spaced (as in Fig. 6iv), each point on the junction confronts opposite values and could intercalate in either direction, leading to 0, 2, 4 or more supernumeraries in *any* positions (French et al, 1976). For other reasons, supernumerary legs may form in particular sites, but their orientations will be constrained : M legs will *always* have claws pointing medially, AM legs will point approximately medially (intercalation as in Fig. 6iv) *or* approximately laterally (intercalation in the opposite direction), and P or PL legs will point either medially or laterally. If spacing of values is uneven, 2 supernumeraries will form in constant opposite positions and constant orientations (French et al, 1976).

In the Boundary Model supernumeraries form where AM, AL and P 'compartments' meet (Fig. 3). After a 180° rotation this occurs at *4* positions : A and P supernumeraries point medially and laterally, and the M and L legs are double structures (because of the presence of 2 compartments) pointing approximately medially (Fig. 6v). A 4 'compartment' (AM, AL, PM, PL) model would also give A,P,M and L legs but their structure and orientation is difficult to predict.

The results of the 180° leg rotation (Fig. 6iii) are not in agreement with the predictions of either model, but further grafts of variously orientated legs into positions other than the normal leg site may decide whether either model adequately describes regeneration in the ventral thorax. Similar experiments will be used to investigate wing primordium and dorsal thorax, and to relate pattern formation in the thoracic segment and the much simpler abdominal segments.

REFERENCES

Abbott L, Karpen G, Schubiger G (1981). Compartmental restrictions and blastema formation during pattern regulation in Drosophila imaginal leg discs. Dev Biol 87: 64.

Anderson H, French V (1985). Cell division during intercalary regeneration in the cockroach leg. J Embryol exp Morph (in press).

Bart A (1971). Morphogenèse surnuméraire au niveau de la patte du Phasme Carausius morosus Br. Wilhelm Roux Arch 166:331.

Bohn H (1965). Analyse der Regenerationsfähigkeit der Insekten-extremität durch Amputations- und Transplantationsversuche an Larven der afrikanschen Schabe Leucophaea maderae Fabr. (Blattaria). II. Mitt. Achsenedetermination. Wilhelm Roux Arch 156:449.

Bohn H (1970). Interkalare Regeneration und segmentale Gradienten bei den Extremitäten von Leucophaea-Larven (Blattaria). I. Femur und Tibia. Wilhelm Roux Arch 165:303.

Bohn H (1974). Extent and properties of the regeneration field in the larval legs of cockroaches (Leucophaea maderae), I. Extirpation experiments. J Embryol exp Morph 31:557.

Bryant P, Simpson P (1984). Intrinsic and extrinsic control of growth in developing organs. Q Rev Biol 59:387.

Bryant SV, French V, Bryant PJ (1981). Distal regeneration and symmetry. Science 212:993.

Bullière D (1970). Interprétation des régénérats multiples chez les Insectes. J Embryol exp Morph 23:337.

Bullière D (1971). Utilisation de la régénération intercalaire pour l'étude de la détermination cellulaire au cours de la morphogenèse chez Blabera craniifer (Insecte Dictyoptère). Dev Biol 25:672.

French V (1978). Intercalary regeneration around the circumference of the cockroach leg. J Embryol exp Morph 47:53.

French V (1980). Positional information around the segments of the cockroach leg. J Embryol exp Morph 59:281.

French V (1984). The structure of supernumerary leg regenerates in the cricket. J Embryol exp Morph 81:185.

French V (1986). Interaction between the leg and surrounding thorax in the beetle, Tenebrio. J Embryol exp Morph (in press).

French V, Bryant PJ, Bryant SV (1976). Pattern regulation in epimorphic fields. Science 193:969.
Haynie J, Bryant PJ (1976). Intercalary regeneration in the imaginal wing disk of Drosophila melanogaster. Nature 259:659.
Kiehle C, Schubiger G (1985). Cell proliferation changes during pattern regulation in imaginal leg discs of Drosophila melanogaster. Dev Biol 109:336.
Meinhardt H (1983). Cell determination boundaries as organizing regions for secondary embryonic fields. Dev Biol 96:375.
Meinhardt H (1984). Models for positional signalling, the threefold subdivision of segments and the pigmentation pattern of molluscs. J Embryol exp Morph 83 Suppl: 289.
O'Brochta D, Bryant PJ (1986). Patterns of growth during the regeneration of Drosophila imaginal wing discs. Dev Biol (in press).
Truby PR (1983). Blastema formation and cell division during cockroach limb regeneration. J Embryol exp Morph 75:151.
Wolpert L (1969). Positional information and the spatial pattern of cellular differentiation. J theor Biol 25:1.

Progress in Developmental Biology, Part A, pages 289–292

KRÜPPEL AND TAILLESS AFFECT COMPLEMENTARY REGIONS OF THE SEGMENTED DROSOPHILA EMBRYO

Teresa R. Strecker and John R. Merriam

Department of Biology, University of California at Los Angeles, Los Angeles, CA 90024

INTRODUCTION

In this report we investigate the proposal that the pattern mutants Krüppel and tailless affect complementary regions of the segmented Drosophila embryo. Based upon their observations of the gap mutants Krüppel, knirps and hunchback, Nüsslein-Volhard and Wieschaus (1980) proposed that an early step in the process of segmentation may be the organization of the egg into large continuous subregions defined by the phenotypes of the gap mutants. The total domain affected by gap mutants, however, does not account for the entire ectodermal segmentation pattern. Instead, another zygotic lethal pattern mutant, tailless, affects the remainder of the embryo left unaffected by gap mutants, namely the anterior head lobes and posterior abdominal segments. Thus, we propose an additional aspect to the original model of Nüsslein-Volhard and Wieschaus: during embryogenesis the anterior and posterior ends of the presumptive segmented embryo are established apart from the central trunk region.

A specific test of this model is the comparison of the segmentation pattern defects in gap mutant and tailless embryos. We predict that there will be no overlap between the domains of the gap and tailless mutants. The following is a report of a comparison of the segmented Kr and tll 9 hr embryos observed using SEM. Krüppel was selected for this first comparison with tailless because it is the most thoroughly studied gap mutant (Gloor, 1950; Wieschaus *et al.*, 1984). Furthermore, a comparison of 9 hr embryos permits clear observation of the anterior head lobes and gnathal segments prior to head involution, and of the posterior-most abdominal segments, A9 and A10, before the formation of the telson (Fig. 1d).

RESULTS

The segmentation pattern (Fig. 1e) of the 9 hr *Kr* embryo is characterized by a large unsegmented thoracic and abdominal region. The only clearly delineated trunk segments in these embryos are A7, A8, A9, and A10. The anterior head lobes and gnathal segments appear normal. The mutant *Kr* phenotype observed at 9 hrs is consistent with that observed in mature *Kr* embryos at 24 hrs (Fig.

Legend: The mature cuticle phenotype observed by dark-field microscopy (22x) of a. wildtype, b. Kr^1/Kr^1 (Gloor, 1950), and c. tll^1/tll^e (Strecker *et al.*, 1985) embryos at 24 hrs. The 9 hr segmentation pattern observed by SEM (230x) of d. wildtype, e. Kr^1/Kr^1, and f. tll^1/tll^1 embryos. (The seventh abdominal segment is marked with an arrow in each picture for comparison. C=clypeolabrum, P=procephalic lobe, G=gnathal segments, T=thoracic segments, A=abdominal segments). Note in mature *Kr* embryos the sixth abdominal segment is characteristically reduced and, anterior to it, there is a large mirror image duplicated denticle belt. In the 9 hr *Kr* embryo the thoracic and abdominal region correspon-ding to T1-A6 is missing and in its place is a large un-segmented region which later develops into the mirror im-age duplicated denticle belt and reduced sixth abdominal belt. The anterior head lobes and gnathal segments of *Kr* embryos do not differ consistently from wildtype. The slightly deviant shape of the clypeolabrum in this *Kr* embryo is probably a fixation artifact. Note in mature *tll* embryos the absence of the eighth denticle belt and telson and abnormal head skeleton. In the 9 hr *tll* embryo, anteriorly, both the clypeolabrum and procephalic lobe are reduced in size, and posteriorly, segments A9-A10 are missing.

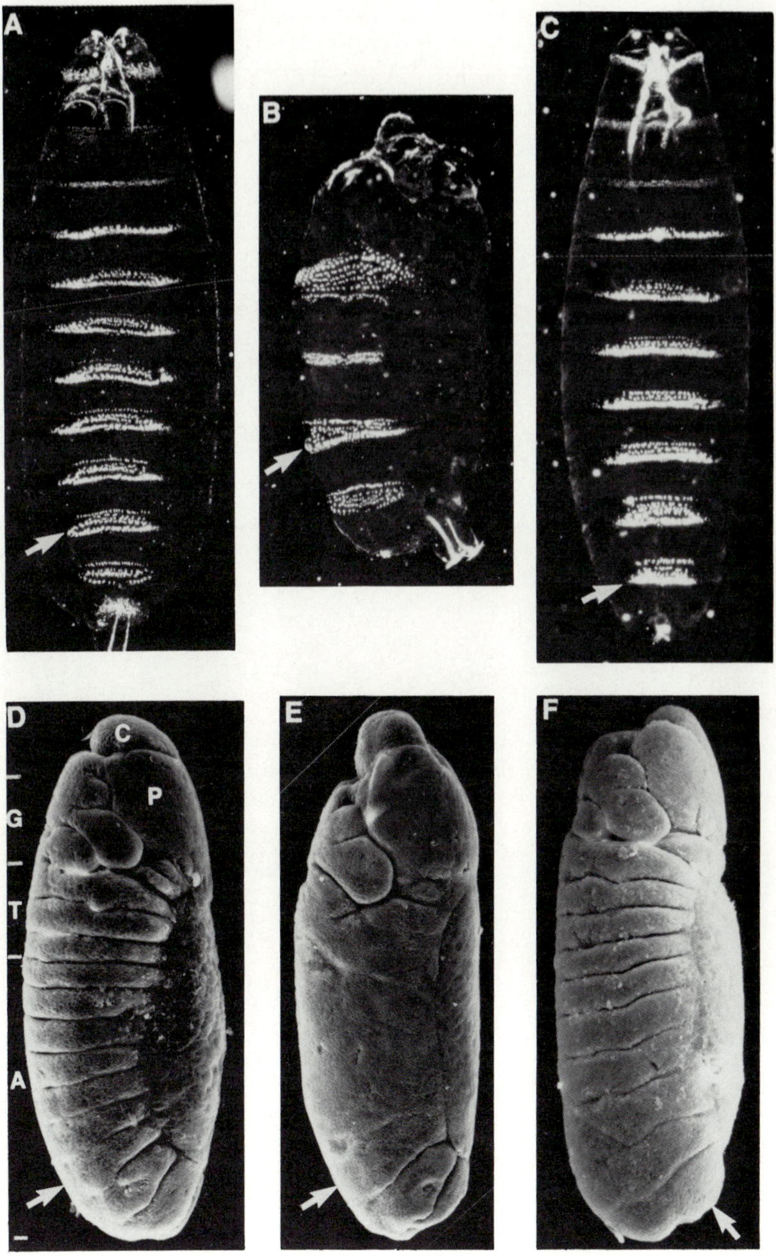
A
B
C
D
C
P
G
T
A
E
F

1b; Gloor, 1950; Wieschaus *et al.*, 1984): a) the cuticular derivatives of the embryonic segments A7-A10, namely the seventh and eighth abdominal denticle belts and telson are normal, and b) we observed in embryos slightly order than 9 hrs the differentiation of the large unsegmented region into the reduced sixth abdominal denticle belt and the mirror image duplicated denticle belt seen in the mature *Kr* embryo.

The segmentation pattern of *Kr* embryos contrasts that observed with 9 hr *tll* embryos (Fig. 1f; Strecker *et al.*, 1985). The *tailless* mutation results in the absence of the posterior abdominal segments A8, A9, and A10, and anteriorly, results in an abnormal clypeolabrum and procephalic lobe. Thus, we observe no overlap in the regions of the segmentation pattern affected by *tailless* and *Krüppel*.

The results of this comparison are consistent with the model that an early step in embryonic pattern formation is the determination of anterior and posterior ends apart from a central trunk region. Further tests of this model, however, include comparisons of the segmentation pattern defects caused by the remaining gap mutants, *knirps* and *hunchback* to those observed in the 9 hr *tailless* embryo.

REFERENCES

Gloor H (1950). Schädigungsmuster eines Letalfaktors (*Kr*) von *Drosophila melanogaster*. Arch Jul Klaus Stiftung 25:38-44.

Nüsslein-Volhard C, Wieschaus E (1980). Mutations affecting segment number and polarity in *Drosophila*. Nature (London) 287:795-801.

Strecker T, Kongsuwan K, Lengyel J, Merriam J (1985). The zygotic mutation *tailless* affects the anterior and posterior regions of the segmented *Drosophila* embryo. (in press, Dev Biol).

Wieschaus E, Nüsslein-Volhard C, Kluding H (1984). *Krüppel*, a gene whose activity is required early in the zygotic genome for normal embryonic segmentation. Dev Biol 104:172-186.

Progress in Developmental Biology, Part A, pages 293–296

EXTRA JOINTS AND MISORIENTED BRISTLES ON *DROSOPHILA* LEGS

Lewis I. Held, Jr., Christine M. Duarte, and Kourosh Derakhshanian

Developmental Biology Center, University of California, Irvine, California 92717

INTRODUCTION

The legs of the fruitfly *Drosophila melanogaster* display three main types of cuticular protuberances: bristles, hairs, and hairlike structures called "bracts" which are only found adjacent to bristle sockets (Hannah-Alava, 1958). Nearly all bristles, hairs, bracts, and "bract-socket vectors" (imaginary vectors pointing from bracts to their associated bristle sockets) point in a distal direction. In various mutants these cell polarities are disturbed (Gubb and Garcia-Bellido, 1982). Since the bristles on wild-type legs are arranged in orderly rows within which they are regularly spaced, such misorientations provide a means of investigating the role of cell polarities in bristle patterning.

RESULTS AND DISCUSSION

We examined the legs of flies from 16 different strains of *D. melanogaster* carrying mutations known to affect bristle or hair polarity in some part of the body (*sple*, *fz*, *dsh*, *in*, *hy*, *al*, *dp*^{ov}, *fy*^{2}, *mwh*, and *pk*) or known to cause partial or missing tarsal joints (*ey*^{D}, *app*, *d*, *fj*, *rn*, and *l(3)1215*). (For descriptions see Lindsley and Grell, 1968, and Bryant *et al*., 1980.) The latter class of mutants was studied because four of the mutants in the former class (*sple*, *fz*, *dsh*, and *in*) were found to possess extra tarsal joints--many of which were partial or otherwise abnormal--so that we broadened the investigation to include an analysis of joint formation as well as cuticular polarity.

1. Cuticular polarity.

In all but three of the 16 strains (al, dp^{ov}, and pk), we observed misorientations of bristles, hairs, bracts, or bract-socket vectors on the legs. From previous reports, abnormal polarites had been expected in two of the mutants known to have defective joints (ey^{D} and l(3)1215), but the finding of misorientations in the other "joint" mutants was surprising, considering that the abnormal polarites were frequently far removed from the defective tarsal joints.

In several of the mutants a correlation between bristle polarity and bristle spacing was discovered: when adjacent bristles in the same row point toward one another, the distance between them is typically larger than normal, whereas when they point away from one another, the interval is typically smaller than normal. We interpret this correlation in terms of a hypothetical mechanism of bristle spacing (Held and Pham, 1983) in which bristle cells repel one another, perhaps via filopodial extensions. We conjecture that the cell extensions are longer in the direction of the bristle shaft than in the opposite direction (Fig. 1).

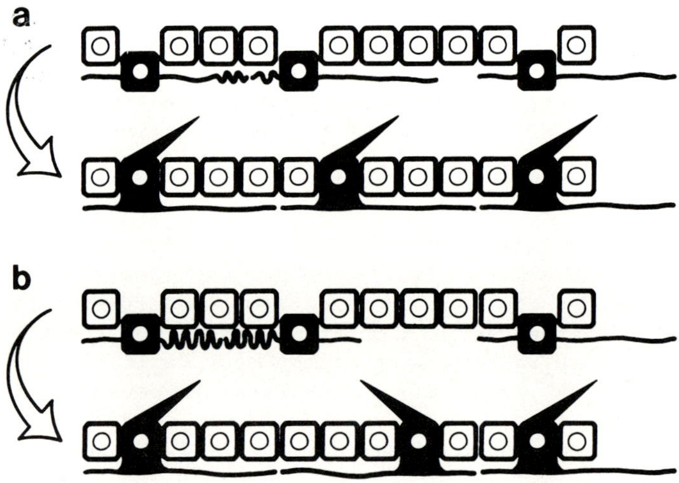

Figure 1. Hypothetical explanation for how uniform bristle spacing is achieved in wild-type flies (a) and for why bristle spacing varies in accordance with bristle polarity in mutant flies (b). Bristle cells are black.

2. Extra joints.

The tarsi of wild-type flies have four joints. Each tarsal joint is a hinge consisting of a condyle (extending from the distal segment) which pivots within a socket (inside the proximal segment). The tarsi of sple flies contain as many as eight joints: four at the normal locations, and an extra joint in each of the proximal four tarsal segments (Fig. 2). Nearly all of the extra joints have inverted polarity. This same phenotype of extra, inverted joints alternating with the normal tarsal joints was also observed to a lesser extent in fz, dsh, and in; and two other mutants --app and d--frequently showed cuticular invaginations or vesicles indicative of partial joint formation at the same ectopic joint sites. The extra joint in the fourth tarsal segment has the same distinctively dark pigmentation as the normal joint between the fourth and fifth tarsal segments, suggesting that each extra joint is a mirror-image copy of the joint at the distal end of its segment. This inference is consistent with the absence of any extra joint in the fifth tarsal segment, since this segment has no distal joint.

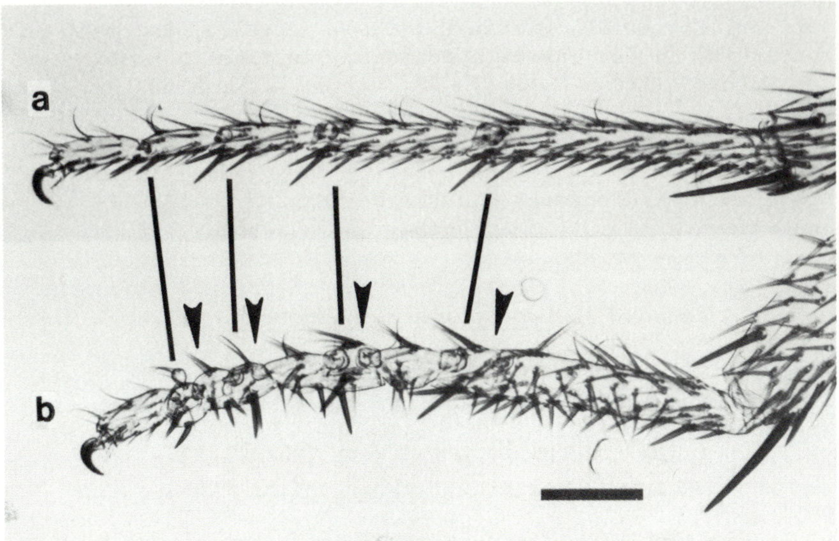

Figure 2. Tarsus of a second leg from a wild type fly (a) and a sple fly (b). Lines connect normal joints. Arrowheads indicate extra joints. Bar length = 0.1 mm.

A similar phenotype of extra, inverted segment boundaries alternating with normal boundaries has been reported for the pattern of body segments in the embryonic-lethal mutant _patch_ (Nusslein-Volhard and Wieschaus, 1980). Interestingly, _sple_ and _patch_ are located within one map unit of one another on the second chromosome, possibly at the same locus, suggesting that leg and body segmentation may share a common developmental or genetic control mechanism. In the bug _Oncopeltus_, an extra, inverted body-segment boundary has been induced surgically by the removal of a transverse strip of epidermis from an abdominal tergite (Wright and Lawrence, 1981). According to the model that the authors propose (and assuming that positional values along the length of a segment are uniformly distributed), segments that contain extra, inverted joints should be shorter than normal--a prediction which we have found not to hold for at least some tarsal segments. The model of Russell (1985) can more easily accomodate the induction of extra, inverted boundaries without a change in segment length.

REFERENCES

Bryant PJ, Girton JR, Martin P (1980). Physical and pattern continuity in the insect epidermis. In "Insect Biology in the Future" (Locke M, Smith DS, eds.), N.Y.: Acad. Pr., pp. 517-542.

Gubb D, Garcia-Bellido A (1982). A genetic analysis of the determination of cuticular polarity during development in _Drosophila melanogaster_. J Embryol Exp Morph 68: 37-57.

Hannah-Alava A (1958). Morphology and chaetotaxy of the legs of _Drosophila melanogaster_. J Morph 103: 281-310.

Held LI Jr., Pham TT (1983). Accuracy of bristle placement on a leg segment in _Drosophila melanogaster_. J Morph 178: 105-110.

Lindsley DL, Grell EH (1968). "Genetic Variations of _Drosophila melanogaster_". Washington, D.C.: Carnegie Inst. of Wash. Publ. No. 627.

Nüsslein-Volhard C, Wieschaus E (1980). Mutations affecting segment number and polarity in _Drosophila_. Nature 287: 795-801.

Russell MA (1985). Positional information in insect segments. Dev Biol 108: 269-283.

Wright DA, Lawrence PA (1981). Regeneration of the segment boundary in _Oncopeltus_. Dev Biol 85: 317-327.

Progress in Developmental Biology, Part A, pages 297–300

GROWTH PATTERNS IN DROSOPHILA IMAGINAL DISCS

Peter J. Bryant and David A. O'Brochta

Developmental Biology Center
University of California, Irvine
Irvine, CA 92717, USA

INTRODUCTION

When *Drosophila* imaginal disc fragments are cultured in adult hosts prior to transfer to larval hosts for metamorphosis, they undergo regeneration or duplication, depending on the size of the starting fragment, during a period of 4-7 days (Bryant, 1975; Haynie and Bryant, 1976; Adler, 1981). Regeneration involves local stimulation of cell proliferation (Dale and Bownes, 1980) and does not occur under conditions which prevent cell proliferation (Schubiger, 1973). It is therefore an example of epimorphic regeneration as defined by Morgan (1901).

During the first one or two days of the culture period, disc fragments undergo wound healing, restoring the continuity of the disc epithelium (Reinhardt *et al.*,1977; Reinhardt and Bryant, 1981). In recent experiments (S. Fraser and P. Bryant, unpubl.) we have marked individual cells in the starting fragments by iontophoretic injection of high molecular-weight fluorescent dyes, and shown that wound healing brings into contact cells which were previously widely separated in the disc. We have suggested (French *et al.*,1976) that the discontinuity in positional information (Wolpert, 1971) created by wound healing is the stimulus for cell proliferation and the associated intercalary regeneration between the confronted cells.

Two classes of models have been suggested in which cell proliferation is closely coupled to pattern regulation during intercalary regeneration. In sequential models the newly generated cells adopt positional values adjacent to those of their neighbors, whereas in averaging models the newly generated cells adopt positional values which are the average of those of their neighbors. Several attempts have been made to distinguish between these possibilities by subjecting regenerating fragments to metamorphosis after varying amounts of time for growth. However, the results are contradictory, probably because of the limited number of pattern elements which can be used as markers, and because of difficulty in scoring incompletely developed markers.

We have tried to distinguish between the two types of model by investigating the patterns of cell proliferation during regeneration of several wing disc fragments (O'Brochta and Bryant, 1985). The sequential model predicts that growth will be highly localized throughout regeneration, that the size of the growth zone will be independent of the extent of the positional information discontinuity, and that the time required for regeneration will depend on the amount of tissue to be regenerated. On the other hand, the averaging model predicts that growth will be localized at first, that the growth zone will expand during regeneration, that late in the regeneration process cell proliferation will be dispersed throughout the regenerate, and that the size of the growth zone will be a function of the extent of the positional information discontinuity.

Fragments of the imaginal wing disc were produced by removing 30°, 90°, or 150° sectors of tissue with the cuts radiating from the center of the wing pouch; thus the fragments all had wounds of similar length although they differed in the amount of tissue removed and in the extent of the assumed positional information discontinuity following wound healing. The fragments were cultured in adult female abdomens for 1,2,3,4 or 5 days, labeled in vitro for 30 minutes with tritiated thymidine, serially sectioned and subjected to autoradiography.

There was negligible thymidine incorporation in unoperated controls and in the undamaged parts of the operated discs, indicating that growth termination was completed during the first day of the culture period. Almost 100% of the fragments from which tissue had been removed, as well as controls which were simply cut without removing any tissue, showed a highly localized zone of intense thymidine incorporation ("blastema") at the position of the wound, even after only one day of culture. The blastemas in control discs were short-lived, with over 50% of these discs having completed growth by the third day of culture. Blastemas in discs from which tissue was removed were larger and more persistent; 50% of the -30° fragments had completed growth by the 4th day of culture, 50% of the -90° fragments had completed growth by the 5th day, and all of the -150° fragments were still growing at the end of the 5th day. After one day of culture there was no clear diference in blastema sizes between the different fragments, but on subsequent days the average blastema size was a function of the amount of tissue removed. The number of labeled cells in the blastemas gradually declined after the second day in all experimental fragments except -150% fragments, in which blastema size had not yet declined after five days of culture.

Localized growth stimulation was obvious even in fragments wich had not yet completed healing, and in these cases it was clear that growth was occurring on both sides of the wound. These results, as well as the finding of growth stimulation in control cut fragments, lead to the suggestion that growth can be locally stimulated by contact between the edges of the peripodial membrane and the columnar epithelium (Reinhardt and Bryant, 1981; Reinhardt *et al.*,1977), or by the presence of a free cut edge.

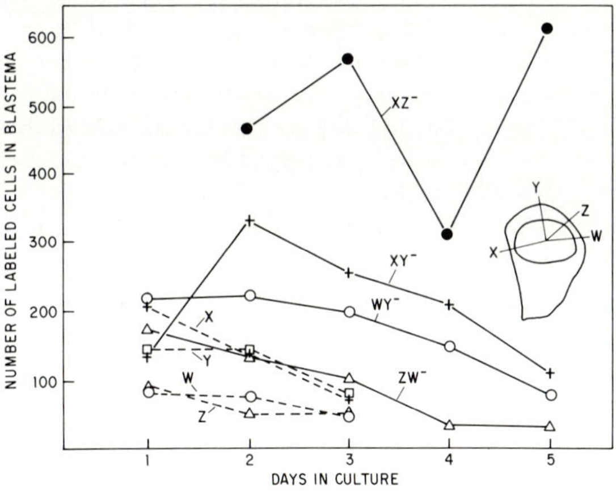

Figure 1. Average number of labeled cells in autoradiographs of regenerating and control fragments of wing discs after various periods of culture *in vivo.* From O'Brochta and Bryant (1985).

The results confirm that the response to wounding includes the local stimulation of cell proliferation, and they tend to support an averaging model rather than a sequential one; the number of labeled cells in the growth zone increases and then declines, and the size of the growth zone is a function of the amount of tissue removed. However, an unexpected feature of the results is that the labeled cells are tightly clustered, rather than dispersed, even late in the regeneration process. This suggests that growth is terminated sequentially from the edges to the center of the growth zone. (Support: NIH grant HD06082).

REFERENCES

Adler PN (1981). Growth during pattern regulation in imaginal discs. Develop Biol 87:356-373.

Bryant PJ (1975). Pattern formation in the imaginal wing disc of *Drosophila melanogaster*: fate map, regeneration and duplication. J exp Zool 193:49-78.

Dale L, Bownes M (1980). Is regeneration in *Drosophila* the result of epimorphic regulation? Wilhelm Roux's Arch 189:91-96.

French V, Bryant PJ, Bryant SV (1976). Pattern regulation in epimorphic fields. Science 193:969-981.

Haynie JL, Bryant PJ (1976). Intercalary regeneration in the imaginal wing disk of *Drosophila melanogaster*. Nature 259: 659-662.

Morgan TH (1901). "Regeneration." London: Macmillan Publishing Company.

O'Brochta DA, Bryant PJ (1985). Patterns of growth during the regeneration of the *Drosophila* imaginal wing disc. Develop Biol (submitted).

Reinhardt CA, Bryant PJ (1981). Wound healing in the imaginal discs of *Drosophila*. II. Transmission electron microscopy of normal and healing wing discs. J exp Zool 216:45-61.

Reinhardt CA, Hodgkin NM, Bryant PJ (1977). Wound healing in the imaginal discs of *Drosophila*. I. Scanning electron microscopy of normal and healing wing discs. Develop Biol 60:238-257.

Schubiger G (1973). Regeneration of *Drosophila melanogaster* male leg disc fragments in sugar fed female hosts. Experientia 29:631.

Wolpert L (1971). Positional information and pattern formation. Curr Topics Develop Biol 6:183-224.

Progress in Developmental Biology, Part A, pages 301–304

EARLY MORPHOGENESIS OF THE METAMERIC QUAIL WING FIELD AND THE DEVELOPMENT OF VASCULAR CHANNELS IN THE SOMATOPLEURE

Stephen Meier and Pamela Burton

Center for Developmental Biology, Department of Zoology, University of Texas, Austin, Texas 78712

INTRODUCTION

The discovery of patterning in the mesoblast layer at earlier times than heretofore realized (Meier, 1979; 1980) is of fundamental importance to our understanding of limb development. From a variety of grafting experiments (Christ *et al.*, 1974a; 1974b; Chevallier *et al.*, 1977), it is known that myogenic precurser cells migrate from the dermamyotomal portion of axially located somites, into the lateral plate limb field, even before there is any external sign of a limb bud (Jacob *et al.*, 1978). Soon after the myogenic cells migrate, neural crest cells move across the somites and enter the limb field (Christ, *et al.*, 1983). Together, the neural crest cells, the myogenic cells of somitic origin, and the local somatic mesoderm of the lateral plate interact to form the skeletal elements, tendons, muscles, and soft connective tissues of the limb. However, so far no one has monitored the developmental preparation of the lateral plate prior to its invasion by axial mesenchyme. We therefore present here some of the salient features of the morphogenesis of the somatic mesoderm that participates in limb development.

RESULTS

In quail, the wing buds eventually initiate outgrowth along the axis adjacent to somites 15-20. Therefore, we monitored development of the lateral plate at this axial level, beginning with the first time the future limb somatic mesoderm can be recognized adjacent to the paraxial segmental

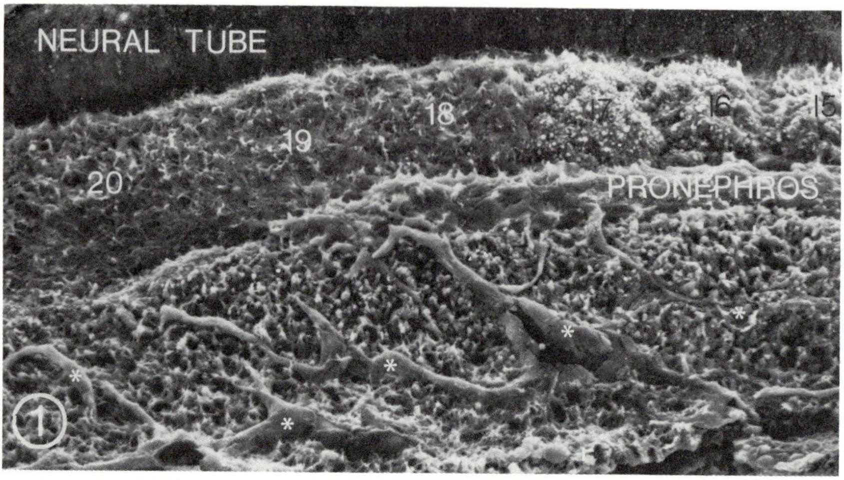

Figure 1. SEM of the right wing (segments 15-20) of a 15 somite embryo, ectoderm removed. Endothelial cells (stars) form a vascular plexus in the somatic mesoderm. X 355.

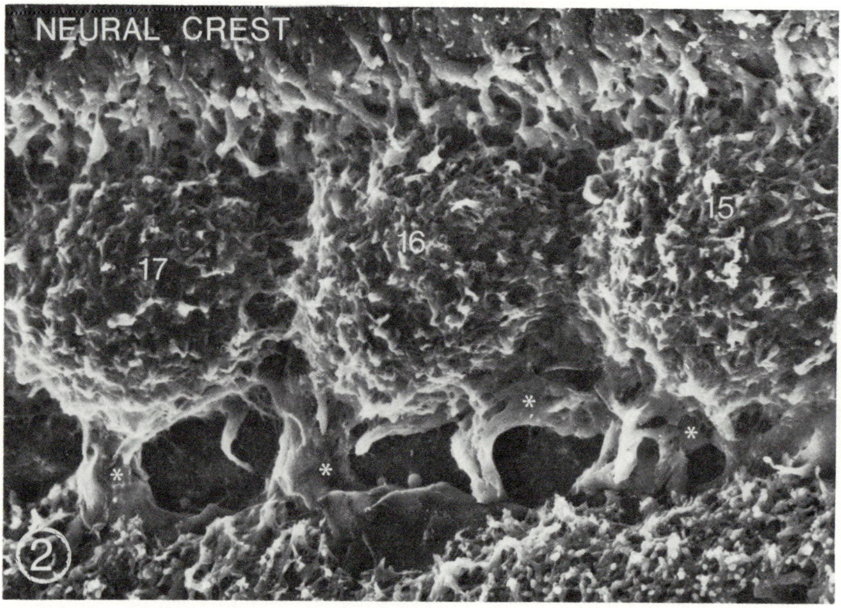

Figure 2. SEM of segments 15-17 of the right wing field of a 20 somite embryo, ectoderm removed. Endothelial cells (stars) anastomose around the somites. X 575.

plate. Initially, the lateral plate is a thin layer of cells that is grooved regularly at intersomitic intervals (Fig. 1). These subectodermal channels are exploited by vascular endothelial cells that anastomose into tubes. Eventually, the vascular complex penetrates deeper into the somatic layer and is no longer obvious from the dorsal surface.

As the lateral body folds raise the medial edge of the somatic mesoderm over the pronephros, to oppose the lateral edge of the somites, numerous vascular endothelial cells from around the somites join with those in the lateral plate to form tubular connections (Fig. 2). By the time the premyogenic cells emerge from the somites to invade the somatic mesoderm (Fig. 3), the lateral plate is a much thicker layer whose interior is highly vascularized.

This morphological study suggests that the premyogenic cells of somitic origin, invade the limb field following already established vascular routes. In theory, the preestablished vascular channels may merely provide a physical

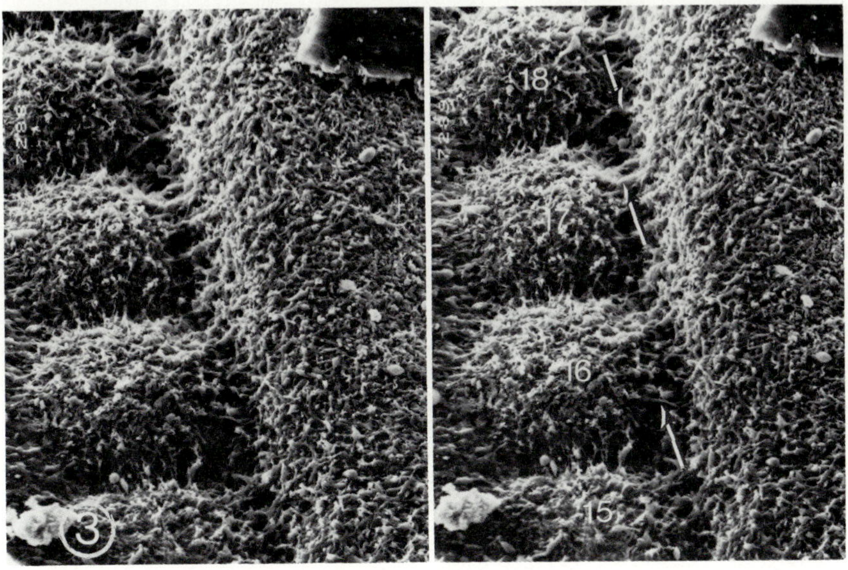

Figure 3. Stereo SEM (tilt angle 10°) of segments 15-18 of the left wing field of a 24 somite embryo, ectoderm removed. Premyogenic cells (arrows) of somitic origin enter the somatic mesoderm, crossing the vascular channels. X 275.

conduit for migrating myogenic cells, or perhaps a chemotactic substance that emanates from the blood vessels (Venkatasubramanian and Solursh, 1984) may lure premyogenic cells along the vascular pathway. Regardless, it seems that the pattern of the vascular plexus which results from exploitation of the segmented lateral plate mesoderm by endothelial cells may be crucial to the proper distribution of future myogenic cells in the wing field.

ACKNOWLEDGMENTS

This work was supported by NSF Grant 8203488 to SM.

REFERENCES

Chevallier A, Kieny M, Mauger A (1977). Limb-somite relationship: origin of the limb musculature. J Embryol Exp Morph 41:245-258.

Christ B, Jacob H, Jacob M (1974a). Uber den Ursprung der Flugelmuskultur. Experimentelle Untersuchungen mit Wachtelund Huhnerembryonin. Experientia 30:1446-1449.

Christ B, Jacob M, Jacob H (1974b). Experimentelle Untersuchungen zur Entwicklung der Brustwand biem Huhnerembryo. Experentia 30:1449-1451.

Christ B, Jacob M, Jacob H (1983). On the origin and development of the ventrolateral abdominal muscles in the avian embryo: an experimental and ultrastructural study. Anat Embryol 166:87-101.

Jacob M, Christ B, Jacob H (1978). On the migration of myogenic stem cells into the prospective wing region of chick embryos. A scanning and transmission electron microscope study. Anat Embryol 153:179-193.

Meier S (1979). Development of the chick embryo mesoblast: formation of the embryonic axis and establishment of metameric pattern. Develop Biol 73:25-45.

Meier S (1980). Development of the chick embryo mesoblast: pronephros, lateral plate and early vasculature. J Embryol Exp Morph 55:291-306.

Venkatasubramanian K, Solursh M (1984). Chemotactic behavior of myoblasts. Develop Biol 104:428-433.

Progress in Developmental Biology, Part A, pages 305–308

EXO UTERO SURVIVAL OF MOUSE EMBRYOS

K. Muneoka, N. Wanek, S. V. Bryant

Developmental Biology Center, University of California, Irvine, CA 92717

INTRODUCTION

Studies on developing mammalian embryos have been limited primarily due to the inaccessibility of the embryo for surgical manipulations. Some experiments have been performed *in utero* on the later embryonic stages of a variety of mammals, and these studies have furthered our understanding of various aspects of mammalian development (see Shatz, 1983). However, experimentation on early embryonic stages is difficult due to the fragility of the embryo coupled with the frequent contractions of the myometrium. While smooth muscle relaxants are somewhat effective in reducing uterine contractions, the teratological effects of these drugs on normal development are for the most part unknown.

Our interest in experimentation on mammalian embryos stems primarily from the finding that in other experimental systems regenerating and developing limbs share many underlying mechanisms for the establishment of pattern (see Bryant & Muneoka, 1985). We are interested in understanding what changes are occurring in the limbs of mammals which may be linked to regenerative failure at later stages in development. We have developed techniques which allow us to surgically manipulate mammalian embryos at stages prior to the overt differentiation of the limb bud. We circumvented the difficulties associated with surgery *in utero* by finding that development of mouse embryos from day 10 onwards will continue outside the confines of the uterus. These *exo utero* embryos develop normally to term.

MATERIALS AND METHODS

All operations were performed on timed-pregnant outbred Swiss Webster mice on days 10 - 14 of gestation (plug day = day 1). Mice were anesthetized with Nembutal (0.06 mg/g body weight) and Ketaset (0.2 mg/ mouse). A midline abdominal incision was made on the pregnant female mouse to expose both horns of the uterus. The number of embryos on each uterine horn was recorded and all embryos on one horn were exposed by making a longitudinal incision in the uterus at a position 180° removed from the placentae. The uterus contracts toward the placentae, thus exposing the embryos (Fig. 1). The contralateral horn of the uterus was left untouched and served as a control for normal survival. Both horns of the uterus were placed back into the abdominal cavity and at term the female mouse was sacrificed and all surviving pups were fixed in Bouin's fixative. The frequency of survival, as well as the size of the pups (crown-rump length and fixed body weight), were compared between control and exposed embryos.

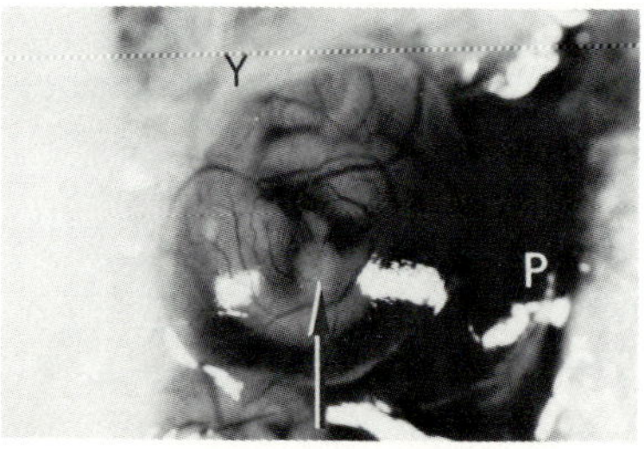

Fig. 1: Exposed day 12 embryo. Arrow indicates right hindlimb bud; P=placenta; Y=yolk sac.

RESULTS

The results of this study indicate that mouse embryos freed from the confines of the uterus, yet still attached via the placenta to the uterus, develop normally to term. Development occurs _exo utero_ and pups at term must be surgically removed. Histological examination of embryos fixed immediately following this operation shows that the embryos are released from the confines of the muscular layers of the uterus without damage to the embryo itself. Exposing mouse embryos on days 10 through 14 of gestation resulted in normal development to term in 81% of the cases (Table 1). Survival _in utero_ to term in this experiment was 84%, indicating that survival of embryos _exo utero_ is essentially identical to survival _in utero_.

Table 1.

Stage		n	No. Survived	% Survival	% of Control Length	% of Control Weight
Day 10	exo utero	18	15	83	97	105
	in utero	22	19	86		
Day 11	exo utero	22	17	77	93	92
	in utero	29	25	86		
Day 12	exo utero	24	17	71	97	93
	in utero	24	16	67		
Day 13	exo utero	14	12	86	92	84
	in utero	10	10	100		
Day 14	exo utero	11	11	100	96	91
	in utero	17	16	94		
Total	exo utero	89	72	81	95	93
	in utero	102	86	84		

Pups which survived exo utero were normal in gross morphological appearance (Fig. 2). When pup sizes were compared between pups which developed exo utero versus pups which developed in utero, there appeared to be a slight decrease in both crown-rump length and fixed body weight (Table 1). In general exo utero pups were 95% of controls in crown-rump length and 93% of controls in fixed body weight.

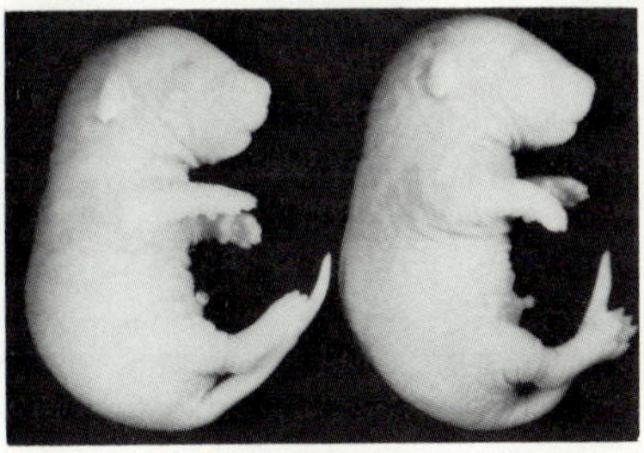

Fig. 2: Right - exo utero pup at term, uterus removed at day 10
Left - in utero control from contralateral horn

DISCUSSION

The development of the exo utero technique has made early staged mouse embryos accessible for experimentation. We have been able to successfully perform a variety of surgical manipulations, including amputation and grafting, on developing mouse limbs. Embryonic surgery is performed by making an incision through the yolk sac and the amnion in the region of the developing limb. Following surgery, the extra-embryonic membranes are sutured closed and the embryo is placed into the abdominal cavity of the female mouse. Survival following exo utero operations is variable and dependent upon the amount of damage to the embryo. While it is clear that this technique is useful for studies concerning limb regeneration in mammals, the technique can be used to explore a wide variety of developmental issues with regard to mammals. Furthermore, our results pose new questions and a new approach to fields quite divergent from developmental biology.

The fact that normal development can occur exo utero sheds new light on the role of the uterus in normal embryogenesis. Muscular contractions of the uterine myometrium are known to occur throughout pregnancy. However, these contractions have been shown to temporarily reduce utero-placental blood flow and are suspected to have a detrimental effect on embryonic and fetal development (see Liebgott, 1984). Local, intrauterine regulation of the immune system is thought to be a major factor in the protection of the fetus from the maternal immune response. The maternal tissue immediately adjacent to the trophoblast and developing embryo is uterine decidual tissue which has been shown to have immunosuppressive activity (Bell, 1983). In the mouse, the decidual tissue around most of the embryo regresses by day 12 of gestation, leaving only the decidua basalis underlying the placenta. The exo utero technique described here leaves intact the important placental-decidual-uterine interface, and thus does not seem to interfere with the suppression of the maternal immune response to the developing fetus.

ACKNOWLEDGEMENTS

Research supported by a gift from the Monsanto Company.

REFERENCES

Bell, SC (1983). Decidualization and associated cell types. Implications for the role of the placental bed in the materno-fetal immunological relationship. J Reprod Immunol 5:185.

Bryant, SV, Muneoka, K (1985). Views of limb development and regeneration. Trends in Genetics (in press).

Liebgott, B (1984). Mesometrial smooth muscle in the mouse: its control of uterine blood flow. Anat Rec 208:365.

Shatz, CJ (1983). The prenatal development of the cat's retinogeniculate pathway. J Neurosci 3:482.

Progress in Developmental Biology, Part A, pages 309–314

THE PRESENCE OF CYTOPLASMIC RETINOIC ACID BINDING PROTEINS IN AMPHIBIAN TISSUES AND THEIR POSSIBLE ROLE IN LIMB REGENERATION.

Stella Keeble and Malcolm Maden

National Institute for Medical Research,
The Ridgeway, Mill Hill, London NW7 1AA U.K.

INTRODUCTION

Retinoids show great potential for contributing to our understanding of pattern formation during limb regeneration as they respecify positional information, resulting in the dose-related production of structures extra to those normally regenerated. Perhaps in discovering the mechanism of this respecification the molecular basis of normal pattern specification will unfold.

The morphological effects of retinoids on regenerating limbs have been well characterised (Maden, 1982; 1983), but neither the route of entry into the blastemal cells nor the site and mechanism of their action have been investigated. The receptor model (Jensen & DeSombre, 1973; O'Malley & Means, 1983) which describes the action of steroid hormones in differentiation is currently being applied to the role of retinoids in many systems (Sporn & Roberts, 1983; Zile & Cullum, 1983). For this model to apply to retinoid action in the limb, the molecule should enter the cell, bind to a specific cytoplasmic receptor and be transported to the nucleus, where gene expression is regulated.

Here we report on recent experiments designed to test the application of this model to the mechanism of retinoid action in producing pattern effects during limb regeneration in the axolotl, Ambystoma mexicanum.

RESULTS

Localisation of Retinoic Acid by Autoradiography

An autoradiographic analysis of sections from cone stage regenerating limbs was performed. These limbs had received blocks of 14C all trans retinoic acid impregnated silastin rubber implanted immediately proximal to the blastema (Maden, Keeble & Cox, 1985) at varying times prior to harvesting. This revealed that retinoic acid (RA) diffuses out from the block in a graded fashion and becomes distributed throughout the tissues of the stump and blastema within 7 hours, favouring no tissue above the others. Silver grains were seen above the muscle, cut end of the cartilage, epidermis, blastemal cells and the extracellular matrix. Within these tissues, grains were present above both the cytoplasm and cell nuclei. This demonstrates that RA is able to enter the cells of the blastema and is not simply diffusing into the blood and extracellular spaces.

Localisation of RA by Subcellular Fractionation

Analysis of regenerating blastemas and non-regenerating limb tissues labelled for 24 hours *in vivo* with 3H all trans RA has consistently produced profiles on 15-60% sucrose density gradients in which label is associated with subcellular components at two points along the gradient. One is with the soluble protein phase at the top of the gradient and the other is with the nuclei pelleted at the bottom. Individually, both the skin and muscle of non-regenerating limbs show this characteristic distribution of label. These two components were therefore investigated in more detail.

Nuclear Associated Label

Electron microscopic analysis revealed that the pellets produced from these gradients contain a high proportion of nuclei interspersed with whole cells, extracellular matrix and cell fragments. To confirm that the label associated with the pellet was incorporated into the nuclei, isolated nuclei were prepared from regeneration blastemas which had been labelled for 24 hours with 3H RA-impregnated silastin implants. The isolation procedure involved exposure to Triton X-100 which has been reported to remove the outer nuclear membrane and some of the nuclear phospholipid (Liau, Ong & Chytil, 1981). After this treatment 15-25% of the label in the homogenate was associated with the final nuclear pellet. Thus RA does enter the nucleus of blastemal cells conforming with the model described above. The specificity of this interaction is currently being investigated.

Cytoplasmic Retinoic Acid Binding Proteins

Crude cytoplasmic preparations from several axolotl tissues have been prepared by standard methods (Sani & Corbett, 1977). Such samples are able to bind 3H RA in vitro (Fig. 1) and produce 1 or 2 peaks on 5-20% sucrose gradients. These represent non-specific binding and specific binding which can be competed out by a 100-fold excess of cold RA. In axolotl skin (Fig. 1A) non-specific binding occurs at a molecular weight corresponding to bovine serum albumin (68K) while specific binding coincides with myoglobin (17K). 100 fold excess of retinol palmitate, retinol acetate or retinol failed to compete out the specific peak. However, the arotinoid RO 13-6298 competed almost as effectively as RA. A specific cRABP is also present in axolotl muscle, but not in

Figure 1. Sucrose gradients of cytoplasmic protein preparations (10mg/ml) incubated for 30 mins at 25°C and 4 hours at 4°C with 40 pmoles of 3H RA in the presence or absence of 100x excess of cold retinoids. Incubates were dialysed overnight and centrifuged at 180,000g for 17 hours on 5-20% sucrose gradients. Myoglobin (2S) and BSA (4.6S) were markers on separate gradients. A. Skin + competition with analogues. Control (●—●), + RA (●- -●), + retinol (0--0), + RO 13-6298 (▲—▲). B. Muscle. C. Liver.

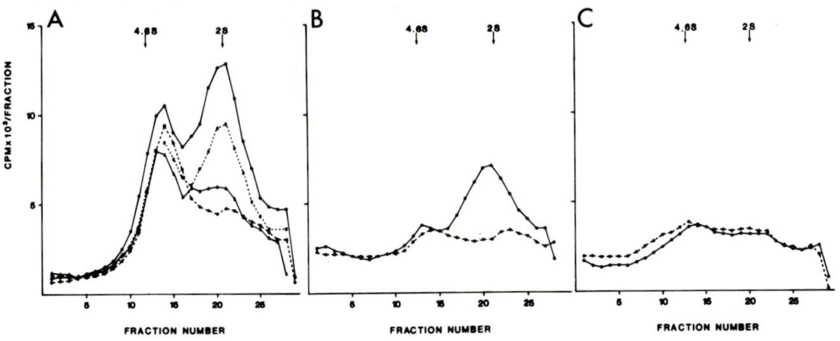

axolotl liver (Fig. 1B,C). In the non-regenerating limb cRABP is present at a concentration of 200 fmoles/mg protein (Fig. 2A). This level rises almost 7-fold in cone stage blastemal preparations (Fig. 2B), falling again to approach normal levels by the late 4-digit stage when regeneration is complete (Fig. 2C). There were no non-specific 4.6S peaks in these limb extracts. The estimated cRABP concentration in each of the six preparations is summarised in Table 1.

Figure 2. Sucrose gradient centrifugation of cytoplasmic protein preparations (3.5mg/ml) incubated with 40pmoles of 3H RA in the presence (●-●) or absence (●—●) of 100x excess cold RA. A. Normal limb. B. Blastema. C. 4-digit stage.

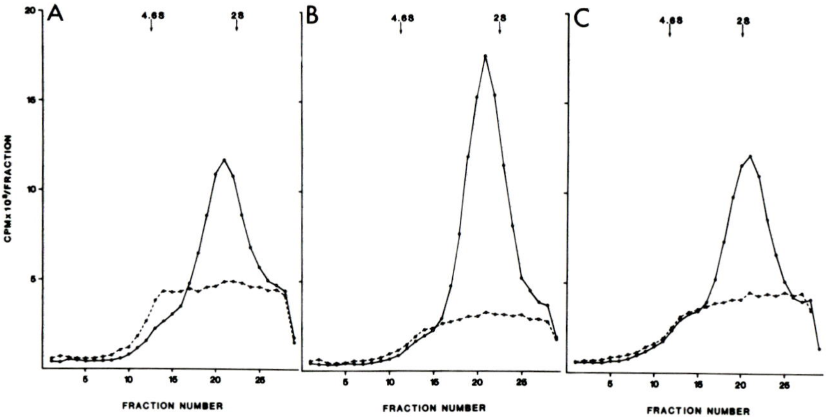

Table 1. Concentrations of cRABP in cytoplasmic protein preparations of axolotl tissues. Estimates are an average of 2-3 experiments.

	Liver	Muscle	Skin	Whole limb	Blastema	4-Digit
cRABP (fmoles/mg)	0	132	240	202	1384	424

DISCUSSION

In the search for the mechanism of retinoid action in limb regeneration, 3H RA was found to localise at two sites, the cell nucleus and a cytoplasmic binding protein. We could not detect other subcellular localisations such as mitochondria (Sherman et al., 1983) or microsomes (Cope, Knox & Hall, 1984).

The cRABP that we have identified was present in skin and muscle, but not in liver. This distribution is similar to that described in mammals (Chytil & Ong, 1984) and the absence of cRABP in the liver reflects the fact that RA is not stored in this organ. Amphibian cRABP reversibly binds RA, shows very low affinity for retinol, retinol palmitate and retinol acetate (not shown), but a high affinity for arotinoid Ro 13-6298. Interestingly, it has been shown that Ro 13-6298 is 50x more potent than RA in altering pattern (Kim & Stocum, 1985). We found a 7-fold increase in cRABP early in regeneration with a fall to normal levels at completion of the limb pattern. These increases agree with similar experiments in the regenerating newt limb (Stocum, unpublished).

The nuclear location of 3H RA in blastemas conforms with data from other systems (Liau, Ong & Chytil, 1981), but it remains to be established whether, in delivering RA, the cRABP enters the nucleus, if the protein itself associates with the DNA or if the RA alone binds. In conclusion, the identification of a cRABP and its apparent ability to translocate RA to the nucleus provides initial evidence that retinoids act on pattern formation in the limb via a similar mechanism to that postulated for the role of steroid hormones in regulating gene expression.

ACKNOWLEDGEMENTS

We would like to thank Hoffmann-La Roche for generous gifts of radiolabelled retinoic acid, R.A. Cox and K. Mustafa for collaboration in this work.

REFERENCES

Chytil F, Ong DE (1984). Cellular retinoid-binding proteins. In Sporn MB, Roberts AB, Goodman DS (eds): "The Retinoids" Academic Press, p 90.

Cope FO, Knox KL, Hall RC (1984). Retinoid binding to nuclei and microsomes of rat testes interstitial cells. I - mediation of retinoid binding by cellular retinoid-binding proteins. Nut Res 4:289.

Kim WS, Stocum, DL (1985). Retinoid-induced pattern duplication in regenerating limbs: comparative efficacy of retinoic acid, etretinate and arotinoid. J exp Zool in press.

Jensen EV, DeSombre ER (1973). Estrogen-receptor interaction Science 182:126.

Liau G, Ong DE, Chytil F (1981). Interaction of the retinol/ cellular retinol-binding protein complex with isolated nuclei and nuclear components. J Cell Biol 91:63.

Maden M (1982). Vitamin A and pattern formation in the regenerating limb. Nature 295:672.

Maden M (1983). The effect of vitamin A on the regenerating axolotl limb. J Embryol exp Morph 77:273.

Maden M, Keeble S, Cox RA (1985). The characteristics of local application of retinoic acid to the regenerating axolotl limb. Roux's Arch Dev Biol 194:228.

O'Malley BW, Means AR (1974). Female steroid hormones and target cell nuclei. Science 183:610.

Sani BP, Corbett TH (1977). Retinoic acid-binding proteins in normal tissues and experimental tumors. Cancer Res 37: 209.

Sherman MI, Paternoster ML, Eglitis MA, McCue PA (1983). Studies on the mechanism by which chemical inducers promote differentiation of embryonal carcinoma cells. In Silver LM, Martin GR, Strickland S (eds): "Teratocarcinoma Stem Cells" Cold Spring Harbor Lab, p 83.

Sporn MB, Roberts AB (1983). Role of retinoids in differentiation and carcinogenesis. Cancer Res 43:3034.

Zile MH, Cullum ME (1983). The function of vitamin A : current concepts. Proc Soc exp Biol Med 172:139.

Progress in Developmental Biology, Part A, pages 315–318

A MODEL FOR THE ROLE OF INTERFACIAL TENSIONS WITHIN CELL SHEETS IN THE EARLY MORPHOGENETIC EVENTS OF AMPHIBIAN DEVELOPMENT

Jay E. Mittenthal

Department of Anatomical Sciences, University of Illinois College of Medicine, 506 S. Mathews Avenue, Urbana, Illinois 61801

INTRODUCTION

Early events of amphibian development (reviewed by Gerhart, 1980) involve the shaping of epithelia. These deformations can be interpreted with a model that extends the differential adhesion hypothesis of Steinberg (1963). The model treats an epithelium as a fluid elastic shell (Mittenthal and McMeeking, 1985). A small multicellular element of the shell resists bending, or isotropic in-plane tension or compression, as does an elastic medium. Its stiffness comes from intracellular filaments and from junctions between cells. The shell is also fluid-like, in that cells in the element can exchange neighbors to relieve shear stress. Such exchanges are necessary to realize the shaping effects of adhesion. Suppose that cells of the same type adhere together preferentially and are segregated in one region of the shell. Interfacial tensions may exist between such regions; these can affect the shape of the shell.

The effect of adhesion on the thickness of the shell depends on the apico-basal distribution of intercellular contacts. If the strongest contacts are apical, the work of adhesion among cells increases with the length of apical perimeters but is independent of cell height. An increase in cohesion will tend to thin and spread the sheet, since spreading increases apical perimeters. If lateral contacts between cells are strong throughout the thickness, cohesion is thickness-proportional. In this case the work of adhesion increases if the shell thickens by columnarization or repacking of cells, to increase the area of lateral contact.

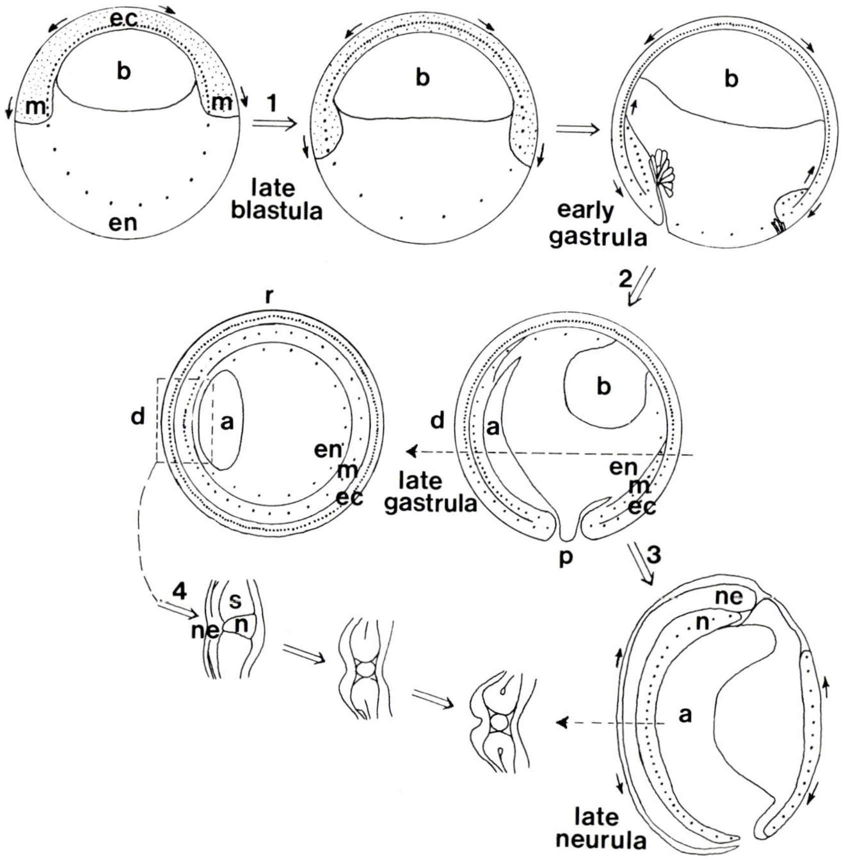

Figure 1. Patterns of intercellular adhesion and processes which these patterns may drive in an amphibian embryo: (1) An increase in thickness-independent cohesion in the blastocoel roof (stippled) may spread it. During gastrulation (2) antero-posterior cohesion waves travel in the mesoderm and ectoderm. These waves may elongate the embryo (3). A dorso-ventral step gradient of cohesion may segregate bands of cells that will form notochord, somites, and pronephros (4). Drawings after Gerhart (1980), Keller (1980), Schroeder (1970). Closer spacing between dots in a row represents higher cohesion. Solid arrow, direction of movement; open arrow, process of development; dashed arrow, plane of section. d, dorsal; p, posterior; r, right. a, archenteron; b, blastocoel; ec, m, en, germ layers; n, notochord; ne, neurectoderm; s, somitic mesoderm.

APPLICATION OF THE MODEL TO MORPHOGENESIS OF XENOPUS

After the mid-blastula transition in Xenopus the blastocoel roof undergoes epiboly, spreading toward the vegetal pole (Fig. 1: 1). Cohesion in the roof increases. The morphology of intercellular junctions suggests that cohesion in the roof is thickness-independent: The superficial cells adhere firmly at an apical junctional complex, but deeper contacts are relatively sparse (Keller, 1980). As the roof spreads the perimeter of apical contacts increases, through spreading of superficial cells and through cell divisions. An increase in thickness-independent cohesion may drive this spreading.

As the roof spreads the leading edge of its deep layer involutes toward the animal pole to form the mesodermal mantle. During this time the relative cohesion in the prospective germ layers is ectoderm > mesoderm > endoderm (Davis, 1984). Measurements of the electrokinetic mobility of cells (Schaeffer et al., 1973) suggest that as a cluster of pre-involution mesoderm cells approaches the site of involution, its cohesion decreases. This occurs first in the prospective head mesoderm and later in prospectively more posterior cells. Involution proceeds at the same speed, so that a time-invariant pattern of cohesion is maintained at the dorsal lip (Fig. 1: 2).

The cohesion of the involuted mesoderm increases as it forms the mesodermal mantle; presumably this increase also occurs as an antero-posterior wave. Later a wave of altered cohesion travels caudad through the mesodermal mantle during the formation of the dorsal organ complex (Poole and Steinberg, 1982).

These waves of altered cohesion may cause the elongation of the embryo which begins during gastrulation (Fig. 1: 3). Non-uniformities of cohesion along a cylindrical tube can elongate it, by favoring an increase in the area of contact between like cells and a decrease in the area of heterotypic contact (Mittenthal and Mazo, 1983). The dorsal midline may extend most because the cohesion waves begin earliest and have largest amplitude there.

The Spemann organizer region -- the prospective notochord -- may induce a dorso-ventral gradient of thickess-proportional cohesion. High cohesion in the notochord may promote repacking of its cells as a stack of disks. High cohesion may also promote the columnarization of cells in the neural plate and somites. If cohesion decreases stepwise from dorsal to ventral, the mismatch in adhesive affinity between steps may constrict and fracture the germ layers (Fig. 1: 4). In the mesoderm such fracture could separate bands of cells that will form notochord, somites, and pronephric duct. Repacking of cells in each band to circularize its cross-section would increase the work of intercellular cohesion. Adhesion-mediated fracture of the ectoderm may separate the neural tube from the epidermis, releasing the neural crest cells.

Thus adhesion may drive epiboly, elongation, and shaping of the dorsal organ complex in amphibian embryos.

REFERENCES

Davis GS (1984). Migration-directing liquid properties of embryonic amphibian tissues. Amer Zool 24:649-655.

Gerhart JC (1980). Mechanisms regulating pattern formation in the amphibian egg and early embryo. In Goldberger RF (ed): "Biological regulation and development", Vol. 2. New York: Plenum, pp 133-316.

Keller RE (1980). The cellular basis of epiboly: An SEM study of deep-cell rearrangement during gastrulation in Xenopus laevis. J Embryol Exp Morph 60:201-234.

Mittenthal JE, Mazo RM (1983). A model for shape generation by strain and cell-cell adhesion in the epithelium of an arthropod leg segment. J Theor Biol 100:443-483.

Mittenthal JE, McMeeking R (1985). Morphogenesis of epithelia: The shaping of a fluid elastic shell. J Theor Biol (submitted).

Poole TJ, Steinberg MS (1982). Evidence for the guidance of pronephric duct migration by a craniocaudally traveling adhesion gradient. Develop. Biol. 92:144-158.

Schaeffer BE, Schaeffer HE, Brick I (1973). Cell electrophoresis of amphibian blastula and gastrula cells: The relationship of surface charge and morphogenetic movement. Develop. Biol. 34:66-76.

Schroeder TE (1970). Neurulation in Xenopus laevis. An analysis and model based upon light and electron microscopy. J. Embryol. Exp. Morphol. 23:427-462.

Steinberg MS (1963). Tissue reconstruction by dissociated cells. Science 141:401-408.

Progress in Developmental Biology, Part A, pages 319–323

THE REACTION-DIFFUSION SYSTEM - A SPATIAL ORGANIZER IN THE VERTEBRATE EPIDERMIS

B.N. Nagorcka

CSIRO, Division of Entomology,
G.P.O. Box 1700,
Canberra 2601, Australia.

A GENERAL VIEW OF PATTERN FORMATION

The translation of genotype into a phenotype involving spatial patterning seems to require positional information (Wolpert 1971). The reaction-diffusion (RD) system, as applied here, does not provide its positional information in the form of a single gradient but rather as a spatial wave-like pattern in the concentration of the reacting diffusing substances which form it (Turing 1952). In the simplest case there are two such substances, denoted here as X and Y. Since their spatial patterns are complementary, only the pattern in X will be illustrated. Provided that the diffusion and reaction rates of X and Y obey certain broad constraints, spatial inhomogeneities (patterns) in the distribution of X and Y arise spontaneously in a homogeneous tissue. No precisely located sources and sinks are required.

An essential pattern forming property of the RD system is the dependence of the patterns on the size and shape of the tissue containing X and Y. This follows since the wavelength, λ (the distance between neighbouring maxima), of the patterns must lie within a narrow range set by the reaction and diffusion rates of X and Y. Hence the RD system will, in general, produce a sequence of patterns during development rather than a single pattern. A general mechanism for pattern formation based on an RD system must, therefore, include a feedback loop consisting of new pattern → proliferation,

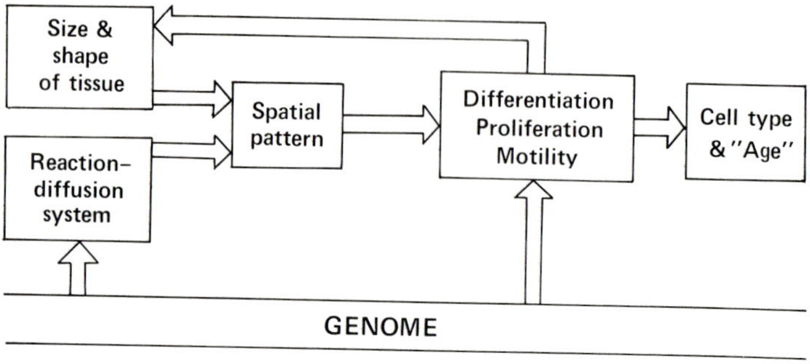

Fig. 1 A view of pattern formation based on an RD system.

differentiation → size and shape → new pattern (Fig. 1). The sequence of patterns produced by an RD system must match the sequence of changes in size, shape, cell type, cell proliferation, etc. observed during development. As an example the sequence of events during hair follicle initiation and development will now be outlined.

HAIR FOLLICLE INITIATION AND DEVELOPMENT

In theory the RD system is switched on (i.e. the interaction between X and Y is modified so that patterns develop spontaneously) by the observed "early dermal message" (Dhouailly 1975). In sheep follicle initiation occurs first of all in the crown and then progressively extends over the rest of the integument (Carter and Hardy 1947). Under these circumstances the RD system produces a series of maxima in [X] (the concentration of X) which are approximately equidistant from each other but otherwise randomly packed (Nagorcka unpublished). The proliferation rate of basal epidermal cells is increased at locations where [X] exceeds a certain threshold causing follicle primordia to appear. These are the early primary central (PCX) follicles.

The follicle primordia develop as cylindrical plugs of cells growing down into the dermis so that the epidermis can no longer be regarded as a two-dimensional sheet of cells. The calculations, of necessity limited to a regular array of follicles (Nagorcka and Mooney 1985), indicate that at a very early stage of follicle development the shape change of the epidermis causes the original pattern of maxima to change. In the new pattern the

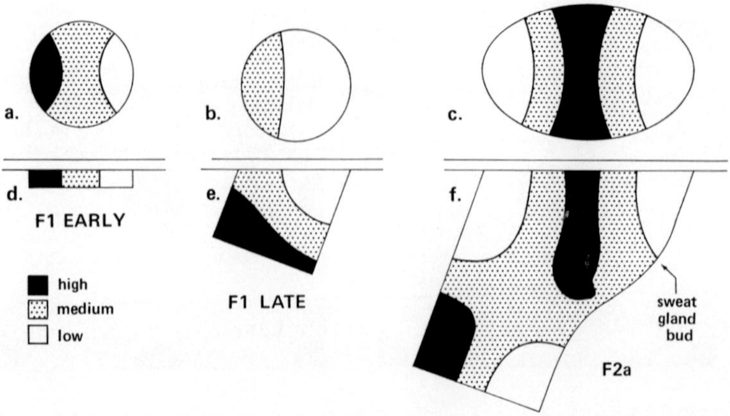

Fig. 2. The spatial patterns of [X] in follicles at stages F1 early, F1 late and F2a, as seen in (a, b, c) cross-sections at the level of the epidermis, and (d, e, f) longitudinal sections through the follicle.

distribution of X across the early primordia is bipolar (Fig 2a,b). Cell proliferation is therefore greater on one side of the follicle than the other and the follicles grow at an angle to the skin surface (Hardy and Lyne 1956) to become late stage F1 follicles (Fig. 2e). At this stage the follicles are large enough so that, as far as the follicles are concerned, the interaction with the epidermis may be ignored. Patterns therefore develop in the follicles independently of the distribution of X in the epidermis. However, the reverse is not true. Indeed the follicles now act as "organizing centres" (Lyne 1966) by dictating the distribution of X at the follicle/ epidermis boundary (Figs. 2 b and c).

When the pattern in the upper part of a late stage F1 follicle (Fig. 2b) is imposed on the epidermis, new maxima appear in between existing PCX follicles (Fig. 3a) initiating a second wave of follicles, called late primary central (PCY) follicles. The location of the minimum in [X] in a late stage F1 follicle (Fig. 2e) is unchanged from that of an early stage F1 follicle and provides the positional information for the initiation of a sweat gland (Mooney and Nagorcka 1985), which subsequently develops as a bud on the side of a stage F2a follicle (Fig. 2f) (Hardy and Lyne 1956).

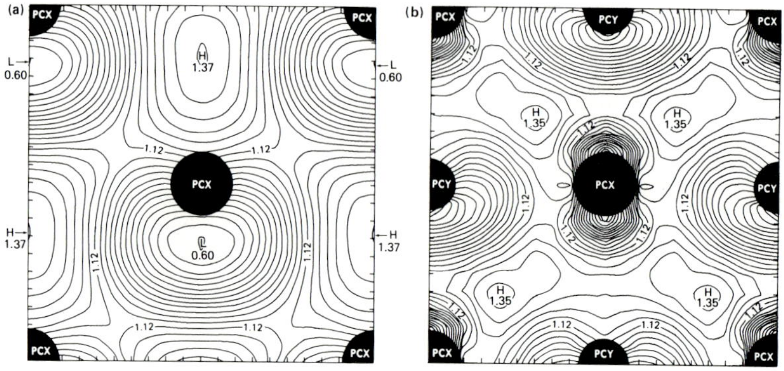

Fig. 3. Spatial patterns of [X] in the epidermis dictated by existing follicles, giving rise to (a) PCY follicles and (b) PLx follicles.

As the sweat gland bud grows and the PCX follicles develop to stage F2a, a new pattern in [X] appears in the PCX follicles. The distribution of X in the upper part of the follicles changes from the bipolar pattern in Fig. 2b to a tetrapolar or bilobed pattern in Fig. 2c. The maxima in [X] appear on opposite sides of the follicle and extend some distance down where they provide the positional information for the initiation of a bilobed sebaceous gland (Mooney and Nagorcka 1985; Hardy and Lyne 1956). At the same time the PCY follicles also advance to late stage F1. The patterns in Fig. 2c and Fig. 2b are imposed at the locations where the PCX and PCY follicles join the epidermis and a new pattern appears in the epidermis with maxima located on either side of the PCX follicles (Fig. 3b). Therefore a third wave of follicles, called early primary lateral (PLx) follicles, initiates forming the trio groups commonly found in mammals. The trio in Fig. 3b forms a triangle; if the the amplitude of the variation of [X] in Fig. 2c is reduced, the trio forms a more linear arrangement (Nagorcka and Mooney 1985).

The sequence of events does not stop here. PCY follicles also develop lateral follicles when they reach stage F2a. This is followed by several waves of secondary follicles, i.e. follicles without sweat glands. Even in mature follicle bulbs the same RD system may continue to play a role during fibre formation influencing the fibre cross-sectional shape and the distribution of various cell types within the fibre (Nagorcka and Mooney 1982).

The same RD system is also capable of providing the positional information required for scale and feather follicle initiation and development (Nagorcka unpublished). Once again the positional information is provided in the form of a sequence of patterns with each new pattern arising because the epidermis changes its size and/or shape in response to the previous pattern.

REFERENCES

Carter HB, Hardy MH (1947). Studies in the biology of the skin and fleece of sheep. Bull Coun Scient Ind Res Melb No 215.

Dhouailly D (1975) Formation of cutaneous appendages in demo-epidermal recombinations between reptiles, birds and mammals. Wilhelm Roux's Arch 177:323-340.

Hardy MH, Lyne AG (1956) The pre-natal development of wool follicles in Merino sheep. Aust J Biol Sci 9: 423-441.

Lyne AG (1966) The development of hair follicle Aust J. Sci 28:370-377.

Mooney JR, Nagorcka BN (1985). Spatial patterns produced by a reaction-diffusion system in primary hair follicles. J Theor Biol (to be published).

Nagorcka BN, Mooney JR (1985). The role of a reaction-diffusion system in the initiation of primary hair follicles. J Theor Biol 114: 243-272.

Nagorcka BN, Mooney JR (1982). The role of a reaction-diffusion system in the formation of hair fibres. J Theor Biol 98:575-607.

Turing AM (1952). The chemical basis of morphogenesis. Phil Trans R Soc B 237:37-72.

Wolpert L (1971). Positional information and pattern formation. Curr Top Dev Biol 6:183-224.

Progress in Developmental Biology, Part A, pages 325–329

STRUCTURAL AND FUNCTIONAL INTERRELATIONSHIPS BETWEEN HEPATOCYTES AND SINUSOIDAL ENDOTHELIUM IN THE MATURING PROCESS OF RAT LIVER

E. Barberá-Guillem and F. Vidal-Vanaclocha

Department of Histology and Cell Biology, College of Medicine, University of the Basque Country (Apdo. 22), 48000-Leioa, Vizcaya, Spain.

INTRODUCTION

Cell differentiation is thought to be mediated by specific inductive environments. There are examples of this in the bone marrow hematopoietic differentiation (Weiss, 1980). Endothelial cells seem to be involved in the establishment of particular microenvironments. In fact, there are different types of endothelium for different organs and even, for distinct areas within a single organ (for review, see Simionescu, 1983).

It is well known that the adult liver is not a homogeneous tissue; Rappaport's acinar zones hold specific structural and functional characteristics closely related to their specific topography. For example, zone 1 hepatocytes contain higher activities of several enzymes including the succinate dehydrogenase and the citrate cycle enzymes (Gumucio, 1983); Kupffer cells are more numerous and active in the zone 1 (Sleyster and Knook, 1982); and liver regeneration after partial hepatectomy starts in zone 1. Recently we have proved the existence of two differentiated domains on the sinusoidal endothelium with respect to their fenestration patterns, which appear to coincide with the functional areas 1 and 3 of the liver acinus. There are three main differences between the two endothelial patterns: the No. of fenestrae, higher in zone 3; the pore size, bigger in zone 3; and the pore clustering in "sieve plates", also higher in zone 3. Because the fenestration of endothelial cells has been related to their function in the selective control of transendothelial exchanges, we proposed a relationship between the fenestration patterns and the filtering function of endothelial cells in both zones (Vidal-Vanaclocha and Barberá-Guillem, 1985).

The liver development has a characteristic chronology. Firstly, the endodermo-mesenchymal interactions lead to the establishment of both a network of prehepatocytes and a vascular system. Later, the liver hematopoietic differentiation takes place (Medlock and Haar, 1983); and finally, at the end of gestation, sinusoidal endothelium and parenchymal cells progressively reach their mature functioning. In the liver, the mature status of both hepatocytes and endothelial cells stands out due to their structural and functional heterogeneity within the liver acinus. However, in the developing liver these cell types change from homogeneous populations of differentiating cells to heterogeneous populations of cells distributed within the liver lobule in two segregated domains which are closely related to the extreme acinar areas 1 and 3.

At present it is not clear which of these two liver cell populations are firstly defined in two segregated domains. Neither is it clear whether reciprocal inductions take place between these cell types, and this may well explain the functional diversification of adult liver cells In order to study this problem, we have analyzed the maturing evolution of parenchymal cells by using as functional marker the succinate dehydrogenase (SDH) activity. Simultaneously, we have studied the fenestration pattern development in endothelial cells of the same livers by an interactive analysis of scanning electron microscope (SEM) images.

MATERIALS AND METHODS

Sprague-Dawley rats in the fetal (20 days of gestation), neonatal (1st and 5th days) and adult periods, were used.

Microspectrophotometric measurement of SDH activity in hepatic parenchymal cells. To quantify this enzyme activity, liver samples were frozen in liquid Nitrogen and tissue-sections of 10 µm thickness were cut at -25°C on a cryostat microtome. The SDH reaction medium and the procedure of measurement and evaluation were performed according to descriptions of Pette et al (1979), with a SMP05-Zeiss microspectrophotometric unit.

Assessment of fenestration characteristics in endothelial cells of liver sinusoids. The procedures used for liver tissue preparation , electron microscopic examination and quantitative estimation of pore density, size and clustering have been described previously (Vidal-Vanaclocha and Barberá-Guillem, 1985).

RESULTS AND DISCUSSION

The SDH histochemical method shows a homogeneously active parenchymal cell population in the immature liver in the last days of the fetal period (Fig.1a). By the first day of neonatal life some high SDH activity hepatocytes appear close to portal vessels and it is possible to recognize them as a small but well differentiated subpopulation (Fig. 1b). On the fifth day of neonatal life two hepatocytic populations are clearly defined (Fig. 1c), and their values of SDH activity are similar to that of adult liver cells (Fig. 1d) although the proportion of high SDH-active hepatocytes is less than in adult livers (Fig. 2). Thus, it seems that the segregation process of the two hepatocytic populations starts after birth and becomes quickly established during the first week of postnatal life.

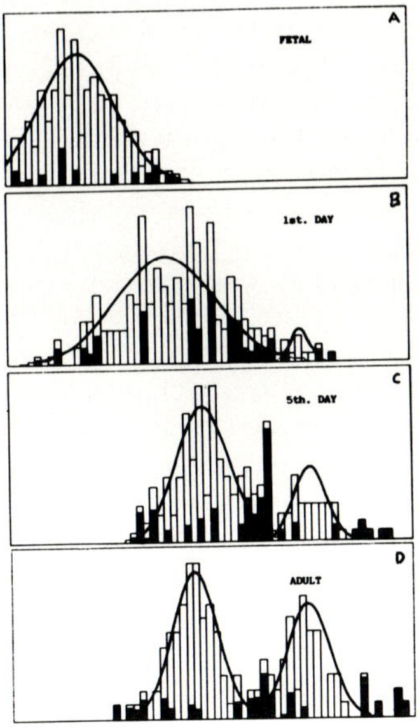

Fig.1 Hepatic parenchymal cell populations discriminated by SDH

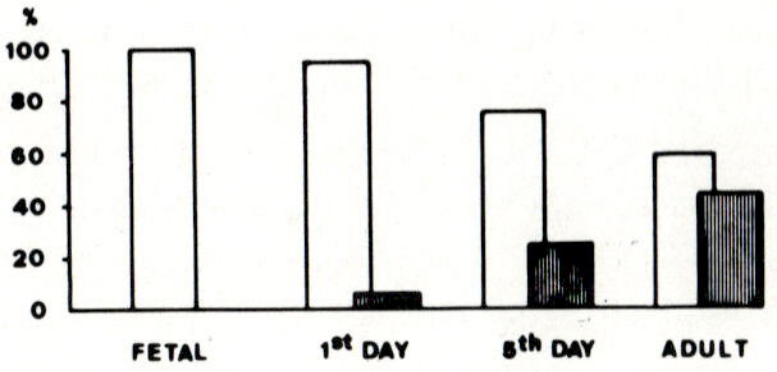

Fig.2 Evolution of hepatic parenchymal cells with respect to the SDH activity. Low active cells ; high active cells

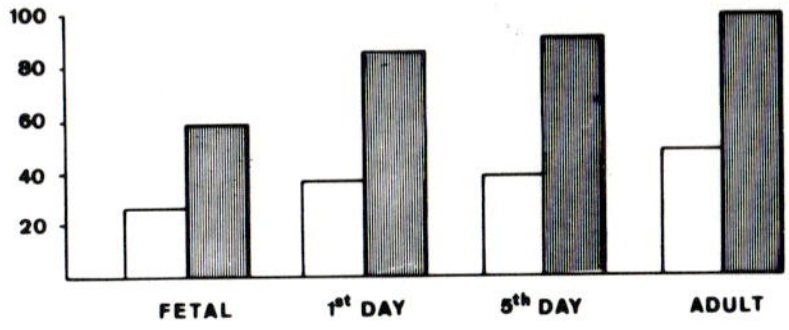

Fig.3 Incidence of sieve plates in the endothelial cells in zone 1 and zone 3 sinusoids.

The study of endothelial fenestrae in a significant number of sinusoids derived from areas 1 and 3 of fetal, neonatal and adult liver acini shows that endothelial cells are provided with dynamic fenestrae which vary in size and distribution according to both their location within the liver acinus and the phase of hepatic development. Before birth, endothelial fenestrae reach an adult-like density per μm^2 of endothelial surface only in 57% of the zone 3 sinusoidal surface, while in zone 1 pore density values are similar to adult ones. Subsequently,in neonatal livers pore density in endothelial cells of zones 1 and 3 turn out to be comparable to that of adult livers, i.e. that in zone 3 being nearly twice the number of that in zone 1 (Vidal-Vanaclocha and Barberá-Guillem, 1985). The size of fenestrae provide a second differential feature. Normal pores of 74 nm in zone 1 and of 96 nm in zone 3 prevail on the other sized pores in endothelial cells of fetal and neonatal livers. However, almost three times as many larger fenestrae (1-2 µm in diameter) are seen in zone 3 endothelium of fetal and 1st day livers. These fenestrae were observed only at these stages and are frequently seen associated to hemopoietic cells, suggesting their possible involvement in the transendothelial passage of hemopoietic cells. Finally, we have recognized early differences in the distribution of sieve plates on the sinusoidal endothelium. As shown in Figure 3, from the fetal period, the incidence of pore clustering during liver development is always greater in zone 3 than in zone 1. To sum up, during the fetal and neonatal periods, endothelial cells form specific fenestration patterns related to their location within the acinus. The segregated domains on endothelial lining defined by fenestration features even before birth, when hepatocytes still constitute a homogeneous population of undifferentiated cells, suggest that the sinusoidal endothelium becomes differentiated before the hepatocytes. This precocious maturation of endothelial cells might indicate their possible involvement in the subsequent definition of the functional heterogeneity of hepatocytes.

REFERENCES

Gumucio JJ (1983). Functional and anatomic heterogeneity in the liver acinus: impact on transport. Am J Physiol 244: G578-G582.

Medlock ES, Haar JL (1983) The liver hemopoietic environment: I. Developing hepatocytes and their role in fetal hemopoiesis. Anat Rec 207: 31-41.

Pette D, Wasmund H, Wimmer M (1979) Principle and method of kinetic microphotometric enzyme activity determination in situ. Histochemistry 64: 1-10.

Simionescu N (1983) Cellular aspects of transcapillary exchange. Physiol Rev 63: 1536-1579

Sleyster EC, Knook DL (1982) Relation between localization and function of rat liver Kupffer cells. Lab Invest 47: 484-490.

Vidal-Vanaclocha F, Barberá-Guillem E (1985) Fenestration patterns in endothelial cells of rat liver sinusoids. J Ultrastruct Res 90: 115-123.

Weiss L (1980) The haematopoietic microenvironment of bone marrow: an ultrastructural study of the interactions of blood cells, stroma and blood vessels. Ciba Found Symp 71: 3-20.

Progress in Developmental Biology, Part A, pages 331–334

CELL-SHAPE MODIFICATION INDUCED BY EXOGENOUS mRNA IN POSTNODAL EXPLANTS OF CHICK EMBRYO BLASTODERM.

F.De Bernardi,C.Sotgia,A.M.Bolzern,
M.Cigada Leonardi,U.Fascio,S.Ranzi.

Department of Biology,University of Milano,
via Celoria 26 20133 Milano,Italy.

Introduction

The postnodal explant of chick embryo is the posterior part of chick embryo,cut at the primitive streak stage 0.6 mm behind the Hensen's node. When cultured on albumen even after 20 hours, it develops in cuboidal ectoderm,entoderm and mesoderm differentiated in mesenchime,blood vessels and blood cells. No further differentiation is possible in these explants without the specific induction of the Hensen's node and of the anterior part of primitive streak both excised in our specimens.

In this paper the results of the treatment of postnodal explants with mRNA specifically coding for myosin heavy chain and for tubulin are summarized. Differentiation of the mesodermal and of the ectodermal cells was obtained in both cases.

Materials and Methods

Chicken embryos (Ross Italia) were explanted at the primitive streak stage (stage 4 of Hamburger and Hamilton, 1951) and were cultured in vitro by New's method (1955).The postnodal explants were cut 0.6 mm behind the Hensen's node and cultured as previously described (Bolzern et al.,1983). The postnodal explants were treated with RNA or with proteins (myosin or tubulin)suspended in Pannet-Compton solution immediatly before use at final concentration of 8-15 μg.in 2 ml.Myosin heavy chain mRNA and myosin were extracted and purified as described (Bolzern et al.,1983). Tubulin mRNA was extracted and purified from chick and from calf brain as described (De Bernardi et al.,1984). Tubulin was purified by two cycles of polymerization and depolymerization (Shelanski et al.,1973).

At the end of the culture period the specimens were fixed,embedded in JB4 medium (Polyscience, Warrington,Pa.) and cut in 1-2 μm sections. The sections were stained with ematoxylin-eosin.

Results

The postnodal explants treated with myosin haevy chain mRNA extracted from chick embryo muscle show modification of the mesodermal cells which became elongated, spindle shaped and organised around a myocoele, to form a true somite (fig.1,2).

The same results were obtained when the explants were treated with myosin heavy chain extracted from chick muscle. Myosin mRNA extracted from other vertebrates (duck,rabbit,trout) induced small somites built up of cells separated by large spaces joined around a large myocoele. Even the myosin mRNA of crayfish muscles induced elongated cells joined at the basal region and connected around a cavity. Heterospecific myosin failed to induce any modification in postnodal explants (Bolzern et al.,1983).

The explants treated with purfied tubulin mRNA,when examined after 20 hours of incubation, showed from one to ten characteristic protuberances from the surfaces irregularly aligned inside the edge of the explant. By light microscopy, the protuberances appeared to consist of a series of lined up ectodermal cells,some of wich appeared to have become elongated. The space defined by the ectodermal cells seemed to be filled by several mesodermal cells, some of which were elongated along the major axis of the protuberance (fig.3). In other cases the ectoderm looked thicker like a pseudostratified epithelium because the cells were slightly prismatic instead of cuboidal, as in control ectoderm. In some explants, some very closely packed elongated mesodermal cells were also observed.

At the electron microscopy some isolated, non-oriented microtubules were present in the ectodermal cells of the control explants. On the contrary, in the elongated ectodermal cells of the treated explants it was possible to observe many microtubules oriented in parallel to the major axis of the cells. The cells at the bases of the protuberances contained many evident bundles of microtubules, in characteristic close association (De Bernardi et al.,1984). By treating the explants with two cycles purified tubulin from chicken brain we obtained

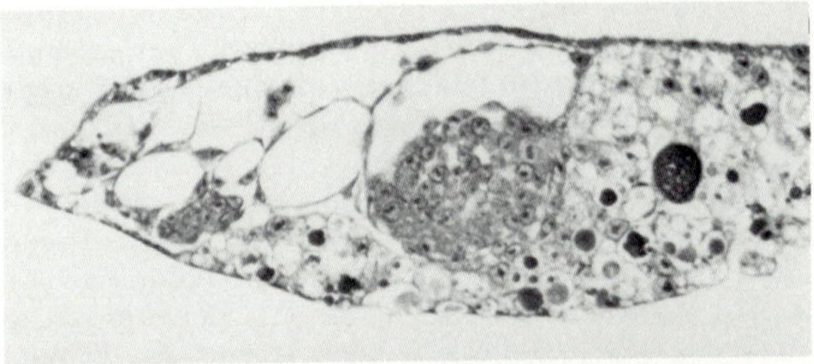

Fig.1. Transverse section of a control explant after 20h of culture (X310)

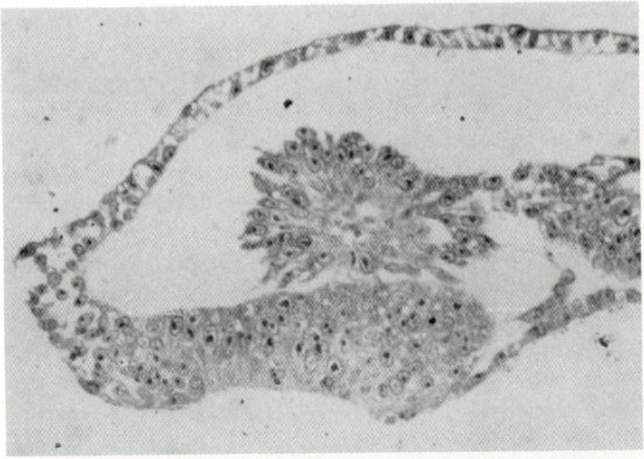

Fig.2. Transverse section of an explant treated for 20h with chick myosin mRNA (X310)

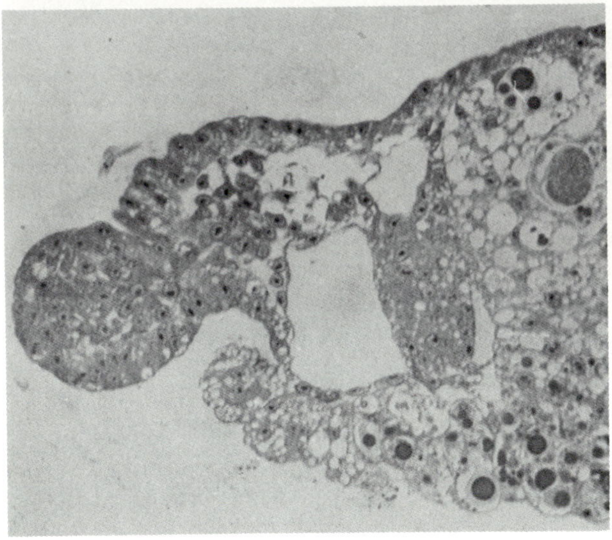

Fig.3. Transverse section of an explant treated for 20h with tubulin mRNA from chick embryo brain (X310)

protuberances that looked very similar to those obtained with tubulin mRNA. The anti-tubulin immunofluorescence showed the presence of cytoplasmatic microtubules inside the elongated ectodermal cells.

Discussion and conclusions

From the experiments carried out with myosin mRNA in presence of transcription and translation inhibitors (Bolzern et al.,1983) we concluded that myosin mRNA can be translated inside the cells and myosin itself promotes the differentiation of mesodermic cells. Therefore we looked to see if the mRNA coding for the major microtubular protein, tubulin, can modify the shape of the cells of the postnodal explants. During cellular differentiation, indeed, the microtubules play a role in determining cell shape.

The protuberances induced by tubulin mRNA in the postnodal explants are completely different from the somites induced by myosin mRNA. These protuberances are present on the surface of the explants and are always in close contact with the ectoderm, while the induced somites were inside the explant, close to the mesodermal layer. Moreover, the mesodermal cells never arranged themselves regularly around a cavity (myocoele) which was characteristic of the somitic cells and of the groups of oriented cells induced by myosin mRNA.

All these results show that the action of myosin mRNA and myosin is particularly evident in the mesodermal cells, producing the elongation and the arrangement of the cells to form a somites. Tubulin mRNA and tubulin can modify the shape of the ectodermal cells increasing the microtubules content.

References

Bolzern A.M.,Cigada Leonardi M.,De Bernardi F.,Fascio U., Maci R.,Ranzi S.,Sotgia C.,(1983). Expl.Cell Biol. 51:228-238.

Bolzern A.M.,Cigada Leonardi M.,De Bernardi F.,Ranzi S., Sotgia C.,(1984). J.Submicrosc.Cytol. 16:101-102.

De Bernardi F.,Saita A.,Sotgia C.,(1984). Acta Embryol. Morphol.Exper.n.s. 5:81-92

Hamburger V.,Hamilton H.L.,(1951). J.Morph. 88:49-67

New D.A.T., (1955). J.Embryol.Exp.Morph. 3:320-331

Shelanski M.L.,Gaskin F.,Cantor C.R.,(1973). Proc.Natl. Acad.Sci. U.S.A. 70:765.

Progress in Developmental Biology, Part A, pages 335–337

FACIAL MORPHOGENESIS AND PATTERN FORMATION

S.E. Wedden and C. Tickle

Dept. of Anatomy and Biology as applied to Medicine, The Middlesex Hospital Medical School, Cleveland St.. London W1P 6DB, U.K.

INTRODUCTION

The development of the face involves the growth and differentiation of 3 main primordia: the frontonasal mass, maxillae and mandibles. In the chick embryo,the frontonasal mass forms most of the upper beak and enlarges and fuses with the maxillae to form the primary palate. The mandibles fuse in the midline to form the lower beak. In both the frontonasal mass and mandibles, outgrowth and differentiation of cartilage occurs in spatially defined patterns. In the upper beak a central rod of cartilage, the prenasal cartilage, develops and a characteristic egg tooth forms at the distal end. In the lower beak, paired rods, Meckel's cartilages form.

Despite the superficial similarities between the morphogenesis of the upper and lower beak, retinoid treatment specifically affects the development of the frontonasal mass. When high concentrations of retinoid are applied to chick embryos on controlled release carriers, severe bilateral clefting of the primary palate occurs (Tamarin et al., 1984). Cartilage differentiation and outgrowth of the frontonasal mass is inhibited, whereas the lower beak is apparently unaffected and cartilage differentiation occurs normally.

Using micromass (high density) cultures (Ahrens et al. 1977), the chondrogenic potential of cells from these two facial primordia have been compared (Wedden et al., 1985). Cultures of cells from the frontonasal mass give rise to a relatively homogeneous sheet of cartilage, whereas in cultures of cells from the mandibles, nodules form. This difference in chondrogenic pattern of the two cell populations may be relevant to the retinoid effect.

To understand the specificity of the retinoid effect we need to know how the frontonasal mass develops. We have investigated how the connective tissue within the frontonasal mass is spatially organised, leading to the differentiation of a cartilage rod in the midline. The developmental potential of small fragments of tissue from different regions of the upper beak, at various stages of development, has been assayed by grafting them into wing buds of host chick embryos. After 6 days of growth, the resulting grafts are examined for the presence of a prenasal cartilage and egg tooth. The results are shown in Table 1.

TABLE 1. The ability of fragments of the frontonasal mass to form midline structures

	No. of grafts with prenasal cartilage +/- egg tooth	
Stage of face development	Fragments abutting Nasal Placodes	Central fragments
15 - 17	1/6 (17%)	4/4 (100%)
18	3/6 (50%)	5/5 (100%)
19	1/3 (33%)	4/4 (100%)
20	0/5 (0%)	6/6 (100%)
21	1/4 (25%)	7/7 (100%)
22 - 23	0/3 (0%)	9/9 (100%)
24	0/1 (0%)	9/9 (100%)
25	-	5/5 (100%)

RESULTS

At early stages, before stage 21 (Hamburger and Hamilton stages), fragments of tissue from the extreme edge of the frontonasal mass, abutting the nasal placodes, give rise to midline structures. At later stages, this ability becomes restricted to a broad central region of the frontonasal mass. However, even when this central region is further sub-divided into one central and two lateral fragments for grafting, midline structures still form from all fragments at stages 20 - 25. In four cases, midline structures were obtained from all three fragments of frontonasal mass and more than one prenasal cartilage and egg tooth may be found in just one graft.

It therefore appears that between stages 21 and 25 the ability to form midline structures has become restricted to more central regions of the frontonasal mass but not exclusively to the midline. Preliminary experiments suggest that in fact the prenasal cartilage is being specified in the centre of the frontonasal mass by stages 29 - 30.

This combined approach of grafting and micromass culture may enable us to probe the cues involved in specifying the pattern in the frontonasal mass. This may then provide an insight into the specifity of retinoid action on upper beak development.

REFERENCES

Ahrens P.B., Solursh M. Reiter R.S. (1977). Stage - related capacity for limb chondrogenesis in cell culture. Dev. Biol. 60:69-82.

Tamarin A., Crawley,A., Lee J., Tickle C. (1984). Analysis of upper beak defects in chicken embryos following treatment with retinoic acid. J. Embryol. exp. Morphol. 84:105 - 123.

Wedden S.E., Lewin-Smith M.R., Tickle C. (1985). The patterns of chondrogenesis of cells from facial primordia of chick embryos in micromass culture. Submitted for publication.

Transmembrane Modulation

Progress in Developmental Biology, Part A, pages 341–344

DIFFERENTIAL KERATIN PHOSPHORYLATION DURING RAT PALATAL EPITHELIAL DIFFERENTIATION

Charles F. Shuler and Stephen A. Schwartz

The Ohio State University College of Dentistry
Columbus, Ohio 43210 (C.F.S.)
The Univ. of Chicago Dept. of Pathology (S.A.S)

INTRODUCTION

The initial development and differentiation of stratified squamous epithelium occurs prenatally. During embryogenesis, the epithelium is derived from a primitive single layer that ultimately develops into the characteristic multilayered tissue of the adult. The morphologic changes that occur during fetal epithelial development resemble those which occur as cells undergo terminal differentiation in a mature stratified squamous epithelium. Palatal epithelium undergoes a highly specialized and unique development. In mammalian palatogenesis epithelial differentiation initially culminates in the adhesion and fusion of the palatal shelves and formation of the secondary palate (Pourtois, 1966). Subsequently, the palatal epithelium develops and differentiates to form the characteristic multilayered stratified squamous epithelium. Different stages of palatal development therefore, serve as accurate indicators of the degree of palatal epithelial development and differentiation. We examined the primary differentiation of palatal epithelium to determine changes in keratin synthesis and post-translational modification which coincide with specific morphologic changes in the epithelium.

The intermediate filament keratins have been shown to be synthesized exclusively by epithelial cells (Moll et al., 1982). An epithelial cell type typically synthesizes 2-8 different keratin proteins, so their characterization is a potentially accurate molecular indicator of the state of differentiation and specialization of epithelial cell types (Fuchs & Green, 1980, Tseng et al., 1982, Eichner et al., 1984). Posttranslational phosphorylation of the keratins has been shown to be the most common modification (Lazarides, 1982). Phosphorylation occurs primarily on the serine residues in the non-alpha helical, glycine-rich variable regions at the amino and carboxy termini of the keratins (Steinert et al., 1983). This modification may be regulated through different keratin kinase pathways (Gilmartin et al., 1984). Importantly, these terminal regions are essential in promoting keratin protein polymerization and intermediate filament formation. Most studies correlating epithelial differentiation and keratin gene expresion have used either mature epidermal tissues or epithelial cells maintained in vitro and have not examined changes in post-translational modification. We have examined the keratin protein composition and phosphorylation in palatal epithelium during its primary differentiation in utero.

RESULTS

The keratin proteins were extracted from the epithelial component of rat palatal mucosa at different stages of development. The proteins were resolved by SDS-PAGE and visualized by silver staining (Figure 1).

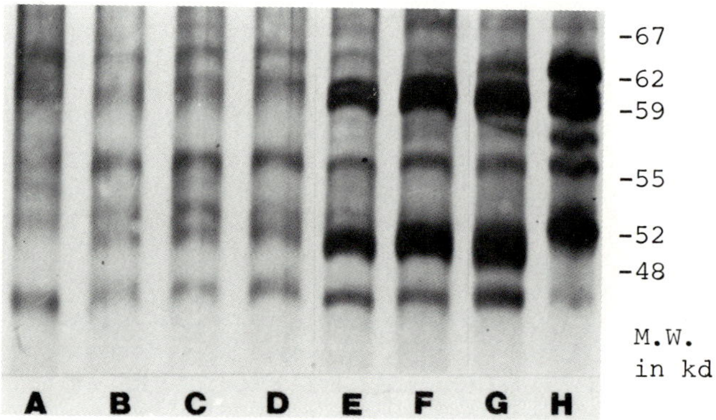

Figure 1. SDS-PAGE of keratins extracted from palatal epithelium at different stages of development. (A-F days in utero) A. 13 B. 14 C. 15 D. 16 E. 18 F. 19 G. 1 day postnatal H. Adult. Samples electrophoresed through 10% polyacrylamide gels, fixed and silver stained. Molecular weight standards, coelectrophoresed. 2ug of protein sample/lane.

The keratin population in palatal epithelium was composed of 9 proteins of molecular weights 48, 50, 52, 55, 57, 59, 60, 62, and 67kd. These polypeptides were absent in the cytoskeletal extracts sampled from underlying mesenchyme. Throughout the course of histodifferentiation from 13 days *in utero* to the adult the same subset of keratins was resolved with no significant qualitative differences. However, significant quantitative changes occurred specifically with the 48, 57, 59, and 62 kd keratins at later stages of development. The adult proteins were identified by reaction with antikeratin antibodies using the immunoblot technique and compared with the fetal proteins by peptide mapping following limited proteolysis with *S. aureus* V8 protease.

The post-translational modification of keratins was examined in epithelium pulse-labelled with ^{32}P orthophosphate immediately following dissection. The pattern of keratin phosphorylation showed distinct changes during palatal epithelial development *in utero* (Figure 2).

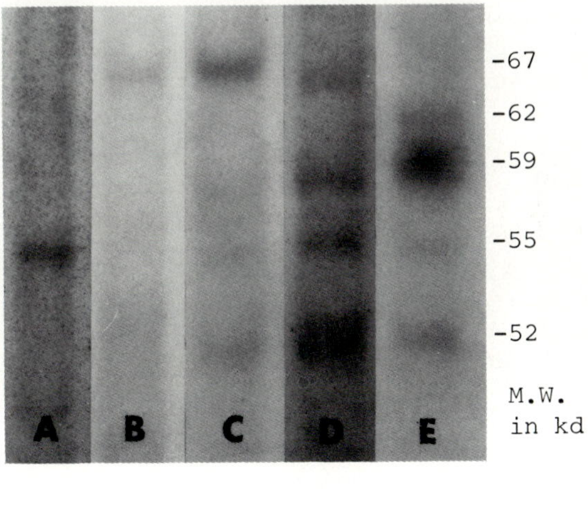

Figure 2. SDS-PAGE of ^{32}P labelled keratins extracted from palatal epithelium at different stages of development. A. 15 days *in utero* B. 17 days *in utero* C. 19 days *in utero* D. 1 day postnatal E. Adult. Samples electrophoresed through 10% polyacrylamide gels, fixed, dried and autoradiographed.

Prior to epithelial stratification at 15 days *in utero* only a single protein was phosphorylated. By 17 days *in utero*, a point coinciding with a significant degree of stratification of the epithelium, all of the keratins resolved by silver staining had incorporated ^{32}P, although to different degrees. The 67kd keratin in particular was disproportionately phosphorylated with respect to its relative concentration. By the 19th day *in utero* and throughout the subsequent developmental stages the pattern of keratin phosphorylation mirrored the previously observed quantitative shifts in keratin biosynthesis.

CONCLUSION

Specific qualitative shifts in the pattern of keratin synthesis were not observed during the primary differentiation of the stratified squamous epithelium of the palate. Changes in the degree and pattern of phosphorylation of these keratins were observed that were closely related to the histologic pattern of epithelial differentiation. These changes in keratin phosphorylation may represent a functional maturation of the proteins that is associated with both the initial and continual stratified differentiation of this epithelium.

REFERENCES

Eichner R, Bonitz P, Sun T-T (1984). J Cell Biology 98:1388-1396.

Fuchs E, Green H (1980). Cell 19:1033-1042.

Gilmartin ME, Culbertson VB, Freedberg IM (1984). J Cell Biology 98:1144-1149.

Lazarides E (1982). Ann Rev Biochem 51:219-250.

Moll R, Franke WW, Schiller DL, Geiger B, Krepler R (1982). Cell 31:11-24.

Pourtois M (1966). J Embryol Exp Morph 16:171-182.

Steinert PM, Rice RH, Roop DR, Trus BL, Steven AC (1983). Nature 302:794-802.

Tseng SCG, Jarvinen MJ, Nelson WG, Huang J-W, Woodcock-Mitchell J, Sun T-T (1982). Cell 30:361-372.

Progress in Developmental Biology, Part A, pages 345–348

AN ULTRASTRUCTURAL STUDY OF PLASMA MEMBRANE LIPID ASSEMBLY DURING BLASTODERM FORMATION IN *DROSOPHILA*

John G. Bluemink*, Willem J. Hage* and Stephan Bilinški☆
*Hubrecht Laboratory, International Embryological Institute, Uppsalalaan 8, 3584 CT Utrecht, The Netherlands. ☆Jagellonian University, Department of Zoology, ul Karasia 6, 30-060 Kraków, Poland

INTRODUCTION

During blastoderm formation in the *Drosophila* embryo some 5000 cells are formed within ca 90 min. Plasma membrane assembly takes place on a tremendous scale (see Fullilove and Jacobson 1971) at a rate of ca 600 $\mu m^2\ s^{-1}$. For the 3.2 x 10^6 μm^2 membrane area needed phospholipid must be synthesized, transported and assembled. During conventional tissue processing for electron microscopy (EM) plasma membrane lipids are completely extracted. Evidence exists (Saffitz *et al.* 1981) that in myeloma cells ca 94% phosphatidyl choline (PC) and dipalmitoyl PC is preserved, and that no significant translocation of labelled PC occurs, when cells are treated with 1.5% tannic acid during fixation for EM. We have used tannic acid to visualize the sites of synthesis and assembly of PC during blastoderm formation in the *Drosophila* embryo.

METHOD

To stabilise PC in aldehyde-fixed embryos tannic acid (A.R. grade code nr.1764 from Mallinckrodt Inc., St.Louis Mo.) was used following a protocol by Kalina and Pease (1977). Early blastoderm stages were processed for ultramicrotomy and freeze fracture EM.

RESULTS

After tannic acid treatment electron dense multilamellar bodies (mlb) resembling liposomes are retained in ultrathin sections. Basal to the nuclei mlb's are numerous (Fig.1) and bound to paired membranes of endoplasmic reticulum (ER). Lateral to the nuclei mlb's are close to microtubules which run apico/basally (Fig.2). Apical to the nuclei mlb's frequently are in contact with the surface of the ingrowing furrows (Fig.3), the inner leaflet of the plasma membrane being continuous with the outermost lamella of the mlb (Fig.3 inset).

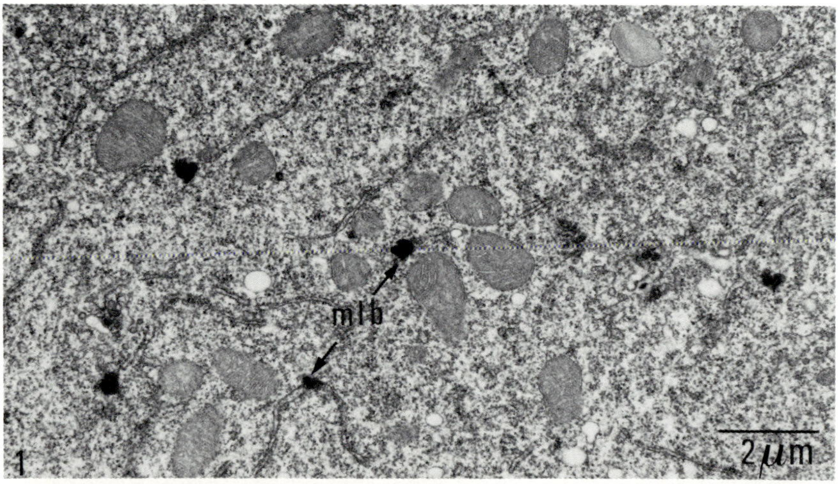

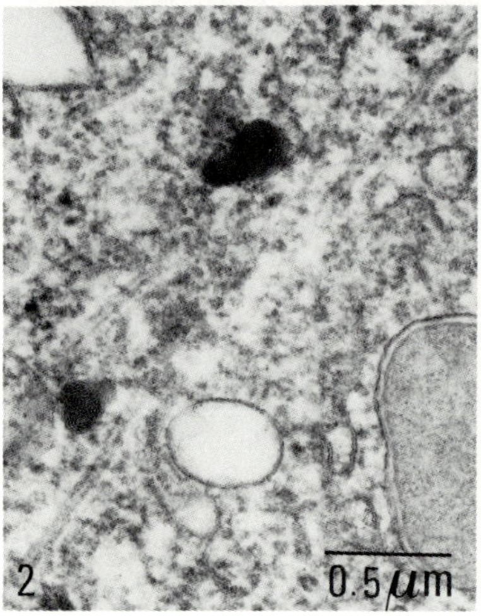

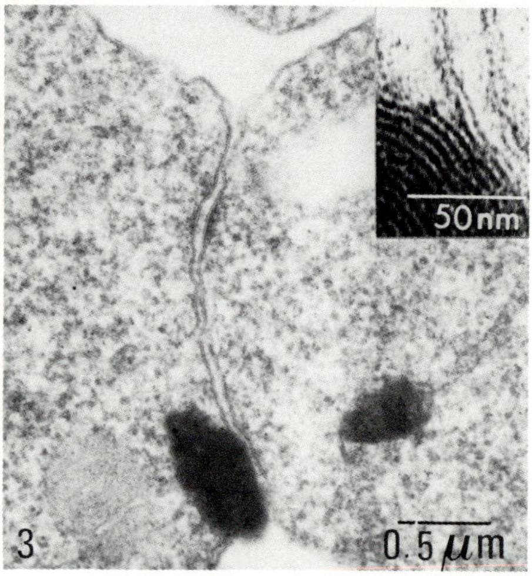

5
0.5 μm
4
1 μm

In freeze fracture replicas the early furrow membrane (Fig.4) is relatively sparsely seeded with intramembrane particles (IMP). In the fractured furrow membrane IMP-free deflections (Fig.5) have concentric rims some of which change stepwise to a next plane of fracturing (arrows, inset Fig.5). In some a single (lipid?) particle is at the top (inset, arrow head). Since the dimensions of these configurations fit those of mlb's, they are taken to arise by fracturing stacked lamellae of mlb's, which are fusing with the furrow membrane.

CONCLUSION

PC lipids synthesized in ER during *Drosophila* blastoderm formation (Heckman *et al.* 1977) are preserved from extraction by tannic acid treatment. Whether PC exists and is transported *in vivo* in the form of mlb's will be investigated in unfixed *Drosophila* blastoderm processed by combining the techniques of high pressure freezing (Moor *et al.* 1980) with freeze substitution (Müller *et al.* 1980) or freeze fracture.

REFERENCES

Fullilove SL, Jacobson AG (1971). Nuclear elongation and cytokinesis in *Drosophila montana*. Devel Biol 26: 560-577.

Heckman GA, Friedman SJ, Shekan PJ, Barrnett RJ (1977). Localization and partial characterization of acyltransferases present during rapid membrane formation in the *Drosophila melanogaster* embryo. Devel Biol 55: 9-32.

Kalina M, Pease DC (1977). The presentation of ultrastructure in saturated phosphatidyl cholines by tannic acid in model systems and type II pneumocytes . J Cell Biol 74: 726-741.

Moor H, Bellin G, Sandri C, Akert K (1980). The influence of high pressure on mammalian nerve tissue. Cell Tissue Res 209: 201-216.

Muller M, Marti T, Kriz S (1980). Improved structural preservation by freeze substitution. Electr Microsc 2: 720-721.

Saffitz JE, Gross RW, Williamson JR, Sobel BE (1981). Autoradiography of phosphatidyl Choline. J Histoch Cytoch 29: 371-378.

Progress in Developmental Biology, Part A, pages 349–352

Assembly of tight junctions in cultured human epithelial cells

Yehuda Ben-Shaul, Esther Cohen, Ortwin Faff, Harald Mückter and Adelbert Bacher

Department of Microbiology, Tel-Aviv University, Israel, and Lehrstuhl für Organische Chemie und Biochemie, Technical University of Munich, Garching, West Germany

In spite of considerable efforts, the molecular structure and the mechanism of formation of tight junctions (TJ) are still largely unknown. The human colon carcinoma cell line HT 29 provides a favourable system for the quantitative analysis of TJ formation. Cells growing in culture have virtually no TJ, but TJ formation proceeds rapidly after brief treatment with various proteases (Polak-Charcon et al., 1976; Talmon et al., 1984; Cohen et al., 1985). Maximum expression of TJ occurs within 15-30 min after protease treatment.

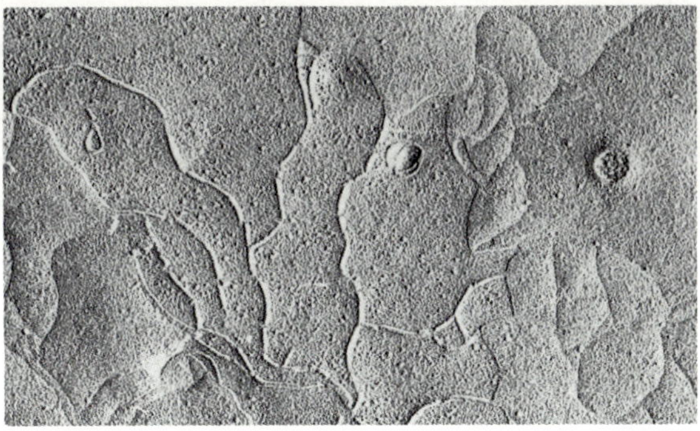

Figure 1. Freeze fractured membrane of HT 29 cells treated with trypsin (1 mg/ml).

TJ induced by proteases are degraded in HT

29 cells during a period of appoximately 2 h (Fig. 2). At the end of this period, the cells are not susceptible to renewed TJ induction unless they are left to recover for a period of several more h.

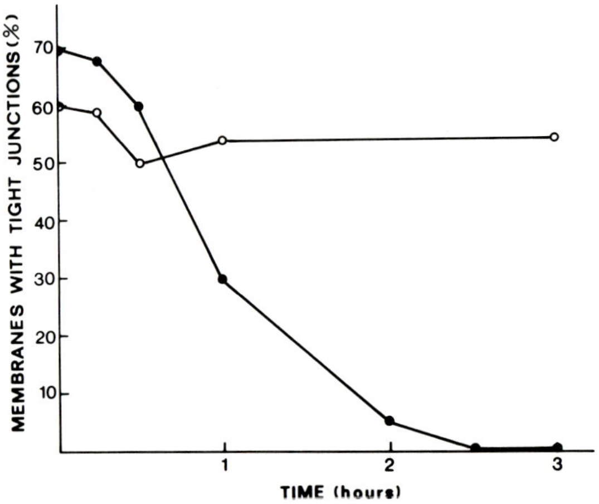

Figure 2. Degradation of trypsin-induced TJ in HT 29 cells. o, Incubation at 3 °C; •, incubation at 37 °C.

When trypsin-treated cells are incubated in the presence of a translation inhibitor (cycloheximide or puromycin), they do not recover the capacity to form TJ when retreated with protease. We conclude that the assembly of TJ requires and consumes a protein component which has to be reexpressed in order to allow the cells to assemble TJ after a first stimulus.

A similar line of evidence was obtained by work with pronase. High concentrations of this protease mixture produce large cell clumps without TJ. These fully viable cells can be stimulated to form TJ only after recovery for several h. The recovery is suppressed entirely by the transcription inhibitor, actinomycin D (Fig. 3).

Treatment of HT 29 cells with 5 mM dinitrophenol in the absence of glucose reduced the ATP level to about 2 % of controls. ATP-depleted cells produce about 20 % TJ when treated with trypsin for 15 min. Treatment with trypsin for

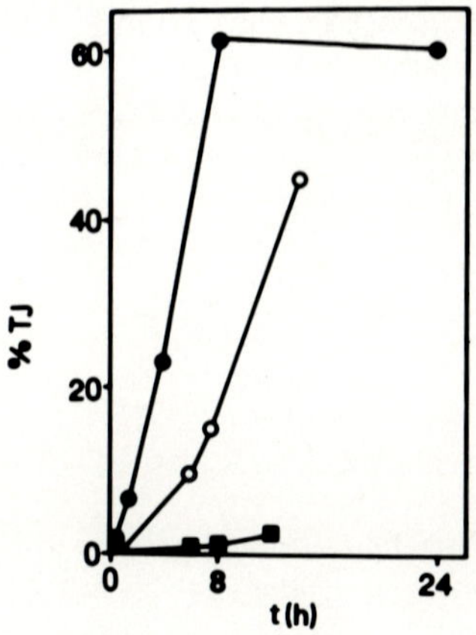

Figure 3. Formation of TJ by trypsin treatment (1 mg/ml, 15 min) after pretreatment of HT 29 cells with pronase (1mg/ml, 15 min). o, actinomycin (2 ug/ml) added after pronase treatment; ■, actinomycin (2 ug/ml) added 1 h prior to pronase treament; ●, no addition.

longer periods can induce the maximum level of TJ found in controls (60 %) (Fig. 4). We conclude that very little or no ATP is required for TJ formation although the rate of formation is reduced in ATP-depleted cells.

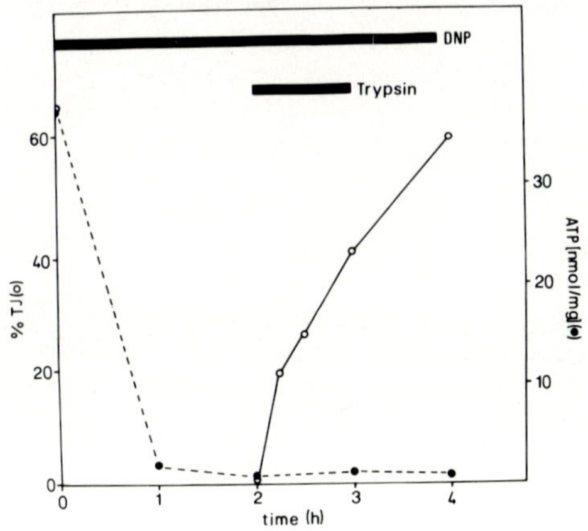

Figure 4. HT 29 cells were treated with 5 mM 2,4-dinitrophenol in the absence of glucose. The ATP-depleted cells were treated with trypsin (1 mg/ml) for the time indicated by the lower bar.

It is a long-standing controversy whether TJ are unstable in the absence of extracellurar Ca^{++} (Stevenson and Goodenough, 1984). We have treated HT 29 cells with trypsin or pronase in the presence of the chelators EGTA or EDTA. In both cases TJ form in the normal amount and at the normal rate. Since TJ can be assembled normally in the presence of either chelator, we conclude that neither Ca^{++} nor other divalent ions are required for their structural integrity.

Chelator	% TJ
none	68
EGTA	68
EDTA	67

Table 1. Induction of TJ by treatment with trypsin (1.5 mg/ml). Chelators were added at a concentration of 3 mM.

TJ can also be induced in HT 29 cells by hypertonic solutions of ammonium sulfate. The effect is ion specific. Sodium chloride at the same osmolarity does not induce any TJ.

This work was supported by grants from the Fonds der Chemischen Industrie and the Ashkenazi Fund for the Biology of Cancer, Tel Aviv University.

REFERENCES

Polak-Charcon S, Shoham J, Ben-Shaul Y (1976). Junction formation in trypsinized cells of human adenocarcinoma cell line. Exp Cell Res 116: 1-13

Talmon A, Cohen E, Bacher A, Ben-Shaul Y (1984). Separation of induction and expression of tight junction formation mediated by proteases. Biochim Biophys Acta 769: 505-507

Cohen E, Talmon A, Faff O, Bacher A, Ben-Shaul Y (1985). Formation of tight junctions in epithelial cells. I. Induction by proteases in a human colon carcinoma cell line. Exp Cell Res 156: 103-116

Stevenson BR, Goodenough DA (1984). Zonulae occludentes in junctional complex-enriched fractions from mouse liver. Preliminary morphological and biochemical characterization. J Cell Biol 98: 1209-1221

Progress in Developmental Biology, Part A, pages 353–356

DIFFERENTIATION OF TRANSMEMBRANE IONIC CURRENTS IN THE EARLY EMBRYONIC CHICK HEART

S. Fujii, R. K. Ayer, Jr. and R. L. DeHaan

Department of Anatomy and Cell Biology, Woodruff Health Science Center, Emory University, Atlanta, Georgia 30322

INTRODUCTION

During the cardiac action potential (AP) at least five potential-sensitive transmembrane currents are activated (Reuter, 1984). Each current crosses the cell membrane through a separate integral membrane protein that serves as an ion-selective channel. The best understood of these conductance mechanisms is the "excitatory sodium current" that underlies the sharp AP upstroke (Fozzard et al, 1985). This current is recognized by its rapid kinetics, and its sensitivity to suppression by a depolarizing prepulse or by nanomolar concentrations of certain poisons such as tetrodotoxin (TTX). In the present work we have focused on its development in the embryonic heart.

Although the structure and function of the sodium channel have been studied extensively and its complete primary amino acid sequence has been determined (Noda, et al 1984), we know little about how this macromolecule is expressed during embryogenesis, or even when these channels first appear and begin to function in normal cardiac development. Some workers have obtained indirect evidence for the presence of sodium channels in the 3d chick heart. However, ventricular tissue from chick embryos aged 2-3 days fires APs that are small in amplitude, have slow upstrokes, and are resistant to TTX. This suggests that sodium channels are lacking from early heart cells, appearing only at later stages (reviewed in DeHaan, 1980). To investigate further the appearance and development of the fast sodium conductance mechanism, we have applied the whole-cell recording patch-electrode technique to ventricle cells isolated from 2-day to 7-day hearts.

METHODS

Single 13-18 μm diameter embryonic chick ventricle cells were freed from the heart by exposure to trypsin, collagenase, or maceration in low-calcium (18 μM) buffer. Cells isolated from embryos after 2, 3, 4, or 7 days of development were transferred to polylysine-coated culture dishes, where they attached without appreciable flattening during 5-20 hrs of incubation at 37°C in an atmosphere of 5% CO_2, 10% O_2, and 85% N_2. Voltage command steps were applied to the cells with patch electrodes in the whole-cell clamp configuration, as described previously (Fischmeister et al, 1984). Since physiological differentiation continues in cardiac cells _in vitro_, the total developmental age of cells at the time of recording was noted, for example, 4d+1, meaning cells taken from the heart of a 4d embryo and maintained _in vitro_ for one additional day.

RESULTS AND DISCUSSION

From a steady membrane potential (V_m) of -80 mV, depolarizing command potentials elicited fast inward currents in a 7d+1 cell (Fig. 1A) that reached a peak in less than 1 ms over most of the voltage range. The same command potentials (Fig. 1C) activated similar currents of smaller magnitudes in 4d+1, 3d+1, and all 2d cells (Fig. 1B) the earliest being a 2d+5h cell, isolated by maceration in low-Ca buffer without enzymes. Examples of the maximal response at each age (Fig. 1D,inset), show that the current increases with age.

The kinetics and voltage range of inactivation of the fast sodium current remained constant throughout development, as did its TTX-sensitivity. At all stages tested, the fast inward current was completely blocked by TTX. The concentration that produced half-maximal inhibition of the current ($K_{1/2}$) in 2d+1, 3d+1, 4d+1 and 7d+1 cells was 0.46, 0.56, 0.63, and 0.85 nM; for the pooled data the $K_{1/2}$ was 0.62 ± 0.11 nM TTX (mean ± s.d.).

The voltage dependence of sodium current activation also remained fairly constant throughout development, as reflected in the I/V curves. However, a systematic developmental change was in its magnitude. This was most readily seen by scaling the currents to measured cell surface, and plotting the I/V curves of different cells in

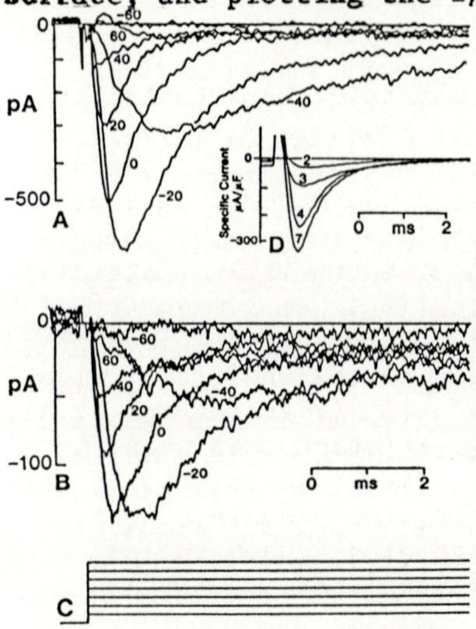

Fig. 1. The excitatory sodium current elicited from a (A) 7d+1 and a (B) 2d+1 cell, by (C) stepwise command potentials from -60 to +60 mV from a holding V_m of -80 mV. (D) Superimposed single currents from a 2d+1, 3d+1, 4d+1 and 7d+1 cell (top to bottom traces) in response to a voltage pulse from -120 to -20 mV.

terms of specific membrane current density (Fig. 2). The increase in current magnitude was the only consistent difference between younger and older cells. This increase in fully saturated current is plotted in Fig. 3.

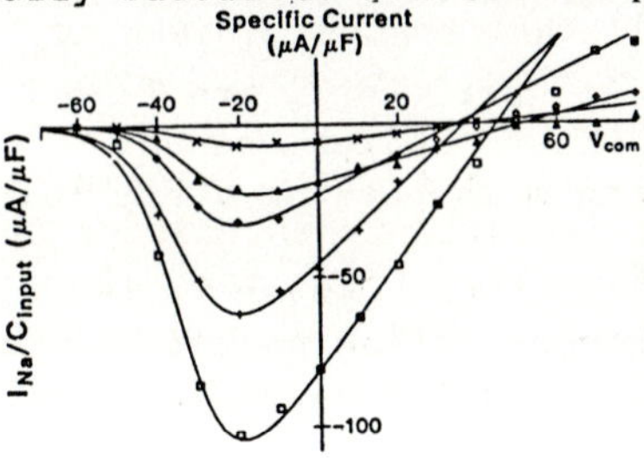

Fig. 2. I/V curves, plotted as specific membrane current density, of net sodium current obtained from typical 2d+5h (X), 2d+1 (▲), 3d+1 (◆), 4d+1 (+) and 7d+1 (□) ventricle cells, from V_m = -80 mV to the command potentials indicated on the V_{com} axis.

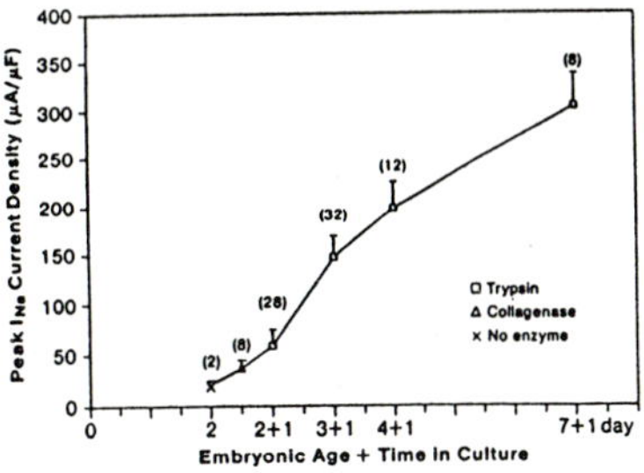

Fig. 3. Maximal sodium current evoked from cells of different developmental ages. Means and standard error bars are shown, with the number of cells (n) in parentheses near each error bar.

These results indicate that a voltage-activatable,

fully functional sodium current is present in the chick ventricle as early as two days of development. The currents were similar at all stages, showing no systematic changes with development in kinetics, voltage dependence, or TTX sensitivity, while current density increased about 17-fold between 2d+5h and 7d+1. From the average 7d cell, the maximal sodium current that could be activated was about 300 uA/cm^2. This value translates into a density of about 5 $channels/um^2$ - which corresponds well to estimates of channel density in the mammalian heart (Doyle et al, 1985) - or about 4000 channels in the membrane of an average 7d cell. A similar cell from a 2d heart would contain only about 300 sodium channels, or fewer than $1/um^2$. The sodium current normally makes no contribution to the 2d AP because its range of activation is more negative than the maximum diastolic potential (about -60 mV) reached by the relatively depolarized cell at that stage. With development, as more channels are inserted, the cells also become more polarized and a sodium-dependent upstroke appears. If the sodium channel protein represents the same gene product in the 2d+5h cell as it does at older stages, transcription of the sodium channel gene must begin before two days.

REFERENCES

DeHaan RL (1980). Differentiation of excitable membranes. Curr Top Dev Biol 16: 117-156.

Doyle DD, Brill DM, Wasserstrom JA, Karrison T, and Page E (1985). Saxitoxin binding and "fast" sodium channel inhibition in sheep heart plasma membrane. Am J Physiol 249: H328-H336.

Fischmeister R, DeFelice LJ, Ayer, Jr. RK, DeHaan RL (1984). Channel currents during spontaneous action potentials in embryonic chick heart cells. The action potential patch clamp. Biophys J 46: 267-272.

Fozzard HA, January CT, Makielski JC (1985). New studies of the excitatory sodium currents in heart muscle. Circ. Res. 56: 475-485.

Noda M, Shimizu S, Tanabe T, Takai T, Kayano T, Ikeda T, Takahashi H, Nakayama H, Kanaoka Y, Minamino N, Kangawa K, Matsuo H, Raftery MA, Hirose T, Inayama S, Hayashida H, Takashi M, Numa S (1984) Primary structure of Electrophorus electricus sodium channel deduced from cDNA sequence. Nature 312: 121-127.

Reuter H (1984) Ion channels in cardiac cell membranes. Ann Rev Physiol 46: 473-484.

Growth Factors and Hormones

Progress in Developmental Biology, Part A, pages 359–364

VERTEBRATE SOMATOTROPIN-LIKE PEPTIDES IN INSECTS

Peter Verhaert, Herman Van Mellaert, Kristien Swinnen and Arnold De Loof

Zoological Institute of the University,
Naamsestraat 59, B-3000 Leuven, Belgium

INTRODUCTION

According to the theory of Evolution it is evident that the comprehensive neuroendocrine control system of vertebrate growth and metabolism must have had its origin in organisms lower on the phylogenetic scale. One could imagine that the basic principles of endocrine regulation might have already been present in the ancient common ancestor of both protostomians and deuterostomians and it would therefore not be impossible that substances similar to the vertebrate hormone(s) promoting growth and various aspects of metabolism may be found as well in invertebrates.

One of these hormones is somatotropin (somatotropic hormone, STH; growth hormone, GH), a polypeptide with about 200 amino acid residues (the number varies depending upon the species) which is known to be synthesized and concentrated within the vertebrate adenohypophysis before its release into the blood stream.

The present paper will concentrate on a study of the potential presence of material resembling vertebrate GHs in insects, comprising a dual approach of this matter, namely an immunohistological and a molecular biological one.

MATERIALS AND METHODS

Periplaneta americana, *Locusta migratoria* and *Sarcophaga bullata* were used in this study, representing Dictyoptera, Orthoptera and Diptera respectively.

Brains and adjacent neuroendocrine structures of adult specimens of these three insect species were microdissected, prepared for immunohistochemistry and processed in the peroxidase antiperoxidase method as earlier described (Verhaert and De Loof, 1985). The primary antisera used were anti-trout GH, a kind donation by Dr. T.F.C. Batten (University of Sheffield, U.K.) and anti human GH, which was purchased from UCB Bioproducts (lot nr. I515/002). Both antisera were controlled by the staining of appropriate vertebrate tissue (anterior pituitary). Serum specificity was judged by the criterium of a consistent distribution of the immunoreactive structures depending upon the primary antibodies only. Method specificity was controlled by the application of two non-immune rabbit sera.

For the nucleic acid approach total cellular RNA was extracted from cerebral tissue of adult roaches and from whole adult flies, following the techniques of Chirgwin et al. (1979) and Ullrich et al. (1977). This RNA was dot blotted onto nitrocellulose (NC) membranes in a serial dilution (starting from 1.6 ug) (Thomas, 1980). Three cloned cDNAs coding either for bovine, human or rat GH were used as probes after radiolabeling by nick translation (Maniatis et al., 1982) upto a specific activity of 5.10^5, $8.5\ 10^5$ and $7.5\ 10^6$ cpm/ug respectively. The NC filters containing the RNA dot blots were prehybridized (minimum 2 hrs, 37°C), hybridized with the radioactive probes (overnight, 37°C), washed (42°C) and exposed to X-ray film (10 ds, -70°C) (Maniatis et al., 1982). Rat and human pituitary RNAs were used as positive controls and poly $(A)^-$ containing RNAs from mouse mammary gland as a negative control.

RESULTS

Immunohistochemistry

In the three insect species under investigation various neurons and neurosecretory cells in different regions of the supra and suboesophageal ganglia as well as neurosecretory nerve endings in the corpora cardiaca were consistently observed to be GH immunopositive (see Fig. 1). It was established that the stainig patterns caused by the human GH antibodies only partially overlap those resulting from treatment with the trout GH. The different distribution and localization of the immunoreactivities of both GH antisera in the insects investigated will be described in more detail elsewhere.

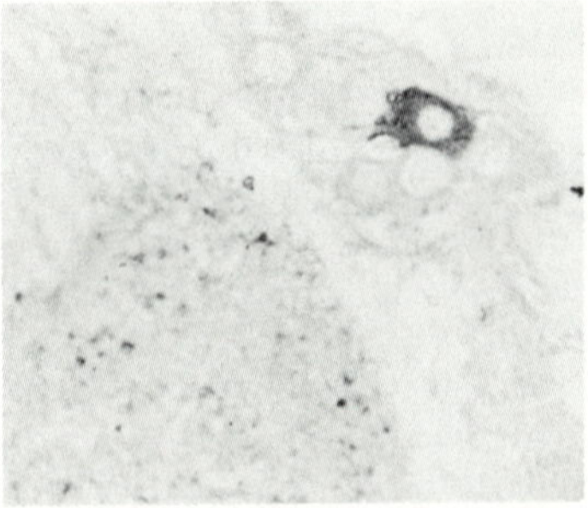

Figure 1. Immunoperoxidase labeling of a large neuronal perikaryon in the suboesophageal ganglion of *Periplaneta americana* by an antiserum to human GH (x280).

Dot Blot Hybridization

All controls gave the expected results as was the case for our immunological staining experiments. Hybridization of the probes to the blotted insect RNAs all resulted in quite obvious signals (see Fig. 2), less intense, however, than the positive controls (not on figure). It was observed that with all three probes the male RNA samples were more intensely labeled than the female RNAs. The signal was also found to be more intense for the cockroach brain RNA than for the equivalent amount of fleshfly whole body RNA (see Fig. 2).

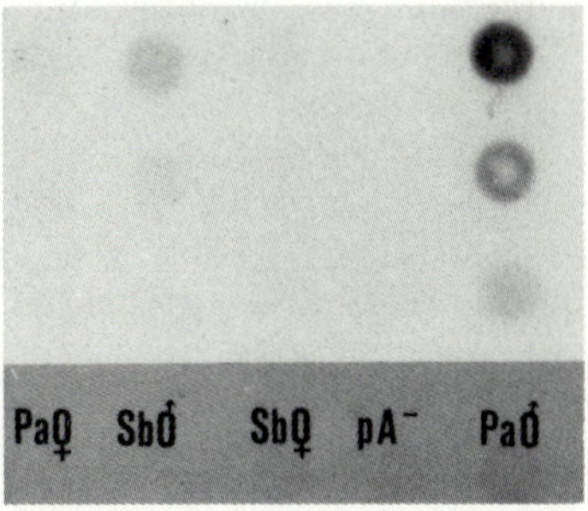

Figure 2. Insect RNA dot blots hybridized with ^{32}P labeled rat GH cDNA probe. Pa, American cockroach brain-retrocerebral complex RNA; Sb, fleshfly total body RNA; pA^-, mouse mammary gland poly(A)$^-$ RNA.

DISCUSSION

Ever since Müller's publications in the early sixties (see Steelman et al., 1971) reporting the presence of a GH-like factor in larvae of a tapeworm, only very few researchers demonstrated GH resembling substances in non vertebrates. Patton and Kuo (1977) showed that a protein with a molecular weight, isoelectric point and electrophoretic migration identical to that of bovine pituitarian GH occurs in insects of several orders and very recently a substance immunochemically related to vertebrate GH was detected in a coleopteran (Veenstra et al., 1985).

In a way the data presented by this study confirm the results of these investigations, for they likewise suggest the presence of GH-like material in some insects. The (partial) difference in localization of the structures immunoreactive to the different GH antibodies within the same species even suggests the occurrence of more than one vertebrate GH-like product. The fact that both GH antisera show different specificities may find its explanation in the polyclonal nature of the antibodies but also in the differences between the two GHs that have served as immunogen. It is indeed well known that quite obvious phylogenetic differences in structure and properties of vertebrate GHs occur (see Bentley, 1982), making it very plausible that an antiserum raised against teleost GH contains different antibody populations than a mammalian GH antiserum.

Nevertheless many endocrinologists are very sceptical about the recent immunochemical observations in invertebrates with antibodies to vertebrate peptides and they emphasize the possibility of accidental fortuitous crossreactions of these antisera with some peculiar invertebrate antigenic determinants, thus without any 'evolutionary' implications. Therefore we decided to try and confirm our immunocytochemical results by non-immunological data, wherever this would be feasible. In this context we already reported some results of a positive LH bioassay confirming LH immunohistochemical results in *Periplaneta* (Verhaert and De Loof, 1986) and in this paper data are provided presenting some evidence indicating that *Periplaneta* and *Sarcophaga* nucleic acids might contain sequences homologous to vertebrate GH cDNA probes. The differences in labeling intensity of the different RNA dots suggest that the concentration of RNA sequences showing some degree of homology to the different GH cDNAs might be higher in the brain (and male) RNA samples than in the total body (and the female) RNA samples. Whether these differences are significant rather than coincidential must be elucidated by further experiments such as more dot blot hybridization tests with newly prepared RNA samples as well as Northern blot hybridizations, which are in progress now. Finally we will try to find a better correlation between our immunohistological and hybridization results. Therefore we started *in situ* hybridizations with our (^{3}H labeled) GH cDNA probes on insect brains at present. However, the present preliminary data lead us to agree with the group of particularly invertebrate neuroendocrinologists who generally postulate the possibility of a phylogenetic relationship between invertebrate and vertebrate neuropeptides, despite their very long evolutionary separation.

As to the potential function of this (these) GH-like compound(s) in insects nothing can be concluded as yet. The distribution of the substance in the corpora cardiaca, however, one of the major insect neurohaemal organs, strongly suggests a neurohormonal role, which is very similar to the vertebrate situation. Nevertheless further characterization of the newly detected 'unknown' substances will be necessary before any physiological experiments can be done. In view of this the preparation of monoclonal antibodies against neuropeptide fractions of locust and roach corpora cardiaca is now in progress, which might lead to a relatively easy purification and isolation of the substances under investigation. On the other hand the present findings indicate that insect RNA sequences can be detected by cDNAs to mammalian GH mRNAs, immediately suggesting a complementary way of isolating the insect (m?)RNAs, which might than lead to the identification of the GH-like substance coded for.

We thank Prof. J. Van Boven for a kind supply of *Periplaneta* as well as Prof. J. Martial, whose lab most of the hybridization experiments were carried out in. The authors are very gratefully indebted to the N.F.W.O. of Belgium for financial support.

REFERENCES

Bentley P J (1982). "Comparative vertebrate endocrinology". 2nd ed., Cambridge University Press, pp 485.

Chirgwin J M, Przybyla A E, MacDonald J and Rutter W J (1979). Isolation of biologically active ribonucleic acid from sources enriched in ribonuclease. Biochemistry 18: 5294-5299.

Maniatis T, Fritsch E F and Sambrook J (1982). "Molecular cloning. A laboratory manual". Cold Spring Harbor Laboratory, New York.

Patton R L and Kuo C-C (1977). The distribution of the STH-like protein in insects. Insect Biochem 7: 487-489.

Steelman S L, Glitzer M S, Ostlind D A and Müller J F (1971). Biological properties of the growth hormone-like factor from the plerocercoid *Spirometra mansonoides*. Rec Prog Horm Res 27: 97-120.

Thomas P (1980). Hybidization of denatured RNA and small DNA fragments transferred to nitrocellulose. Proc Natl Acad Sci

USA 77: 5201-5205.

Ullrich A, Shine J, Chirgwin J, Pictet R, Tischer E, Rutter W J and Goodman H M (1977). Rat insulin genes: Construction of plasmids containing the coding sequences. Science 196: 1313-1319.

Veenstra J A, Romberg-Privee H M , Schooneveld H and Polak J M (1985). Immunocytochemical localization of peptidergic neurons and neurosecretory cells in the neuro-endocrine system of the Colorado potato beetle with antisera to vertebrate regulatory peptides. Histochemistry 82: 9-18.

Verhaert P and De Loof A (1985). Immunocytochemical localization of a methionine-enkephalin resembling neuropeptide in the central nervous system of the American cockroach, *Periplaneta americana* L. J Comp Neurol (in press).

Verhaert P and De Loof A (1986) Substances resembling peptides of the vertebrate gonadotropin system occur in the central nervous system of *Periplaneta americana* L. Immunocytological and some biological evidence. Insect Biochem (in press).

Progress in Developmental Biology, Part A, pages 365–369

CONTROL OF LENS CELL DIFFERENTIATION AND ION FLUXES BY GROWTH FACTORS

David Beebe, Karla Snellings, Michael Silver and Judson Van Wyk
Dept. of Anatomy, USUHS, Bethesda, MD 20814-4799 (D.B., K.S., M.S.) and Dept. Pediatrics, Univ. of N. Carolina, Chapel Hill, NC 27514 (J.V.W.)

INTRODUCTION

Throughout life lens fibers differentiate from the epithelial cells that cover the anterior surface of the lens. Fiber differentiation involves extensive cell elongation (Kuwabara, 1975; Piatigorsky, 1975), the synthesis and accumulation of large amounts of lens crystallins (Papaconstantinou, 1967; Zwaan and Ikeda, 1968; Piatigorsky, 1981), the cessation of DNA synthesis and cell division (Hanna, 1965, Modak, *et al.*, 1968) and the loss of most membrane-bound organelles, including the nucleus (Cohen, 1965; Kuwabara and Imaizumi, 1974). Embryonic chicken lens epithelial cells form lens fibers when cultured in medium containing serum (Philpott and Coulombre, 1965), insulin (Piatigorsky, 1973), or a factor in vitreous humor, called lentropin (Beebe, *et al.*, 1980).

GROWTH FACTORS AND HORMONES

Lentropin was unstable during purification and useable amounts of the purified factor have not been obtained. Therefore, we tested the ability of other purified growth factors and hormones to stimulate fiber cell differentiation in cultured chicken embryo lens epithelia. The insulin-like growth factors, somatomedin C (Sm-C/IGF-I) and multiplication-stimulating activity (MSA/IGF-II), were the only factors, other than insulin, to promote cell elongation and increased crystallin synthesis in these cells (Beebe, *et al.*, submitted). Sm-C/IGF-I was about thirty times more active in causing lens cell elongation than insulin or MSA/IGF-II. A monoclonal antibody, sm-1.2, which inhibits the action of human Sm-C/IGF-I (Russell, et al., 1984), blocked the ability of lentropin, serum and Sm-C/IGF-I to stimulate fiber formation from cultured embryonic lens epithelia. Sm 1.2 does not cross react with insulin (Russell, et al., 1984) and did not prevent insulin from stimulating fiber differentiation. These results suggested that lentropin is an insulin-like growth factor and that insulin causes fiber formation by cross reacting with Sm-C/IGF-I receptors on lens epithelial cells.

ION FLUXES, CELL VOLUME AND CELL ELONGATION

Lens fiber cell elongation *in vitro* and *in vivo* is associated with a proportional increase in cell volume (Beebe, et al., 1981, 1982). We have suggested that increasing volume drives fiber cell elongation. In order to establish whether the increase in volume during fiber formation was due to increased osmotic pressure from the accumulation of ions within the cytoplasm, we have developed a method for measuring ion transport into and out of single six-day-old chicken lens central epithelial cells (Beebe, et al., in preparation). The epithelium was dissected from the lens and the central region, containing cuboidal epithelial cells, was isolated (Piatigorsky, 1975). Groups of 4-6 central explants were cultured in a defined basal medium (Ham's F-10; Ham, 1963) or exposed to vitreous humor for 1 hr (Beebe and Feagans, 1981). Single control or stimulated explants were transferred to microfuge tubes and suspended in 10 µl of basal medium supplemented with 0.1 µCi of either $^{22}Na^+$ or $^{86}Rb^+$, a tracer often used to measure potassium fluxes. Tritium-labelled inulin (0.1 µCi) was also included in the labelling medium to determine extracellular space. After incubation for times ranging from 20 sec to 15 min the medium was overlayered with a silicone oil/hexane mixture having a density of 1.01. This mixture did not immediately displace the culture medium because of adhesion between the medium and the plastic tube. The tubes were then centrifuged for 1 min. This caused rapid separation of the explant, which pelleted at the bottom of the tube, and the medium, which floated at the top of the oil mixture. By comparing the isotope ratios ($^3H/^{22}Na^+$ or $^{86}Rb^+$) in the medium with those in the explant, accurate estimates of isotope uptake could be calculated.

Sodium accumulation reached steady-state within 1 min after exposure to the isotope. Steady state sodium levels were 0.677±0.077 nMoles per explant in unstimulated epithelia and 0.455±0.071 nMoles per explant in epithelia exposed to vitreous humor for 1 hr ($p<.005$; Student's t-test; Beebe, *et al.*, in preparation). Because stimulated explants begin to elongate and increase in volume within the first hour of culture (Beebe and Feagans, 1981), the lower sodium accumulation in the stimulated cells reflected an even lower cytoplasmic sodium concentration. Using volumes estimated by measuring ^3H_2O uptake in a similar manner to that used for sodium uptake (18±1 nL and 25±2 nL/explant for unstimulated and vitreous-humor treated epithelia, respectively), the intracellular sodium concentrations were calculated to be 38 and 18 mM in the unstimulated and vitreous-humor treated epithelia, respectively. Clearly, sodium accumulation was not directly responsible for the increase in cell volume.

Rubidium uptake increased continuously during incubation of explants for at least 15 min. Rubidium accumulation after 15 min was 1.00±0.458 nMoles/explant in unstimulated epithelia and 1.97±0.713 nMoles/explant in epithelia exposed to vitreous humor ($p<.005$). When these values were corrected for cell volume, the potassium concentration was estimated to be 56 and 79 mM respectively. Since it is not certain that 15 min is sufficient time for the Rb^+ uptake to reach steady state, these values are minimum estimates of the potassium concentration within the cells. The increased Rb^+ content and concentration in the elongating epithelia offset and exceeded the decrease in sodium concentration and suggests that potassium accumulation may cause the increase in cell volume seen during differentiation. Experiments are in progress to test this conclusion using EM microprobe analysis.

In these same experiments, Rb^+ accumulation after 1 min was indistinguishable in control and stimulated epithelia (0.224 and 0.240 nM/explant; $p<.30$). This suggests that the rate of potassium influx remains constant during fiber formation and that the increased accumulation seen in stimulated explants may have been due to decreased efflux of the label. This possibility was tested by incubating explants in labelled medium for 15 min, transferring them to unlabelled medium and measuring the percentage of total Rb^+ lost to the medium during a 5 min incubation. In these experiments unstimulated, control explants lost an average of 72% of their rubidium pool, whereas stimulated epithelia lost only 39%. These data support the possibility that, when lens epithelial cells are stimulated to differentiate, decreased potassium efflux and continued potassium influx result in increased potassium accumulation, increased intracellular osmolarity, increased cell volume and cell elongation. This conclusion is being tested by determining the rate constants for potassium influx and efflux and by patch clamp assay of changes in membrane channel conductivity and selectivity during fiber differentiation _in vitro_.

SUMMARY

The results of these studies suggest that an insulin-like growth factor, lentropin, present in the vitreous humor of the eye, is responsible for stimulating lens epithelial cells to differentiate into lens fiber cells during normal lens growth. Lentropin has been shown to stimulate embryonic lens epithelial cells to elongate, specialize for the synthesis of delta crystallin, the major protein synthesized in chicken embryo fiber cells, and to cease DNA synthesis and mitosis. Lentropin appears to decrease the permeability of the plasma membrane to potassium and, by this mechanism, may cause lens fiber cell elongation.

ACKNOWLEDGEMENTS

Supported by NIH Grant EY04853. The opinions or assertions contained herein are the private ones of the authors and are not to be construed as official or reflecting the views of the DoD or the USUHS. Experiments were conducted according to the principles set forth in the _Guide for Care and Use of Laboratory Animals_, ILAR, Nat'l Research Council, DHEW Pub.# (NIH) 78-23.

REFERENCES

Beebe DC, Feagans DE (1981). A tissue culture system for studying lens cell differentiation. Vision Res 21:113

Beebe DC, Feagans DE, Jebens HAH (1980). Lentropin: a factor in vitreous humor which promotes lens fiber differentiation. Proc Nat'l Acad Sci USA 77:490

Beebe DC, Compart PJ, Johnson MC, Feagans DE, Feinberg RN (1982). The mechanism of cell elongation during lens fiber cell differentiation. Develop Biol 92:54.

Beebe DC, Johnson MC, Feagans DE, Compart PJ (1981). The mechanism of cell elongation during lens fiber cell differentiation. In Hilfer SR, Sheffield J (eds): "Ocular Size and Shape During Development," New York: Springer, p. 79.

Cohen AI (1965). The electron microscopy of the normal human lens. Invest Ophthalmol 4:433

Ham RG (1963). An improved nutrient solution for diploid Chinese hamster and human cell lines. Exp Cell Res 29:515.

Hanna C (1965). Changes in DNA, RNA, and protein synthesis in the developing lens. Invest Ophthalmol 4:480

Kuwabara T (1975). The maturation of the lens cell: a morphological study. Exp Eye Res 20:427.

Kuwabara T, Imaizumi M (1974). Denucleation process of the lens. Invest Ophthalmol 13:973.

Modak SP, Morris G, Yamada T (1968). DNA synthesis and mitotic activity during early development of chick lens. Develop Biol 17:544.

Papaconstantinou J (1967). Molecular aspects of lens cell differentiation. Science 156:338.

Philpott GW, Coulombre AJ (1965). Lens development. II. The differentiation of embryonic chick lens epithelial cells in vivo and in vitro. Exp Cell Res 38:635.

Piatigorsky J (1973). Insulin initiation of lens fiber differentiation in culture: elongation of embryonic lens epithelial cells. Develop Biol 30:214.

Piatigorsky J (1975). Lens cell elongation in vitro and microtubules. Ann NY Acad Sci 253:333.

Piatigorsky J (1981). Lens differentiation in vertebrates. A review of cellular and molecular features. Differentiation 19:134.

Russell WE, Van Wyk JJ, Pledger WJ (1984). Inhibition of the mitogenic effects of plasma by a monoclonal antibody to somatomedin C. Proc Nat'l Acad Sci USA 81:2389.

Zwaan J, Ikeda A (1968). Macromolecular events during differentiation of the chicken lens. Exp Eye Res 7:301.

Progress in Developmental Biology, Part A, pages 371–374

GLUCOCORTICOID RECEPTORS IN DEVELOPING MOUSE VISCERAL YOLK SAC AND LIVER

Glen K. Andrews

Department of Biochemistry, University of Kansas Medical Center, Kansas City, Kansas 66103

INTRODUCTION

Glucocorticoid receptors are present in a number of embryonic and fetal tissues including lung (Giannopoulous, 1980), neural retina (Lippman et al., 1974) and secondary palate (Salomon and Pratt, 1979) and glucocorticoids have been shown to effect differention of these tissues. We are interested in possible glucocorticoid effects on development of mouse visceral yolk sac (VYS) and fetal liver. In this study we concentrated on the analysis of glucocorticoid receptors in these developing tissues.

Glucocorticoid responsive target tissues can be distinguished from non-target tissues by virtue of the presence of specific cytosolic receptors which bind glucocorticoids with high-affinity and limited capacity (Munck and Leung, 1977). Cytosolic glucocorticoid receptor complexes are thought to go through a temperature dependent "activation" process which results in their nuclear binding due to increased affinity of specific DNA sequences. Nuclear binding of hormone-receptor complexes in turn results in altered gene transcription (Schmidt and Litwack, 1982; Groner et al., 1984).

RESULTS AND DISCUSSION

Specific triamcinolone acetonide (TA) binding sites in developing VYS and liver cytosol were determined after a 4 h incubation at 4°C with 30 nM TA in molybdate containing buffer (Fig.1). The number of TA binding sites/cell in VYS increased dramatically between day 10 and 14 of gestation, after which it leveled off (Fig. 1A).

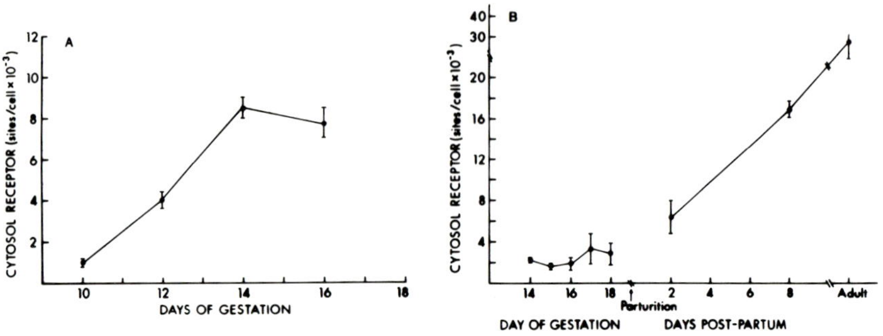

Figure 1. Changes in cytosolic triamcinolone acetonide binding capacity during visceral yolk sac (A) and liver (B) development

The level of receptor in day 10 VYS was near the limit of detection of the assay employed (1 x 10^3 sites/cell). Changes in cytosolic TA binding sites in liver (Fig. 1B) showed a different developmental pattern than those in VYS. The number of receptor sites/cell in fetal liver were low throughout gestation, but increased rapidly after birth, reaching a level of about 1.7 X 10^4 sites/cell by day 9 post-partum. The specific binding capacity of various mixtures of fetal and adult liver cytosol, made before incubation with TA, was predictably additive (not shown) indicating that fetal liver cytosol did not contain an "inhibitor" of TA binding (i.e., free corticosterone or protease).

Scatchard analysis of binding data indicated the presence of a single class of high affinity binding sites in VYS, fetal and neonatal liver (Fig.2). Similar results were obtained with adult liver cytosol (Kd = 3.1 nM; not shown). In each case, the apparent dissociation constant (K_d) for the hormone receptor interaction at 4°C ranged between 2 nM and 4 nM.

Bound (^{3}H)-TA in cytosol from fetal liver and VYS was further characterized by DEAE-Sephadex A-50 column chromatography (Fig. 3). Bound (^{3}H)-TA in cytosol maintained at 4°C ("unactivated") eluted as two peaks during chromatography (Fig. 3A, C). The initial peak in the prewash fractions (0.02 M potassium phosphate buffer) represented non-specifically bound TA as this binding was not displaced by a 100-fold excess of unlabeled TA (Fig. 3C). The second peak which eluted at approximately

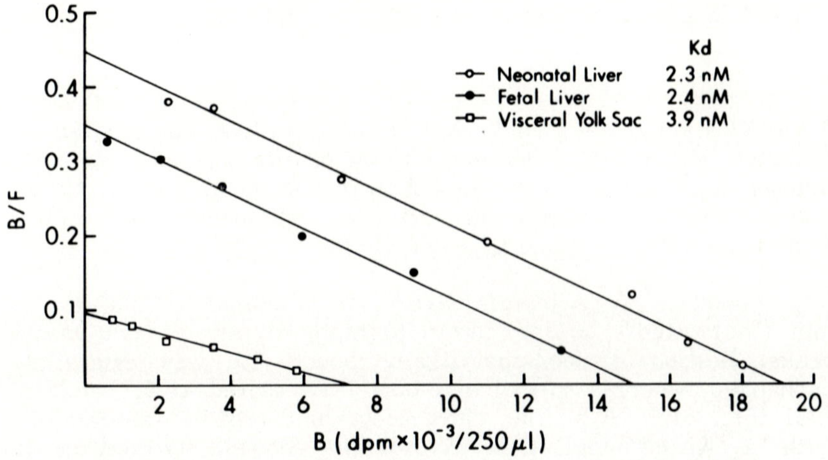

Figure 2. Scatchard analysis of triancinolone acetonide binding to cytosol.

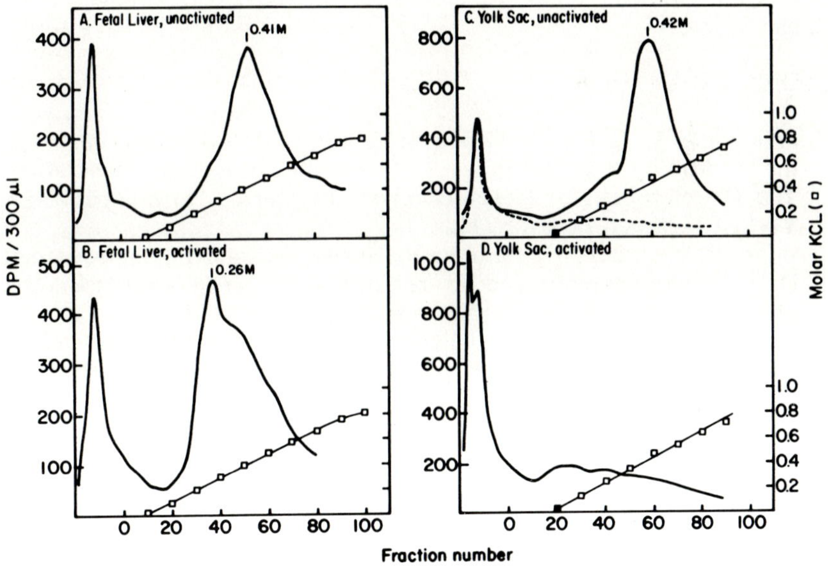

Figure 3. DEAE-Sephadex A-50 chromatogram of bound

(^{3}H)-TA in fetal liver and visceral yolk sac cytosol.

0.4 M KCl in cytosol from both VYS and fetal liver represented TA receptor complexes. A similar elution profile was obtained with cytosol from adult liver (not shown) and is characteristic of "unactivated", glucocorticoid receptor complexes from several adult tissues (Schmidt and Litwack, (1982).

Bound (^{3}H)-TA in fetal liver cytosol heated to 25°C for 30 min ("activated") before chromatography eluted as two major peaks. Elution of non-specifically bound TA was essentially unchanged, whereas specifically bound TA eluted at 0.22-0.26 M KCl (Fig. 3B). A similar chromatographic profile was obtained with "activated" adult liver cytosol (not shown). In contrast to results obtained with liver cytosols, the bound (^{3}H)-TA in heat "activated" cytosol from VYS consistently eluted in the prewash fractions (Fig. 3D). These results indicate that the VYS and fetal liver are target tissues for glucocorticoid hormones. Furthermore, the evidence suggests that quantitative but not qualitative changes in glucocorticoid receptors occur duing liver development and that the VYS glucocorticoid receptors differ physiochemically from those in liver.

REFERENCES

Giannopoulous, G (1980) Glucocorticoids and fetal lung development. In Kay AM and Kaye M (eds.): "Advances in Bioscience: The Development of Responsiveness to Steroid Hormones," New York: Pergamon Press, p241.

Groner, B, Kennedy, N, Skroch, P, Hynes, NE and Porta, H (1984) DNA sequences involved in regulation of gene expression by glucocorticoid hormones. Biochem Biophys Acta 781:1.

Lippman, M, Wiggert, B, Chader, G and Thompson, E (1974) Glucocorticoid receptor characteristics, specificity, and ontogenesis in the embryonic chick neural retina. J Biol Chem 219:5916.

Munck, A and Leung K (1977) Glucocorticoid receptors and mechanisms of action. In Pasqualin: JR (ed): "Receptors and Mechanisms of Action of Steroid Hormones," New York: Dekker, p311.

Salomon, DS and Pratt RM (1979) Involvement of glucocorticoids in the development of the secondary palate. Differentiation 13:141.

Schmidt TJ and Litwack G (1982) Activation of the glucocorticoid receptor complex. Physiol Rev 62:1131.

Progress in Developmental Biology, Part A, pages 375–379

RESPONSIVENESS OF DENTAL EPITHELIAL AND MESENCHYMAL CELLS TO EPIDERMAL GROWTH FACTOR

Anna-Maija Partanen and Irma Thesleff

Institute of Dentistry, University of Helsinki, SF-00280 Helsinki, Finland

INTRODUCTION

During tooth development the multistep process of morphogenesis, dental cell differentiation and secretion of mineralizing matrices of dentin and enamel take place in a specific temporal and spatial pattern, which makes the developing tooth a good model system for studies on embryonic organogenesis and cell differentiation. Several studies have confirmed that tooth morphogenesis and differentiation of the mesenchymal odontoblasts and epithelial ameloblasts are tightly regulated by epithelio-mesenchymal tissue interactions (see: Thesleff and Hurmerinta, 1981), and the role of extracellular matrix molecules in these processes has been emphasized. Our recent studies have shown that also circulating molecules can affect tooth development, but only at the early stages of development. Transferrin, the iron-carrying serum protein, is a necessary growth factor for the development of bud-staged mouse embryonic tooth germ in organ culture, and it is the only growth factor required for the morphogenesis and differentiation of tooth at early cap stage of development (Partanen et al., 1984).

The transferrin-containing chemically defined culture medium that supports early tooth morphogenesis made it possible for us to study the effects of other growth factors on morphogenesis, proliferation and differentiation of mouse embryonic tooth in controlled conditions. Epidermal growth factor (EGF) which is a potent mitogen for both epithelial and mesenchymal cells in monolayer cultures (Carpenter and Cohen, 1979) has been shown to stimulate cell proliferation also during embryonic development of lung and palate (Sundell et al., 1975; Hassell, 1975). In our studies EGF also stimulated the proliferation of tooth explants during culture, but inhibited early morphogenesis and differentiation (Partanen et al., 1985). Epidermal growth factor exerts its biological effects by binding to a specific cell surface receptor, a membrane glycoprotein with intrinsic tyrosine-kinase activity (Schlessinger et al., 1983). Since EGF and functional EGF-receptors are found in mouse embryonic tissues from 10-11 days of gestation onwards (Nexo et al., 1980; Hortsch et al., 1983; Adamson and Meek, 1984), EGF or EGF-like molecules probably have a physiological role in embryonic development. We have studied the binding of EGF in tooth germs of 14 day old mouse embryos to find out if the response of dental tissues to EGF is related to the distribution of EGF-receptors in the tooth organ.

MATERIALS AND METHODS

Organ Culture

Mandibular first molars from 14 days old mouse embryos were dissected free from surrounding tissues under a stereo microscope. The explants were cultured in a Trowell-type organ culture, where they were supported by a Nuclepore filter on a metal grid. The chemically defined culture medium was Improved Minimum Essential Medium , I-MEM supplemented with 50 ug/ml of transferrin. Glutamine (4 mM) was added to the medium prior to culture and ascorbic acid (0.9 mM) after 2 days of culture.

Epidermal growth factor was from Collaborative Research. For light microscopy the explants were fixed in Zenker´s solution, embedded in paraffin, serially sectioned at 5 um and stained with hematoxylin and eosin.

125 I-EGF Binding Assay

The freshly dissected day 14 first mandibular molar teeth were incubated with 5 ng/ml of 125 I-EGF (from Collaborative Research) in MEM containing 20 mM Hepes and 0.1 % bovine serum albumin. To control the specificity of 125 I-EGF binding the control teeth were incubated with 5 ug/ml of unlabelled EGF added to the 125 I-EGF-containing binding medium. After several washes the teeth were fixed with 3 % glutaraldehyde and prepared for histology as described above. For autoradiography the deparaffinized tissue sections were dipped in Ilford's Nuclear Emulsion (G5). After 5 weeks' exposure they were developed, fixed and stained with hematoxylin and eosin.

RESULTS

The results are presented in figures 1 and 2.

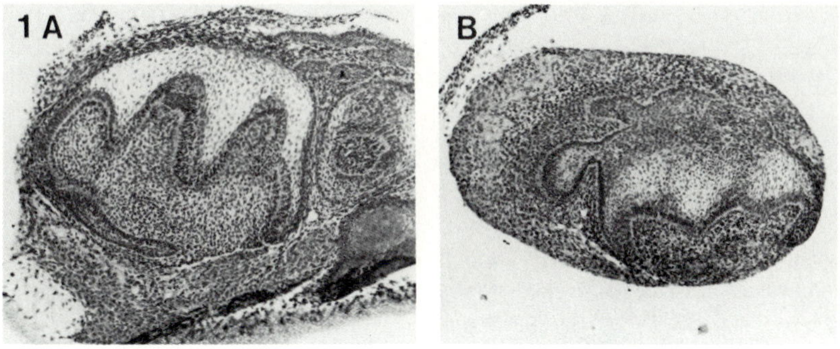

Figure 1. When day 14 first mandibular molar teeth which were at the early cap satge of development at the onset of culture (see Fig. 2) were grown in I-MEM supplemented with transferrin (50 ug/ml) for 6 days, they had undergone cuspal morphogenesis (A). The presence of epidermal growth factor (20 ng/ml) in the culture medium inhibited tooth morphogenesis, but the proliferation of the surrounding non-dental me-

senchymal tissue had been stimulated (B).

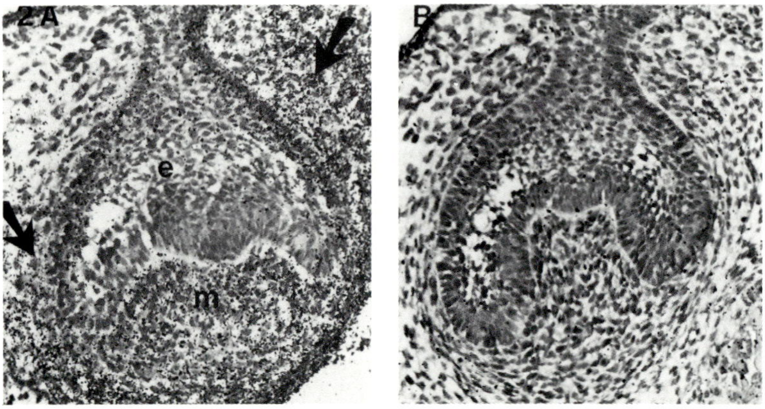

Figure 2. Autoradiographs of freshly dissected day 14 first mandibular molar tooth germs incubated with 125 I-EGF (5 ng/ml) A. Large numbers of EGF-receptors were found in the nondental mesenchyme surrounding the tooth germ (arrows) while there were significantly less EGF-receptors in the tooth. In tooth epithelium (e) only the outer epithelial cell layer bound EGF, but there was some EGF binding all over the dental mesenchyme (m). B. In the presence of 1000-fold excess unlabelled EGF only slight background labelling was seen which confirmed the specificity of EGF binding.

DISCUSSION

The inhibition of tooth morphogenesis caused by EGF probably resulted from inhibited proliferation of the dental mesenchyme (Partanen et al., 1985). At the same time EGF stimulated cell proliferation in the dental epithelium. The surrounding non-dental mesenchyme also responded to EGF by stimulation of cell proliferation. This tissue was now shown to bind EGF abundantly. The distribution and amount of EGF binding in the dental epithelium and dental mesenchyme were, however, not similarly correlated to stimulated cell proliferation. We suggest that epithelio-mesenchymal tissue interactions control the responsiveness of dental epithelial and mesenchymal cells to EGF.

REFERENCES

Adamson E D, Meek J (1984). The ontogeny of epidermal growth factor receptors during mouse development. Dev Biol 103: 62-70.

Carpenter G, Cohen S (1979). Epidermal growth factor. Ann Rev Biochem 48: 193-216.

Hassell J R (1975). The development of rat palatal shelves in vitro and ultrastructural analysis of the inhibition of epithelial cell death and palate fusion by the epidermal growth factor. Dev Biol 45: 90-102.

Hortsch M, Schlessinger J, Gootwine E, Webb C G (1983). Appearance of functional EGF receptor kinase during rodent embryogenesis. EMBO J 2: 1937-1941.

Nexo E, Hollenberg M D, Figueroa A, Pratt R M (1980). Detection of epidermal growth factor-urogastrone and its receptor during fetal mouse development. Proc Nat Acad Sci USA 77: 2782-2785.

Partanen A M, Thesleff I, Ekblom P (1984). Transferrin is required for early tooth morphogenesis. Differentiation 27: 59-66.

Partanen A M, Ekblom P, Thesleff I (1985). Epidermal growth factor inhibits morphogenesis and cell differentiation in cultured mouse embryonic teeth. Dev Biol 101 (in press).

Schlessinger J, Scheiber A B, Levi A, Lax I, Liberman T, Yarden Y (1983). Regulation of cell proliferation by epidermal growth factor. Brit Rev Biochem 14: 93-111.

Sundell H, Serenius F, Barthe T, Friedman C, Kanarek K, Escobedo M, Orth D M, Stahlman M T (1975). Effect of epidermal growth factor on fetal lamb lung maturation. Pediatr Res 9: 371-376.

Thesleff I, Hurmerinta K (1981). Tissue interactions in tooth development. Differentiation 18: 75-88.

Progress in Developmental Biology, Part A, pages 381–384

EMBRYONIC MOUSE LUNG MORPHOGENESIS IN SERUMLESS, CHEMICALLY-DEFINED MEDIUM *IN VITRO*

Tina F. Jaskoll, Randall Johnson,
Grace Don, and Harold C. Slavkin

Laboratory for Developmental Biology, School of Dentistry, University of Southern California
Los Angeles, California 90089-0191

INTRODUCTION

Numerous investigations have been conducted concerning which factors might influence lung morphogenesis and development (see review Ballard, 1982; Kikkawa et al, 1971; Smith and Sabry, 1983). Epithelial-mesenchymal interactions appear to influence specific epithelial branching patterns (Spooner and Wessells, 1970; Masters, 1976) and subsequent epithelial cell differentiation (Smith and Sabry, 1983; Jaskoll and Slavkin, 1984). Hormones, for example glucocorticosteroids and thyroxin, can stimulate and accelerate lung development (see review Ballard, 1982; Kikkawa et al, 1971; Honig et al, 1984). Cortisol stimulates fetal lung fibroblasts to release a factor which enhances alveolar type II cell differentiation and the subsequent synthesis and secretion of pulmonary surfactant (Smith, 1979). Glucocorticoid and thyroxin may act synergistically to promote fibroblast-epithelial cell interactions in the fetal lung (Smith and Sabry, 1983). Due to the complex nature of inductive events during lung organogenesis, a complete understanding of these developmental events has eluded researchers.

Serumless, chemically-defined culture techniques provide a possible experimental approach for simplifying the developmental events in this complex organ system. This paper describes the *in vitro* morphogenesis and differentiation of the embryonic *mouse lung* in serumless, chemically-defined medium. In the absence of exogenous serum or humoral factors, endodermally-derived lung epithelia differentiated into alveolar type II cells.

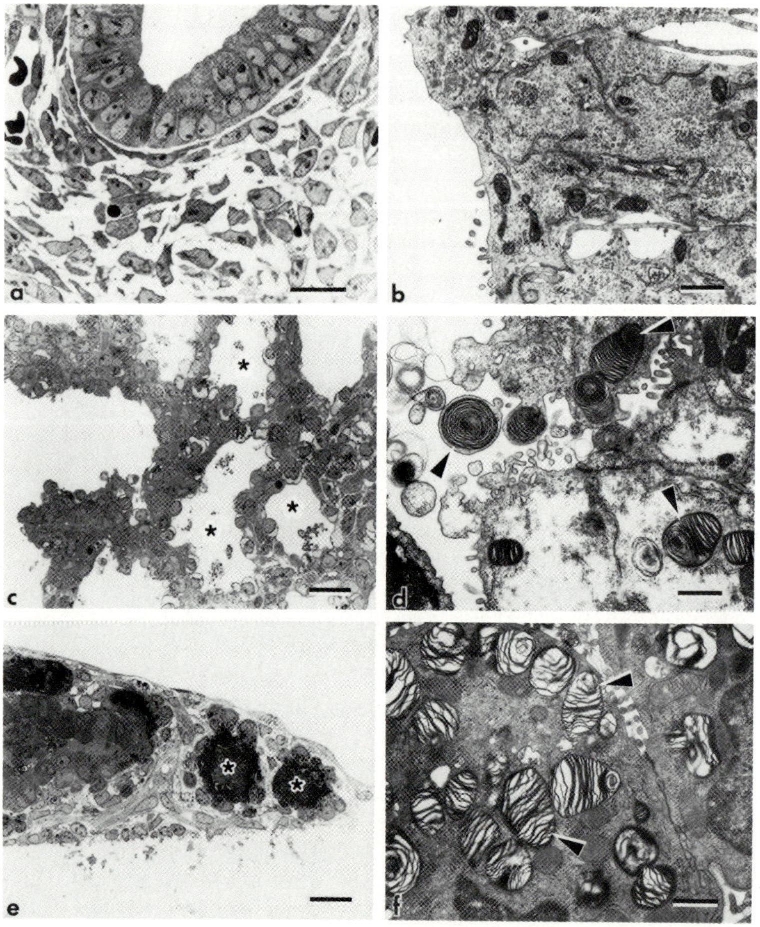

Fig.1 Mouse lung development. (a) Section (1μm) of Theiler stage 20 (12 days gestation) embryonic lung primordia stained with toluidine blue (bar=20μm). (b) Transmission electron micrograph (TEM) as in (a) shows the ultrastructure of the undifferentiated epithelium (bar=1μm). (c) Section (1μm) of fetal lung in situ (Theiler stage 26; 18 days gestation) stained with toluidine blue. Air sacs (*) are associated with extensive epithelial morphogenesis (bar=20μm). (d) TEM as in (c). Alveolar type II cells can be identified by glycogen deposits and multiple lamellar bodies (arrows) within the cytoplasm. Note the presence of lamellar bodies in the lumen of the air sacs (bar=1μm). (e) Section (1μm) of lung explant cultured for 27 days in vitro stained with toluidine blue. Epithelial morphogenesis and air sac (*) formation are delayed relative to the in situ lung (bar=20μm). (f) TEM as in (e). Type II cells can be identified by the presence of multiple lamellar bodies (arrows)(bar=1μm).

Development of Lung Primordia in Organ Culture

Embryonic mouse lung primordia (Fig 1), Theiler stage 20 (Theiler, 1972),have been cultured in serumless, chemically-defined medium for up to 27 days in vitro. Using a modified Trowell method (Trowell, 1954; Yamada et al, 1980), lung primordia were placed on millipore (Bedford, MA) AABP filter discs which were then placed upon stainless steel grids over 70µl medium within Grobstein Falcon Dishes (Becton-Dickinson, Oxnard, CA) under optimal humidity conditions. The medium consisted of BGJ_B (GIBCO, Grand Island, NY) supplemented with 0.2mg Ascorbic Acid/ml and 50u Pen/Strep (Flow Laboratories, McLean, VA),pH 7.2, and was changed every two days. The explants were evaluated by morphology, biochemistry, and immunochemistry.

Cytologically, alveolar type II cells can be identified with ultrastructure by the presence of multiple and discrete lamellar bodies and glycogen deposits (Fig. 1d). Lamellar bodies are released into the lumen of the air sacs (Fig. 1d) and represent organelles containing pulmonary surfactant (Mason et al, 1977). In the absence of serum or hormonal supplementation, morphogenesis and cytodifferentiation were observed (Figs. 1e and f). Type II cells were identified by the presence of multiple lamellar bodies within the cytoplasm (Fig. 1f); Clara cells and ciliated cells were also observed (data not shown). Preliminary investigations using Western transfer techniques and indirect immunofluorescent microscopy demonstrated the synthesis and secretion of pulmonary surfactant proteins after 24 days in vitro (unpublished data). The rate of expression of the type II cell phenotype is significantly slower in vitro than in situ (17 days in situ vs. 36 days in vitro). These results indicate that the developmental program for type II cell cytodifferentiation and pulmonary surfactant production is not under hormonal control.

CONCLUSION

Exogenous factors may play an important role during morphogenesis and differentiation of embryonic organ systems. The embryonic mouse lung is capable of expressing its sophisticated developmental program in serumless, chemically-defined culture conditions. The developmental program for alveolar type II cell phenotype appears to be intrinsic in the developing lung organ and its expression is not dependent on extrinsic serum or humoral factors.

ACKNOWLEDGEMENT

We wish to thank Mr. Scott Trowbridge for help in preparing this manuscript. These studies were supported in part by research grant HL-28325 and training grant DE-07006, National Institute of Health.

REFERENCES

Honig LS, Smith BT, Slavkin HC, Donahue HG (1984). Influence of the major histocompatibility complex (H2) on glucocorticoid-stimulated pulmonary surfactant synthesis in two congenic mouse strains. Proc Soc Exp Biol Med 176:419-425.

Jaskoll TF, Slavkin HC (1984). Ultrastructural and immunofluorescence studies of basal lamina alterations during mouse lung morphogenesis. Differentiation 28:36-48.

Kikkawa Y, Kaibara M, Motoyama EK, Orazalesi MM, Cook CD (1971). Morphologic development of fetal rabbit lung and its acceleration with cortisol. Amer J Pathol 64:423-442.

Mason RJ, Dobbs LG, Greenleaf RD, Williams MC (1977). Alveolar type II cells. Fed Proc 36:2697-2702

Masters JRW (1976). Epithelial-mesenchymal interaction during lung development: The effect of mesenchymal mass. Dev Biol 51:98-108.

Smith BT (19/9). Lung maturation in the fetal rat: Acceleration by the injection of fibroblast-pneumocyte factor. Science 204:1094-1095.

Smith BT, Sabry K (1983). Glucocorticoid-thyroid synergism in lung maturation: A mechanism involving epithelial-mesenchymal interactions. Proc Natl Acad Sci USA 80:1951-1954.

Spooner BS, Wessells NK (1970). Mammalian lung development: Interactions in primordium formation and branchial morphogenesis. J Exp Zool 175:445-454.

Theiler K (1972). "The House Mouse." New York: Springer-Verlag.

Trowell OA (1954). A modified technique for organ culture in vitro. Exp Cell Res 6:246-258.

Yamada M, Bringas P, Grodin M, MacDougall M, Slavkin HC (1980). Chemically-defined organ culture of embryonic mouse tooth organs: Morphogenesis, dentinogenesis and amelogenesis. J Biol Buccale 8:127-139.

Progress in Developmental Biology, Part A, pages 385–388

ONTOGENY OF A NERVE GROWTH FACTOR-LIKE SUBSTANCE IN CULTURED MOUSE ATRIUM

B B Rawdon

Department of General Anatomy, School of Dentistry, University of the Witwatersrand, Johannesburg 2001, South Africa.

INTRODUCTION

In organ culture, neurites from sympathetic ganglia from neonatal mice extend preferentially towards cardiac or smooth muscle of the same age; in the case of cardiac muscle extension appears to be under the control of nerve growth factor (NGF), but in smooth muscle NGF and another factor, immunochemically distinct from NGF, are involved (Rawdon and Dockray, 1983). The object of the present study is to extend the observations made on heart of neonatal mice to embryonic stages.

MATERIALS AND METHODS

A Swiss-Webster strain of mouse was used for this experiment. Pieces of atrium (0.6 x 0.4 mm) were dissected from embryos aged 10, 11, 14 and 17 days (E10, E11, E14 and E17) and from neonates: small pieces of superior cervical and of coeliac ganglion were obtained from neonates. Explants were set in 1mℓ of a nutrient collagen gel (see Rawdon and Dockray, 1983) in a 35mm culture dish with the two ganglia positioned on opposite sides 1mm from a centrally placed piece of atrium. Pairs of identical cultures, in which the explants were provided by the same embryos and neonates, were prepared; to one of these (control) 10μℓ of normal rabbit serum (NRS) was added, to the other (experimental) 10μℓ of anti-NGF (Collaborative Research). The ability of the anti-NGF to block the effect of exogenous NGF was tested on both types of ganglia: these cultures received 5, 10 or 20ng/mℓ 2.5S NGF (Collaborative Research) and 10µℓ NRS (control) or 10µℓ anti-NGF (experimental). Cultures were incubated in air at 37°C for 48 hours and fixed in 10% formalin.

The outgrowth of neurites from ganglia was assessed in two ways: (i) the length of the 10 longest neurites on the side of the ganglion facing towards the heart explant (explant (E) side) and on the side facing away from the explant (non-explant (NE) side) was determined and the mean obtained; (ii) the number of neurites exceeding 0.5mm in length was counted on the E and NE sides of ganglia. Data were analysed by means of Student's t-tests for paired and unpaired samples.

RESULTS

Addition of NGF at 5, 10 or 20 ng/mℓ to cultures of both types of ganglia stimulated outgrowth of a dense halo of neurites from the explants; this was totally blocked by anti-NGF. In control cultures of superior cervical and coeliac ganglia and atrium from embryonic or neonatal mice the mean length of neurites on the E side always exceeded that on the NE side (Fig. 1); with increase in age of atrial targets from E10 through to E17 there was a gradual increase in the mean length of neurites on the E side; this was paralleled by an increase in the number of fibres exceeding 0.5mm in length (Fig. 1; Table 1). Growth of neurites towards the older atrial targets was blocked by anti-NGF, but with the younger targets there was a tendency for preferential growth of neurites towards the explant to persist in its presence (Fig. 1; Table 1).

DISCUSSION

This experiment has demonstrated that with advance in embryonic age of atrium there is an increase in both length and number of neurites extending towards heart explants; this is totally blocked, or markedly reduced by antiserum to NGF. This indicates that in culture, mouse atrium produces NGF and that with increase in embryonic age, production rises.

In cultures uitilizing atrium from younger embryos, there is evidence that the extension of neurites towards the target is not fully suppressed by anti-NGF. As this antiserum was shown to block the effects of addition of 20 ng/mℓ NGF to control cultures (a level unlikely to be exceeded; see Barth,

Table 1. Co-culture of neonatal superior cervical ganglion with embryonic or neonatal atrium. Mean number of neurites (±SEM) crossing a hairline positioned 0.5mm from the ganglion on the E side. With few exceptions, neurites on the NE side did not attain 0.5mm in length.

Group	Neurite number ± SEM				
	E10	E11	E14	E17	neonate
Control	4.7 ± 2.0	10.2 ± 2.0	14.3 ± 3.6	23.0 ± 3.3	16.0 ± 5.5
Anti-NGF	0.2 ± 0.2	3.0 ± 1.4	0.0	0.0	0.0

Figure 1. Mean neurite length (±SEM) on the E and NE sides of neonatal superior cervical ganglion co-cultured with embryonic or neonatal atrium in control (light screen) and in anti-NGF treated (dark screen) cultures.

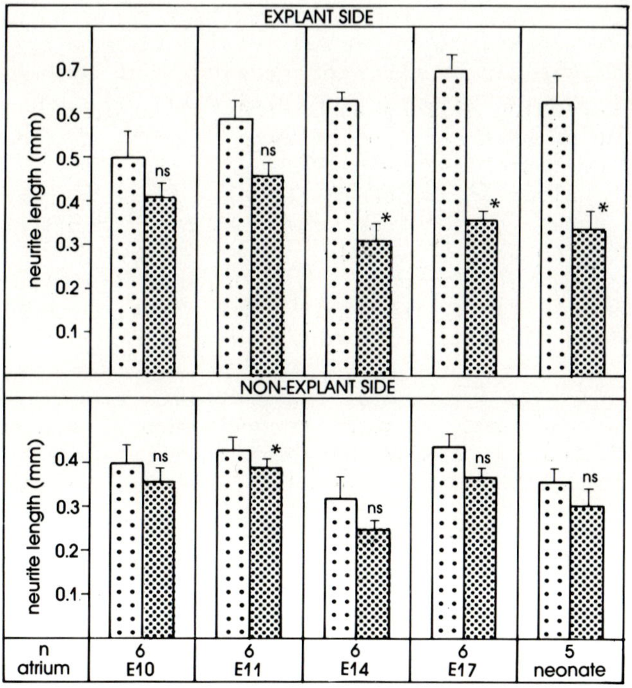

*Upper panel: significantly different from the E side of control cultures ($P < 0.05$).

*Lower panel: significantly different from the E side of anti-NGF treated cultures ($P < 0.05$). ns not significant.

Korsching and Thoenen, 1984), it seems probable that embryonic atrium produces a small amount of another neuronotrophic factor as well as NGF. Indeed, there is a considerable body of evidence supporting the view that in vitro vertebrate heart muscle produces such factors; for instance, extracts of embryonic chick heart and murine heart conditioned medium both support neurite growth in the presence of anti-NGF (Ebendal et al, 1979; Coughlin, Bloom and Black, 1981). While heart conditioned medium appears to contain large amounts of a factor distinct from NGF, the present results show that in explants of mouse atrium NGF production predominates and other factors are found only in small amount.

The support of the South African Medical Research Council and the Council Research Committee of the University of the Witwatersrand, Johannesburg is gratefully acknowledged.

REFERENCES

Barth EM, Korsching S, Thoenen H (1984). Regulation of nerve growth factor synthesis and release in organ cultures of rat iris. J Cell Biol 99:839-844.

Coughlin MD, Bloom EM, Black IB (1981). Characterization of a neuronal growth factor from mouse heart-cell-conditioned medium. Develop Biol 82:56-68.

Ebendal T, Belew M, Jacobson C-O, Porath J (1979). Neurite outgrowth elicited by embryonic chick heart: partial purification of the active factor. Neuroscience Letters 14: 91-95.

Rawdon BB, Dockray GJ (1983). Directional growth of sympathetic nerve fibres in vitro towards enteric smooth muscle and heart from mice with congenital aganglionic colon and their normal littermates. Develop Brain Res 7:53-59.

Progress in Developmental Biology, Part A, pages 389–392

CHARACTERIZATION OF A GROWTH FACTOR SECRETED BY RAT SERTOLI CELLS IN CULTURE

Stephen D. Holmes, Gerald Spotts and Roy G. Smith

Dept. of Urology, Baylor College of Medicine

Houston, TX 77030

INTRODUCTION

The endocrine control of spermatogenesis is a complex process requiring the presence of FSH and high intratesticular levels of testosterone. In the absence of these hormones spermatogenesis does not proceed beyond meiotic prophase (Steinberger, 1971; Weissenberg et al., 1982). There is no definite evidence demonstrating that these hormones act directly on the germ cells rather that they have some indirect effect. Germ cells in the early portion of spermatogenesis have direct access to regulatory factors in the extratubular milieu, any regulation of the germ cells in the latter half of spermatogenesis must be mediated indirectly through the Sertoli cells (Steinberger, 1971). This study examines the preliminary characterization of a mitogen secreted by rat Sertoli cells in culture that may be involved in the regulation of spermatogenesis.

RESULTS AND DISCUSSION

It has been demonstrated that Sertoli cell conditioned medium contains a mitogen, Sertoli cell secreted growth factor (SCSGF) that is capable of stimulating proliferation of epithelial, fibroblast and smooth muscle cells in culture (Holmes et al., 1985). Transferrin, another known growth factor secreted by Sertoli cells could not mimic this effect (Holmes et al., 1985). The secretion of this mitogenic activity was increased by insulin, testosterone, retinol and FSH while LH or TSH had no effect (Table 1; Holmes et al., 1985). The secretion of SCSGF is significantly stimulated by FSH in a dose dependent manner from 5 to 200 ng/ml (Table 1). The secretion of the mitogen parallels the secretion of transferrin while androgen binding protein (ABP) does not follow this pattern (Table 1). Holmes et al. (1983) reported that transferrin binds specifically to pachytene spermatocytes and the SCSGF may act in conjunction with transferrin to promote differentiation of this germ cell type.

Table 1. Effect of FSH on the secretion of SCSGF, transferrin and androgen binding protein (ABP) from rat Sertoli cells. Sertoli cells from 18-20 day old rats were isolated by the method of Tung et al. (1984). The cells were incubated for 2 days in serum free Dulbecco's modified Eagles medium (DME) and then incubated for a further 3 days in DME with insulin (5 ug/ml) with or without FSH. An 0.5 ml aliquot of Sertoli cell conditioned media with 1.0 ml DME was added to A431 cells (8K) and incubated for 5d with a media change on day 3. Control cells consisted of DME with insulin and FSH (40 ± 2 K). Values are mean ± SD. Transferrin and ABP were measured by RIA.

	SCSGF Cell number x 10^3	Transferrin ng/d/10^6cells	ABP fmoles/d/10^6cells
Control	120 ± 5	500 ± 5	207 ± 3
FSH, 5 ng/ml	137 ± 2	518 ± 4	275 ± 10
FSH, 20 ng/ml	151 ± 2	579 ± 11	318 ± 1
FSH, 200 ng/ml	177 ± 6	683 ± 18	301 ± 5
FSH, 500 ng/ml	182 ± 5	715 ± 56	245 ± 11

The ability of conditioned medium from Sertoli cells to displace ^{125}I-EGF binding to its receptor was also investigated. Using Sertoli cell conditioned medium at a final concentration of 20% to 90% produced a slope that was not significantly different from EGF. Sertoli cells incubated with insulin contained 1.28 ng equivalents of EGF/ml while FSH (200 ng/ml) and GH (100 ng/ml) in the presence of insulin further increased the secretion of the EGF competing activity to 1.86 and 2.40 ng equivalents EGF/ml respectively. The amount of EGF competing activity was also positively correlated with mitogenic activity. Although EGF displacement activity was present in Sertoli cell conditioned medium the medium was negative for any

EGF immunoreactivity (Holmes et al., 1985).

The mitogenic and EGF competing activities of Sertoli cell conditioned medium also have other parallel properties: heat sensitive (10-20% activity remaining after 10 min at 100C); unaffected by reducing agents (50 mM dithiothreitol, 18h at 4C); and inactivated by trypsin. The SCSGF also does not bind to Concanavalin A sepharose.

Size exclusion HPLC of Sertoli cell conditioned medium (Fig. 1) revealed two peaks of EGF competing activity that were coincident with peaks of mitogenic activity. There was also some mitogenic activity eluting at the void volume of the column which can probably be attributed to the transferrin in the sample. The apparent molecular weights of the activity eluting at fractions 14 to 15 and 20 are difficult to estimate because the standards used did not produce a linear plot.

Figure 1. Size exclusion HPLC of Sertoli cell conditioned medium (SCCM). Proteins in SCCM were resuspended in 0.5M ammonium acetate pH 7 and chromatographed on a TSK G2000 SW size exclusion column with guard column equilibrated with 0.5M ammonium acetate pH 7. The flow rate was 0.5 ml/min and 1 ml fractions were collected, lyophilized and resuspended in DME media. EGF competing activity was expressed as percent displacement from the zero point. Thymidine incorporation was expressed as percent incorporation above control (3000 dpm).

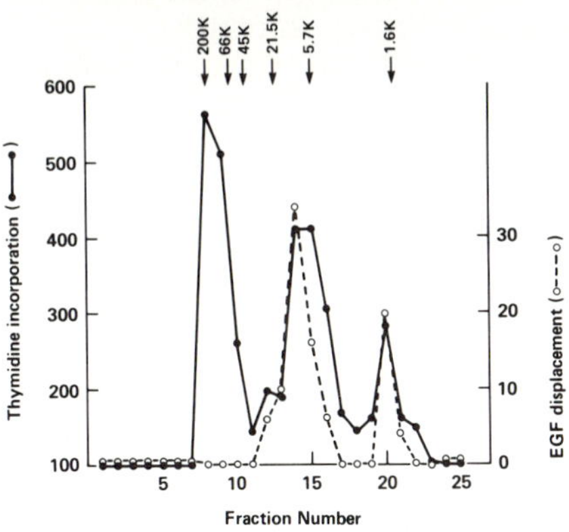

Other growth factors reported as products of the seminiferous epithelium differ in some characteristics to SCSGF (Brown et al., 1982; Hall et al., 1983; Feig et al., 1980). The function of SCSGF is speculative but it may regulate the proliferation rate of prepubertal Sertoli cells, or it may be targeted for other testicular cells. There is the possibility that SCSGF may regulate spermiogenesis or be involved in stem cell proliferation.

Acknowledgements: This work was supported by NIH grants AA 06167 and DA 03431.

REFERENCES

Brown KD, Blakeley DM, Henville A, Setchell BP (1982) Rete testis fluid contains a growth factor for cultured fibroblasts. Biochem Biophys Res Commun 105:391.

Feig LA, Bellve AR, Horbach-Erickson N, Klagsbrun M (1980) Sertoli cells contain a mitogenic polypeptide. Proc Natl Acad Sci USA 77:4774.

Hall K, Ritzen EM, Johnsonbaugh RE, Parvinen M (1983) Pubertal rise of immunoreactive somatomedin and its eventual source. In Spencer EM (ed.): Insulin-like growth factors/somatomedins, New York: Walter de Gruyter Co., p 611.

Holmes SD, Spotts G and Smith RG (1985) Rat Sertoli cells secrete a growth factor that blocks EGF binding to its receptor. Submitted to J Biol Chem.

Holmes SD, Bucci LR, Lipshultz LI and Smith RG (1983) Transferrin binds specifically to pachytene spermatocytes. Endocrinology 113:1916.

Steinberger E (1971) Hormonal control of mammalian spermatogenesis. Physiol Rev 51:1.

Weissenberg R, Eshkol A and Lunenfeld B (1982) Hormonal dependence of the first spermatogenic wave in the mouse. Archives Andrology 9:135.

Progress in Developmental Biology, Part A, pages 393–396

ISOLATION AND CHARACTERIZATION OF A GROWTH REGULATORY FACTOR FROM 3T3 CELLS

John L. Wang and Yen-Ming Hsu

Department of Biochemistry
Michigan State University
East Lansing, MI 48824 USA

Treatment of sparse, proliferating cultures of 3T3 cells with medium conditioned by exposure to density-inhibited 3T3 cultures (CM) resulted in an inhibition of growth and division in the target cells when compared with similar treatment with unconditioned medium (UCM) (Steck et al., 1979). Gel filtration of the ammonium sulfate precipitate of CM on a column of Sephadex G-50 partitioned the inhibitory activity into two major components. The material eluting from the Sephadex G-50 column at a position corresponding to polypeptides in the molecular weight range of 10,000- 15,000 showed enrichment in specific biological activity (growth inhibition assay) and was designated FGR-s (Fibroblast Growth Regulator - soluble form) (Steck et al., 1982). Isoelectric focusing and polyacrylamide gel electrophoretic analysis showed that the principal components of the FGR-s fraction were polypeptides of M_r 10,000 (pI ~ 7.5) and M_r 13,000 (pI ~ 10) (Steck et al., 1982; Wang et al., 1982). Using similar procedures, Wells and Mallucci (1983) have shown that secondary cultures of mouse embryo fibroblasts release into the medium a growth inhibitory activity whose polypeptide composition and physicochemical behavior closely parallel those of FGR-s.

DEAE-cellulose chromatography of the FGR-s fraction resulted in further fractionation of the components. However, only the material not bound by the ion-exchange column exhibited growth inhibitory activity; the sum of the growth inhibitory activity in these unbound fractions accounted for ~80% of the total activity applied to the column. There was a six-fold enrichment in terms of specific activity in this fractionation step (Hsu and Wang, 1985). Polyacrylamide gel electrophoretic analysis of this active material yielded a single polypeptide migrating at a position corresponding to a molecular weight of 13,000. Identical results were obtained irrespective of whether the polyacrylamide gels were electrophoresed under reducing or nonreducing conditions (Figure 1, inset, lanes 1 and 2). This material is hereafter designated FGR-s (13 K).

Using the partially purified fraction FGR-s as the immunogen, we have carried out in vitro immunization of rat splenocytes and have generated hybridoma lines secreting monoclonal antibodies directed against components of the FGR-s preparation. One such antibody, designated 2A4, specifically bound the M_r 13,000 polypeptide (Hsu et al., 1984). When the growth inhibitory activity of FGR-s was assayed in the presence of Antibody 2A4, the activity of the partially purified fraction was neutralized. Moreover, the growth inhibitory activity of a highly purified preparation of FGR-s (13 K) was depleted when the material was passed through an affinity column containing Antibody 2A4 (Hsu and Wang, 1985). These results strongly suggest that FGR-s (13 K) is directly responsible for the observed growth inhibitory activity.

The growth inhibitory effect of FGR-s (13 K) on target cells was dependent on the concentration of ligand added (Figure 1). The concentration of inhibitor required for 50% inhibition was ~3 ng/ml, corresponding to ~0.23 nM. The inhibition was reversible within 20 hours after the removal of FGR-s (13 K) (Hsu and Wang, 1985).

If FGR-s (13 K) plays a physiologically significant role in density-dependent inhibition of growth, one would expect that addition of Antibody 2A4 to dense cultures of 3T3 cells should at least partially neutralize the activity of endogenous FGR-s (13 K) molecules in the culture and reverse the effect of density inhibition. Indeed, the addition of Antibody 2A4 resulted in a higher level of DNA synthesis in 3T3 cultures without any exogenously-added FGR-s. In contrast, the addition of Antibody 104, which is not reactive with FGR-s, failed to yield the same effect (Wang and Hsu, 1984).

Antibody 2A4 binds directly to live or unfixed 3T3 cells. The binding is saturable, suggesting that there are

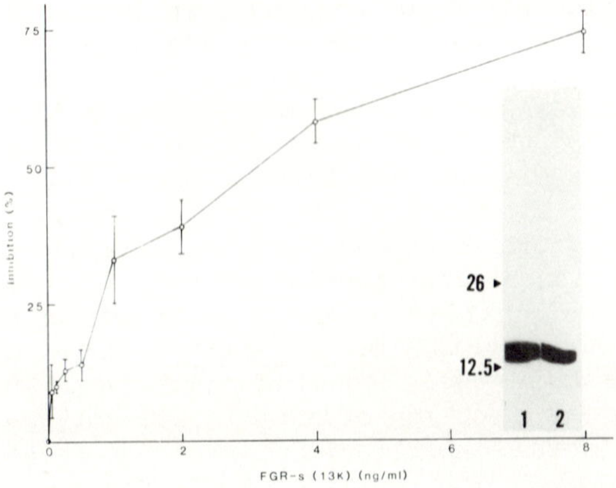

Figure 1. Dose-response curve of growth inhibitory activity of FGR-s (13 K) on target 3T3 cells (2 x 10^4 cells/cm^2). The data represent the averages of triplicate determinations and are expressed as percent inhibition relative to control cultures. The inset shows the results of polyacrylamide gel electrophoresis in the presence of sodium dodecyl sulfate of a ^{35}S-methionine labeled preparation of FGR-s (13 K) under non-reducing (lane 1) and reducing conditions (lane 2).

only a finite number of antigenic targets exposed at the cell surface. This is consistent with the finding that radioactively-labeled preparation of FGR-s can bind to or exchange with components of the cell surface (Voss et al., 1982). We infer that FGR-s (13 K) or a cross-reactive precursor on the membrane may be functioning in the normal mechanism of density-dependent inhibition of growth in 3T3 cells.

Our results on the purification of FGR-s (13 K) suggest that it may join Transforming Growth Factor-β (Tucker et al., 1984; Roberts et al., 1985) as another example of a growth regulator that functions in an autocrine pathway. Perhaps a family of growth inhibitors may be defined when structural information becomes available for these negative regulators of cell growth.

Acknowledgements: This work was supported by grants

GM-27203 and GM-32310 from the National Institutes of Health. J.L. Wang was supported by Faculty Research Award FRA-221 from the American Cancer Society.

REFERENCES

Hsu YM, Barry JM, Wang JL (1984). Growth control in cultured 3T3 fibroblasts: Neutralization and identification of a growth-inhibitory factor by a monoclonal antibody. Proc Natl Acad Sci USA 81:2107-2111.

Hsu YM, Wang JL (1985). Growth control in cultured fibroblasts. V. Purification of a M_r 13,000 polypeptide responsible for growth inhibitory activity. in preparation.

Roberts AB, Anzano MA, Wakefield LM, Roche NS, Stern DF, Sporn MB (1985). Type β transforming growth fractor: a bifunctional regulator of cellular growth. Proc Natl Acad Sci USA 82:119-123.

Steck PA, Voss PG, Wang JL (1979). Growth control in cultured 3T3 fibroblasts. Assays of cell proliferation and demonstration of a growth inhibitory activity. J Cell Biol 83:562-575.

Steck PA, Blenis J, Voss PG, Wang JL (1982). Growth control in cultured 3T3 fibroblasts. II. Molecular properties of a fraction enriched in growth inhibitory activity. J Cell Biol 92:523-530.

Tucker RF, Shipley GD, Moses HL, Holley RW (1984). Growth inhibitor from BSC-1 cells closely related to platelet type β transforming growth factor. Science 226:705-707.

Voss PG, Steck PA, Calamia JC, Wang JL (1982). Growth control in cultured 3T3 fibroblasts. III. Binding interactions of a growth inhibitory activity with target cells. Exp Cell Res 138:397-407.

Wang JL, Steck PA, Kurtz JW (1982). Growth control in cultured 3T3 fibroblast: Molecular properties of a growth regulatory factor isolated from conditioned medium. In Sato GH, Pardee AB, Sirbasku DA (eds): "Growth of Cells in Hormonally Defined Media", Cold Spring Harbor: Cold Spring Harbor Laboratory Press, pp 305-317.

Wang JL, Hsu YM (1984). Isolation and characterization of a growth regulatory factor from 3T3 cells. In Skehan P, Friedman SJ (eds): "Growth, Cancer, and the Cell Cycle", Clifton: Humana Press, pp 177-186.

Wells V, Mallucci L (1983). Properties of a cell growth inhibitor produced by mouse embryo fibroblasts. J Cell Physiol 117:148-154.

Progress in Developmental Biology, Part A, pages 397–400

MOLECULAR CLONING OF DNA COMPLEMENTARY TO BOVINE PITUITARY mRNA

Binhui Ni, Dexiu Song and Yinghsien Shih

Institute of Developmental Biology,
Academia Sinica, Beijing, CHINA

INTRODUCTION

Growth hormone,along with at least two other polypeptide hormone,chorionic somatomam motropin and prolactin, forms a set of proteins with amino acid sequance homology and to some extent overlapping biological activities(T.A.Bewley et al, 1972). Since the gene of this set of proteins probably have a common ancestral origin(T.A.Bewley, 1972),they constitute an excellent model to study the evolution, structure and differential regulation of related genes. In addition, bovine growth hormone is of considerable use in animal husbandary (Machlin L.J, 1973).

We now report the synthesis and cloning of DNA complementary to bovine pituitary mRNA that appear to have the greatest aboundence of bGH mRNA. The double-stranded cDNA were inserted into HindIII site of pBR322 by addition of HindIII Linkers. Difference sizes of inserts were founded in agrose electrophoresis by EB stain.

RESULTS

Poly A mRNA was isolated from bovine pituitary collected shortly after killing at a local abattoir and were frozen immediotely in liquid N_2 Poly A RNA was prepared by Sucrose Gradient Centrifugation, phenol extraction and ethanol precipitation. Total mRNA was prepared by oligo(dT)-cellulose chromatography. Integrity of the Poly A RNA was assessed by wheat germ cell-free system, which resulted in stimulation of ^{14}C-leucine incorporation when Poly A mRNA was added(Robert B.E, 1973).

Total Poly A mRNA from the bovine pituitary were used as template for the synthesis of cDNA with reverse transcriptase using Oligo(dT) primers for first strand synthesis and self-primer for the second strand with E.coli DNA polymerseI . Efficiencies of reverse trancription were generally 10% for the first strand. We have examined the enzymatic reactions in double-stranded DNA synthesis with the aim of obtaining full length duplex copies in good yield using relatively small amounts of the messengers. Reverse transcriptase reactions were stoped by chilling on ice and centrifuged briefly. Tubes were then placed in a boiling water bath for 3 min to seperate mRNA-first strand hybirds. DNA Pol.I reactions were carried out sequentially in the same reaction vessel with very little manipulation of the reaction products. The amount of DNA Pol.I and the incubation time of the reaction were determined by carrying out a set of pilot-scale reaction(Marvin P.Wickens, 1978).

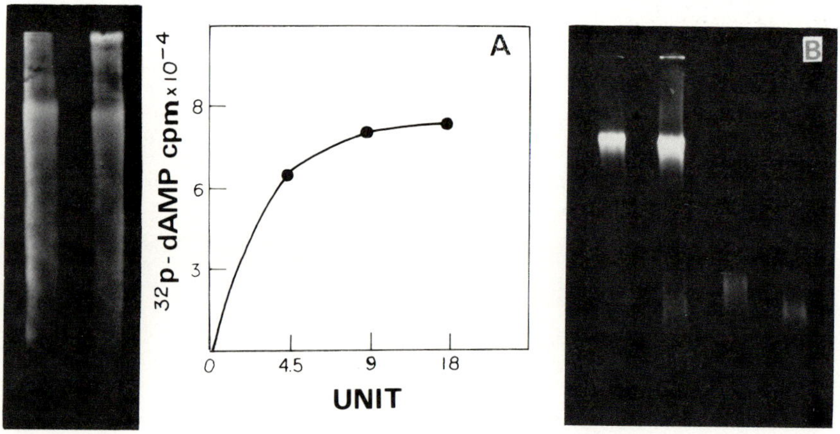

Figure 1. Gel electrophoresis of double-stranded cDNA products formed with different amount of DNA PolI(A)and differet incubation time(B)

We found that amount of DNA Pol.I and incubation time have marked effects on the size distribution of DNA PoI I products. Second strand cDNA made in 9 units of Pol I display slightly larger than that made in 18 units.

The double-stranded cDNA was then trimmed with S_1. HindIII linkers were added to each end of the cDNA and cohesive ends were generated by digestion with HindIII(Fig 2). The resulting cDNA was ligated to HindIII-cut pBR322 plasmid DNAs. About 1200 colonies selected by antibiotic resistance ($Tet^s Amp^r$) were grown, and the plasmid DNAs were isolated.The DNA was digested with HindIII and analyzed by gel electrophoresis. Several clones generated inserts of 600-400bp (Fig 3).

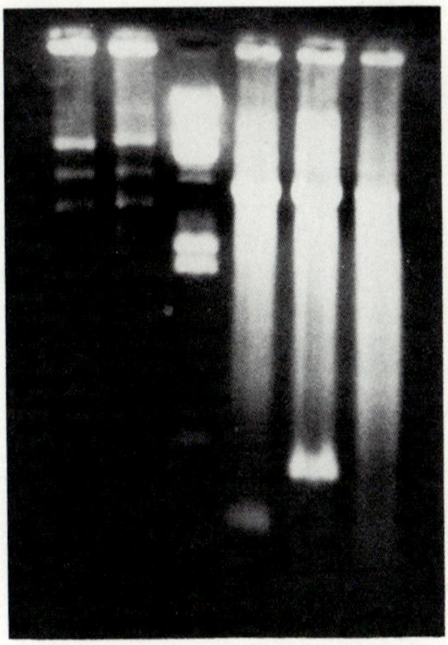

Figure 2. Autoradiogram of cDNA with labled HindIII linkers in agrose gel.

Figure 3. Several recombinant plasmids digested by HindIII generated inserts by gel electrophoresis.

The electron micrographs of pBR322 and recomplasmids showed that circumference of recombinant plasmid was longer than that of normal pBR322 plasmid by measurement(Qian Biao et al, 1984)(Fig.4).

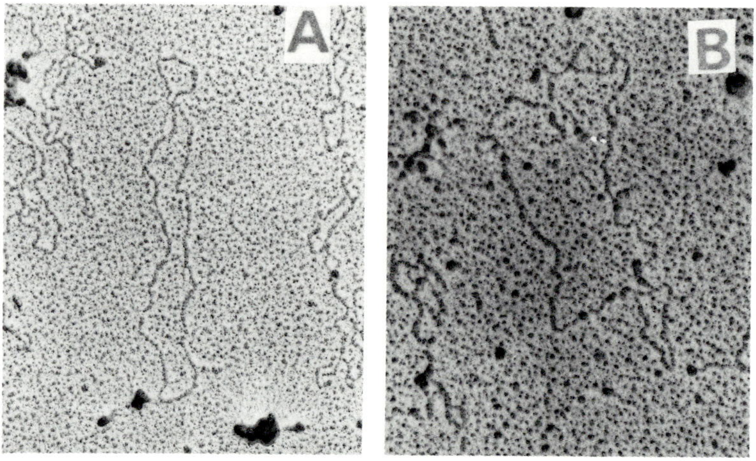

Figure 4. Electron micrographs of pBR322 plasmid (×30,000)(A) and recombinant plasmid(×30,000)(B).

These suggested that these clones did contain cDNA complementary to bovine pituitary mRNA.

REFERENCES

Machlin L.J. (1973) J.Dairy Science 56:575-585.
Marvin P.Wickens, Gary N. Buell and Robert T.Schimke (1978). J. Bio. Chem. 253:2483-2495.
Qian Biao, Wang Changcai (1984). HEREDITAS 6:6-10.
Robert B.E. (1973). Proc. Natl. Acad. Sci. USA. 70: 2330-2334.
T. A. Bewley, J.S. Dixon and C. H. Li (1972). Int. J. Pept. Protein Res. 4:281-283.

Early Events During Embryogenesis

Progress in Developmental Biology, Part A, pages 403–406

INVOLVEMENT OF THE CYTOSKELETON IN EARLY GRAY CRESCENT FORMATION IN AXOLOTL OOCYTE: RELATIONSHIPS WITH GERMINAL VESICLE BREAKDOWN.

Jean Gautier and Jean-Claude Beetschen

Laboratoire de Biologie Générale, UA CNRS n° 675
Université P. Sabatier, 118 route de Narbonne
31062 Toulouse. FRANCE.

INTRODUCTION

The gray crescent of amphibian egg is a morphological marker of the future dorsal side of the embryo. Its formation during the first cell cycle, between fertilization and first cleavage, provides the first evidence for dorso-ventral polarity in the egg.

We have shown (Gautier and Beetschen, 1985) that it is possible to induce early gray crescent formation in the axolotl oocyte(*Ambystoma mexicanum*), at least three conditions being required:

-progesterone-induced cytoplasmic biochemical changes (until the tenth hour of maturation),

-interaction between mature cytoplasm and nuclear factors issued from the oocyte nucleus after germinal vesicle breakdown.

-protein synthesis inhibition (PSI), around the first polar body formation period.

In fertilized eggs, or after experimental induction in oocytes, gray crescent results in the asymmetrical distribution of cortical and deep cytoplasmic constituents.

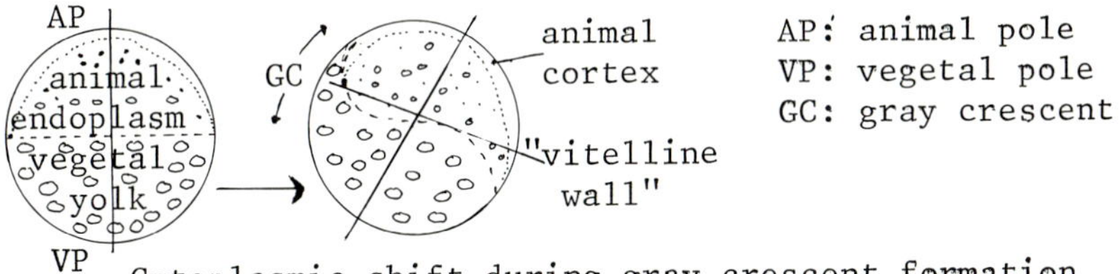

Cytoplasmic shift during gray crescent formation.

It is now becoming clear that the oocyte cytoskeleton, particularly microtubules (MT), must play a key role in the establishment of dorso-ventral polarity (Scharf and Gerhart 198 , Ubbels et al 198 , Scharf and Cande 1985).

We performed two types of experiments in the Axolotl oocyte model system, which provide further evidence for MT partcicipation in establishing asymmetry :

1. Interference with the cytoskeleton was obtained by microinjecting cytoskeleton inhibitors or antibodies against the main cytoskeletal proteins.

2. Nuclear substitution was performed in order to investigate the role of nuclear components in the generation of asymmetry.

RESULTS

1.

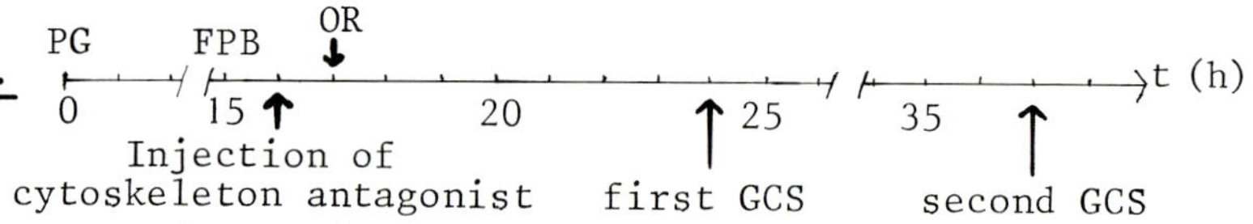

Time course of experiments . PG : time of progesterone addition ; FPB : first polar body formation ; OR : oocytes orientation (80° rotation) ; GCS : gray crescent scores.

Table 1: Effect of cytoskeleton antagonists on gray crescent formation in rotated axolotl oocytes.

Injected solution	Final concentration in the oocytes	Total number of injected oocytes	GC % at 1st GCS	GC % at 2nd GCS
Anti-actin	800 ug/ml	40	0	0
Anti-tubulin	800 ug/ml	164	21	59
Anti-keratin	15 ug/ml	20	0	cytolysis
Anti-vimentin	50 ug/ml	50	pigment shift not a GC	
Control	800 ug/ml	30	0	0
Colcemid	10^{-4}M	20	30	80
Colcemid	10^{-5}M	20	25	60
Control (Buffer)	–	45	0	2
Vinbastin sulfate	10^{-4}M	55	3	9
Nocodazole	10^{-4}M	65	18	69
Nocodazole	10^{-5}M	40	20	70
Taxol	10^{-4}M	30	3	3
Cytochalasin	10^{-5}M	30	0	3
Control (DMSO)	–	55	0	3

Microtubules inhibitors can induce early gray crescent formation in maturing oocytes; this reaction was slow : only a few GC were scored at the first GCS. Only those inhibitors which reversibly depolymerize MT were efficient.

2. Timing of experiments

As previously shown (Gautier and Beetschen 1985), enucleated oocytes never form a GC after injection of a protein synthesis inhibitor (or after injection of a cytoskeleton antagonist : data not shown).

In order to determine the nuclear contribution to cytoskeleton modifications in EGC formation, nuclear substitution experiments were performed

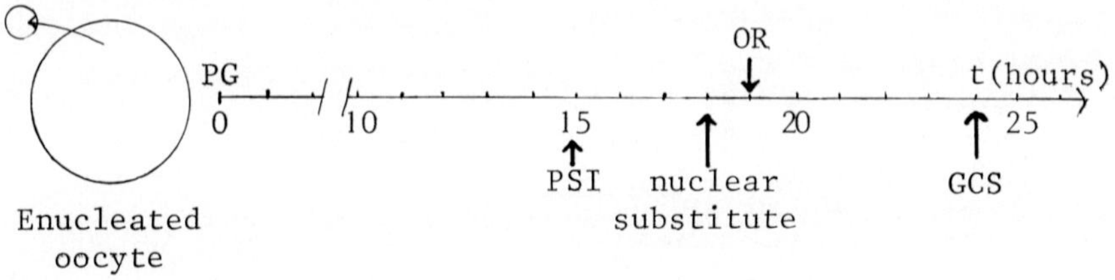

PSI : Protein synthesis inhibitor injection.

Injection of Taxol or microtubule associated proteins (MAPs) into enucleated oocytes, restores the ability of early gray crescent formation (after protein synthesis inhibition). However Taxol (33% GC) was not as potent as MAPs (70% GC).

CONCLUSION

Early gray crescent formation should transform a cytoplasmic steady state into another stable anisotropic cytoplasmic arrangement. We suggest, as a provisional hypothesis, that these changes require at least three steps :

- microtubule depolymerization experiments with antagonists, which allows:
- cytoplasmic shift;
- microtubule repolymerization (nuclear substitution experiments)

Germinal vesicle contribution could consist of regulatory factors, allowing microtubule polymerization.

The contribution of germinal vesicle components for microtubule organization during Xenopus oocyte maturation has already been shown (Huchon et al 1981; Jessus et al 1984). Polymeric tubulin cycling has been studied by Elinson (1985) in Xenopus egg during physiological gray crescent formation, and is under investigation in the axolotl oocyte model system.

We thank Dr D.Paulin (Institut Pasteur, Paris) for anticytoskeleton antibodies gift, Dr C.Jessus (Université Paris VI) for providing the MAPs. This work was supported by the Centre National de la Recherche Scientifique and the Ministry of Education.

REFERENCES

Elinson RP (1985). Changes in levels of polymeric tubulin associated with activation and dorsoventral polarization of the frog egg. Dev Biol 109:224-233.

Gautier J, Beetschen J-C (1985). A three-step scheme for gray crescent formation in the rotated axolotl oocyte. Dev Biol 110:192-199.

Huchon D, Crozet N, Cantenot N, Ozon R (1981). Germinal vesicle breakdown in Xenopus laevis oocyte: description of a transient microtubular structure. Reprod Nutr Dev 21:135-148.

Jessus C, Freiderich E, Francon J, Ozon R (1984). In vitro inhibition of tubulin assembly by a ribonucleoprotein complex associated with the free ribosome fraction isolated from Xenopus laevis oocytes: effect at the level of microtubule- associated proteins. Cell Differ 14:179-187.

Scharf SR, Cande WZ (1985). A role for microtubules in frog egg axis determination. Cell Differ 16S:156S (Abstract).

Scharf SR, Gerhart JC (1983). Axis determination in eggs of Xenopus laevis: a critical period before first cleavage identified by the common effect of cold, pressure and ultraviolet irradiation. Dev Biol 99:75-87.

Ubbels GA, Hara K, Koster CH, Kirschner MW (1983). Evidence for a functional role of the cytoskeleton in determination of the dorsoventral axis in Xenopus laevis eggs. J Embryo Exp Morphol 77:15-37.

Progress in Developmental Biology, Part A, pages 407–410

CYCLIC AMP SHORTENS MITOTIC PHASE OF SEA URCHIN EGG CLEAVAGE CYCLE

Akira Suzuki*, Ikuo Yasumasu* and Katsumi Ishida**

*Department of Biology, School of Education, Waseda University and **Department of Urology, Teikyo University School of Medicine

INTRODUCTION

It has been well known that adenosine 3',5'-monophosphate (cAMP) has various effects on cell growth and proliferation (Hsie and Puck, 1971; Sheppard, 1971; Johnson and Pastan, 1972; Whitfield et al., 1980; Mayer, 1982). We have already reported that the cleavage cycle of sea urchin eggs was shortened by extracellular cAMP (Ishida and Yasumasu, 1982). The intracellular concentration of cAMP also changed periodically with close relation to the cleavage cycle (Yasumasu et al., 1973; Ishida and Yasumasu, 1982). These suggest that periodical change in cAMP content induced to accelerate or protract the pace of cleavage cycle. Indeed, caffeine, which is a potent inhibitor of cAMP breakdown, apparently delayed the cleavage pace (Ishida and Yasumasu, 1982). However, it has been unknown how the extracellular cAMP shortens the cleavage pace. The present paper deals with what phase of the cleavage cycle is influenced by the extracellular cAMP.

Gametes of sea urchins, Hemicentrotus pulcherrimus and Strongylocentrotus nudus, and oocytes of starfish, Asterina pectinifera were used as materials.

Eggs and sperm of the sea urchins were obtained by 0.54 M KCI induced spawning. Eggs were washed several times with filtered sea water before insemination. Sperm was kept on ice until insemination. After fertilization was confirmed, 2-3 fertilized eggs were developed using a small glass chamber where fresh sea water was continuously supplied. The temperature of the surrounding sea water was kept at 18°C (H. pulcherrimus) or 22°C (S. nudus) during the development. After fertilization, the chamber was placed under an interference difference microscope equipped with Nomarsky prism and the process of cleavage cycle of eggs was monitored and recorded with a video tape recorder.

In order to estmate DNA synthesis during the cleavage cycle, the eggs were developed in sea water containing 1mM cAMP and 0.1μCi/ml of H-thymidine at a density of 20,000 eggs/l at 18°C (H. pulcherrimus) or at 22°C (S. nudus). Adequate volume of the egg suspension was collected at every 5 min. after fertilization. After DNA fraction was extracted from the eggs, the radioactivity of the incorporated radioactive thymidine into DNA was determined.

Protein estimation was carried out by the method of Lowry et al. using bovine serum albumin as a standard (1951).

Ovaries of the starfish were removed from the starfish by a forccpt. The ovaries were dipped in Ca^{2+}-free sea water for 1 hour and oocytes were discharged from the ovaries by addition of 0.54M KCl to Ca^{2+}-free sea water. The oocytes thus obtained were placed in a small chamber on a slide glass and treated with sea water containing 1 μM 1-methyladenine and 1 mM cAMP. Process of the first and the second polar body extrusion was monitored and recorded by an interference difference microscope equipped with Nomarsky prism at 22°C.

All chemicals were of reagent grade obtained from Sigma Chemical Co.

RESULTS

Figure 1 shows DNA synthesis velocity after fertilization of H. pulcherrimus. It is clear that cAMP does not enhance DNA synthesis. Oocytes of starfish are able to be matured without DNA synthesis. This suggest that the extrusion of polor body is highly considered to be accelerated by cAMP. This consideration was confirmed by an observation that the extrusion of polor bodies from 1-methyladenine treated starfish oocytes was apparently shortened by extracellular cAMP (Figure 2). Therefore, the accelerative effect of cAMP on the cleavage cycle of the sea urchin eggs

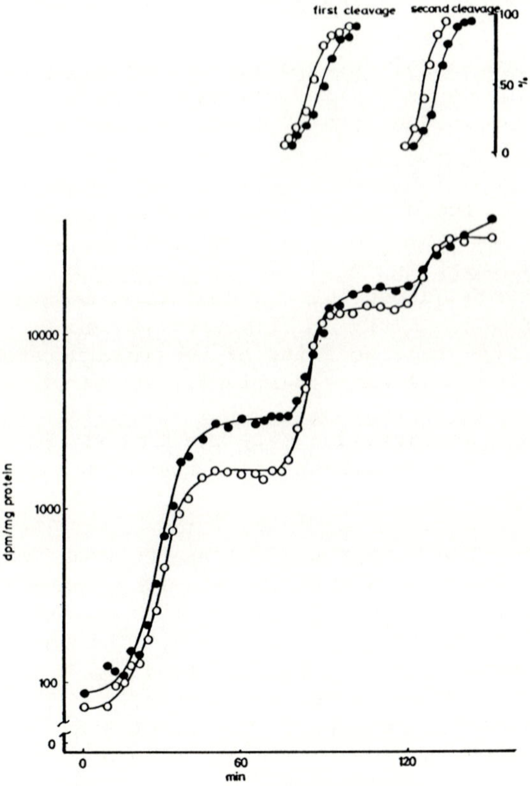

Figure 1.Abscissa : Time in minutes after fertilization at 22°C. Ordinate : Radioactivity of labeled thymidine incorporated into DNA fraction. Solid circle : Control. Open circle : Eggs developed in the sea water containing 1 mM cAMP. Ordinate of upper small figure: Cleavage rate.

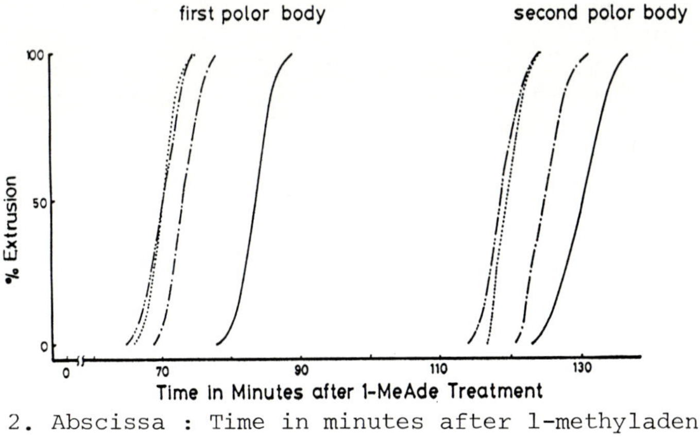

Figure 2. Abscissa : Time in minutes after 1-methyladenine treatment. Ordinate: Percentage of polor body extrusion. —— : control. - --: +mM cAMP--------: + 2.5 mM cAMP - -: +5mM cAMP.

was not related to DNA synthesis.

Analysis of the cleavage cycle of the sea urchin eggs by using a video tape recorder revealed that the acceleration of cleavage cycle by cAMP was due to a reduction of interval in a phase post DNA synthesis. The intervals between fertilization and diaster formation were 57'22"±3'52" (n=7) for control eggs and 57'32"±4'54" (n=16) for eggs developed in condition containing 1 mM cAMP, respectively. Though these intervals, of course, include DNA synthesis period, the intervals between the diaster formation and furrowing were 24'42"±3'21" in the control eggs and 16'52" ±2'57" in the experimental eggs, respectively. Therefore, about 33% reduction of the interval in the post DNA synthesis phase was calculated. From the continuous microscopical observation, it makes clear that this reduction was mainly caused by stimulating elongation of astral rays of mitotic apparatus.

REFERENCES

Hsie AW, Puck T (1971). Proc Natl Acad Sci USA 68:358-378.

Ishida K, Yasumasu I (1982). Acta Embryol Morphol Exp 2: 115-119.

Ishida K, Yasumasu I (1982). Biochim Biophys Acta 720:266-273.

Johnson GH, Pastan I (1972). J Natl Cancer Inst 48:1377-1387.

Lowry OH, Rosebrough NJ, Farr AL, Randall RJ (1951). J Biol Chem 193:265-275.

Mayer TC (1982). Deveop Biol 94:509-514.

Sheppard JR (1971). Proc Natl Acad Sci USA 68:1316-1320.

Whitfield JF, Boynton AL, MacManus JP, Sikorska M, Tsang BK (1979). Mol Cell Biochem 27:155-179.

Whitfield JF, Boynton AL, MacManus JP, Rixon RH, Sikorska M, Tsang B, Walker PR (1980). Ann New York Acad Sci 339: 216-240.

Yasumasu I, Fujiwara A, Ishida K (1973). Biochem Biophys Res Comm 54:628-632.

Progress in Developmental Biology, Part A, pages 411–418

CYTODIFFERENTIATION vs. ORGANOGENESIS IN KIDNEY DEVELOPMENT

Eero Lehtonen and Lauri Saxén

Department of Pathology, University of Helsinki, Finland

INTRODUCTION

The development of the permanent mammalian kidney, the metanephros, has proved to be a suitable model-system for analysis of early organogenesis. Its development involves differentiation and strict spatial assembly of three cell lineages: (1) the epithelial cells of the Wolffian-duct-derived ureter bud, giving rise to the collecting system; (2) the mesenchymal cells of the blastema to be converted into the epithelium of the excretory tubules; and (3) the endothelial cells of outside origin constructing the kidney vasculature including the glomerular loop. Through complex interactions the three cell populations become organized in spatial and temporal synchrony to form the complete nephrons consisting of cells of several types. Expression of these phenotypes during the development of the nephron is characterized by changing morphology and polarization of the cells and by appearance of new proteins of which some are secreted into the extracellular matrix (ECM). These changes have been well described in whole embryonic kidneys in vivo and in three-dimensional, organotypic cultures of the tissue components (Wartiovaara, 1966; Grobstein, 1967; Saxén et al., 1968, 1981; Bernstein et al., 1981; Ekblom , 1984; Lehtonen and Saxén, 1985).

In order to understand the complex process of kidney tubulogenesis, it seems important to separate the various events at the cellular and tissue levels and to explore to which extent the expression of new cellular phenotypes is a consequence of the organogenetic assembly of the cells as opposed to an independent differentiation of single cells. An experimental set-up recently modified by us from the classic transfilter technique (Grobstein, 1956) might allow us to distinguish between these two types of differentiation (Lehtonen et al., 1985).

THE EXPERIMENTAL MODEL-SYSTEM

The original technique was based on the finding that differentiation of the mesenchymal cell population of the nephric blastema requires a morphogenetic stimulus from the ingrowing ureter, and that this "inductor" can be replaced experimentally by certain heterotypic tissues including embryonic spinal cord. Accordingly, an isolated metanephrogenic mesenchyme can be exposed to the inductor by bringing the two in contact in an organotypic culture. This can also be implemented through a porous filter membrane (Grobstein, 1956), provided that the filter pores are large enough to allow penetration of cytoplasmic processes and establishment of actual cell-cell contacts between the interactants (Lehtonen, 1976). Kinetic studies of the process have shown that induction in these conditions is time-consuming and completed after some 24 to 26 h of transfilter contact (Saxén and Lehtonen, 1978). Our modification of the technique is based on the finding that at the time of completed induction, the mesenchymal target cells are still undifferentiated and show no signs of epithelial transformation (Wartiovaara, 1966; Saxén et al., 1968). These undifferentiated though determined cells can be separated from the inductor and subcultured either in three-dimensional organotypic explants or as monolayers.

When kept in three-dimensional conditions, the induced mesenchyme follows the differentiation program set during the initial 24-h induction period and develops into advanced nephric structures (Ekblom et al., 1981b; Lehtonen et al., 1983). In the experiments to be described here, both induced and non-induced mesenchymal cells were cultivated as monolayers with no obvious signs of tubule formation. Thus, we could follow the early steps of epithelial transformation not dependent on three-dimensional spatial assembly of the cells. So far, the analysis has been restricted to immunohistology, and antibodies against four proteins have been used: fibronectin and laminin of the ECM and vimentin and cytokeratin type intermediate filaments (IFs).

BACKGROUND

The choice of the four antigenic markers was based on our earlier findings in vivo and in organotypic kidney cultures. The first morphological sign of the onset of tubulogenesis is a condensation of the mesenchymal cells in close proximity to the inductor (Saxén and Wartiovaara, 1966). As shown by Ekblom et al. (1981a), this aggregation is accompanied by changes in the ECM: fibronectin, initially expressed uniformly by the uninduced mesenchyme is lost while laminin can now be visualized in the aggregated zone. The undifferentiated mesenchymal cells still express vimentin at this stage, but after transformation into epithelial cells these filaments become replaced by cytokeratin-positive IFs (Holthöfer et al., 1984; Lehtonen et al., 1985). Hence, changes in the ECM are spatially and temporally correlated to an early aggregation of the mesenchyme, whereas shift in the IF-type is associated with subsequent epithelization of the mesenchymal cells and their polarization within the tubules. It is therefore interesting to follow these events in the monolayer cell cultures, where these organogenetic events are prevented from occurring.

RESULTS

Our main observations can be summarized as follows:

1. Subcultures of uninduced mesenchymal cells regularly expressed both ECM markers (fibronectin and laminin) and vimentin, but not cytokeratin.
2. During prolonged subcultivation, induced mesenchymal monolayers showed cells with cytokeratin-positive filaments, many of which had lost their vimentin. However, co-existence of the two types of IFs in the same cell was common (Fig. 1).
3. Induced mesenchymal cells in monolayers continue to synthesize and deposit fibronectin. This is frequently detected in cytokeratin-positive cells as well (Fig. 2).
4. In similar cultures, laminin is expressed by both vimentin- and cytokeratin-positive cells (Figs. 3 and 4).
5. The four marker proteins show no two-dimensional polarization in the cells of the monolayer cultures.

These results diverge to some extent from those obtained in whole kidneys in vivo and in organotypic culture conditions: determination and epithelial transformation of the

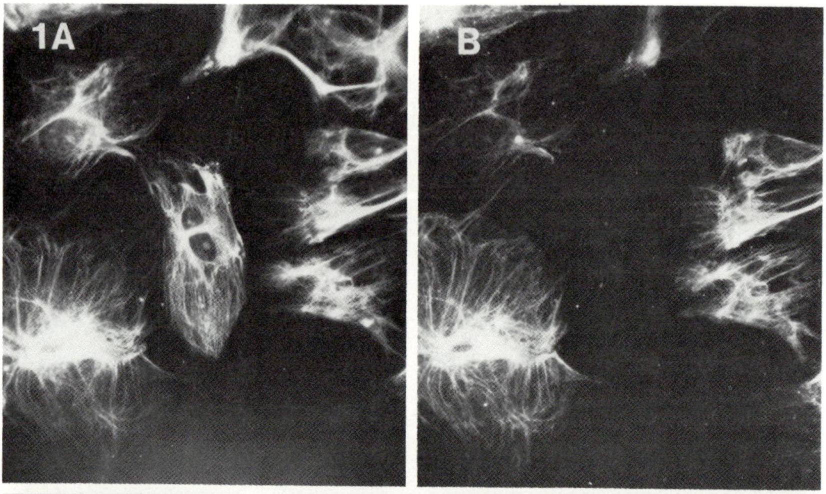

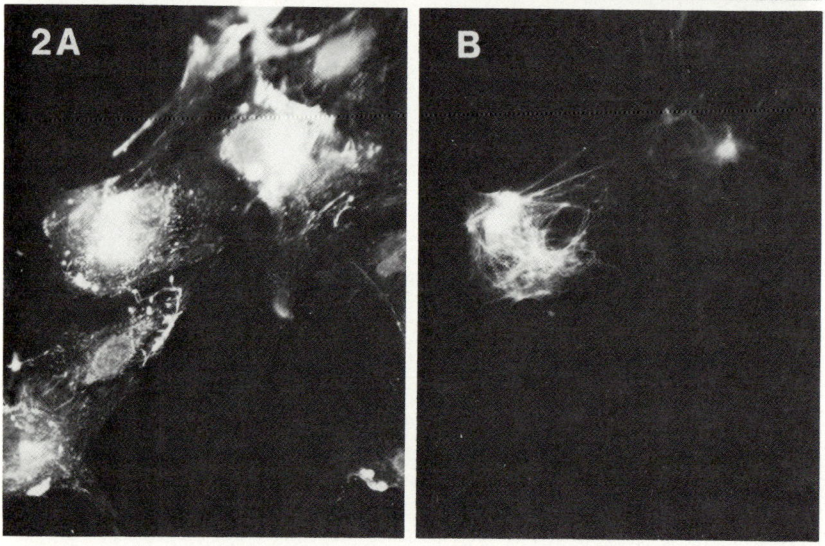

Figure 1. Fluorescence micrographs of a monolayer culture of metanephric mesenchyme cells initially induced for 24 h and subcultured for 7 days. Double immunostaining with vimentin antibodies (A) and cytokeratin antibodies (B)(From Lehtonen et al., 1985).

Figure 2. Fluorescence micrographs of a monolayer culture similar to that in Fig. 1. Double staining for fibronectin (A) and cytokeratin (B).

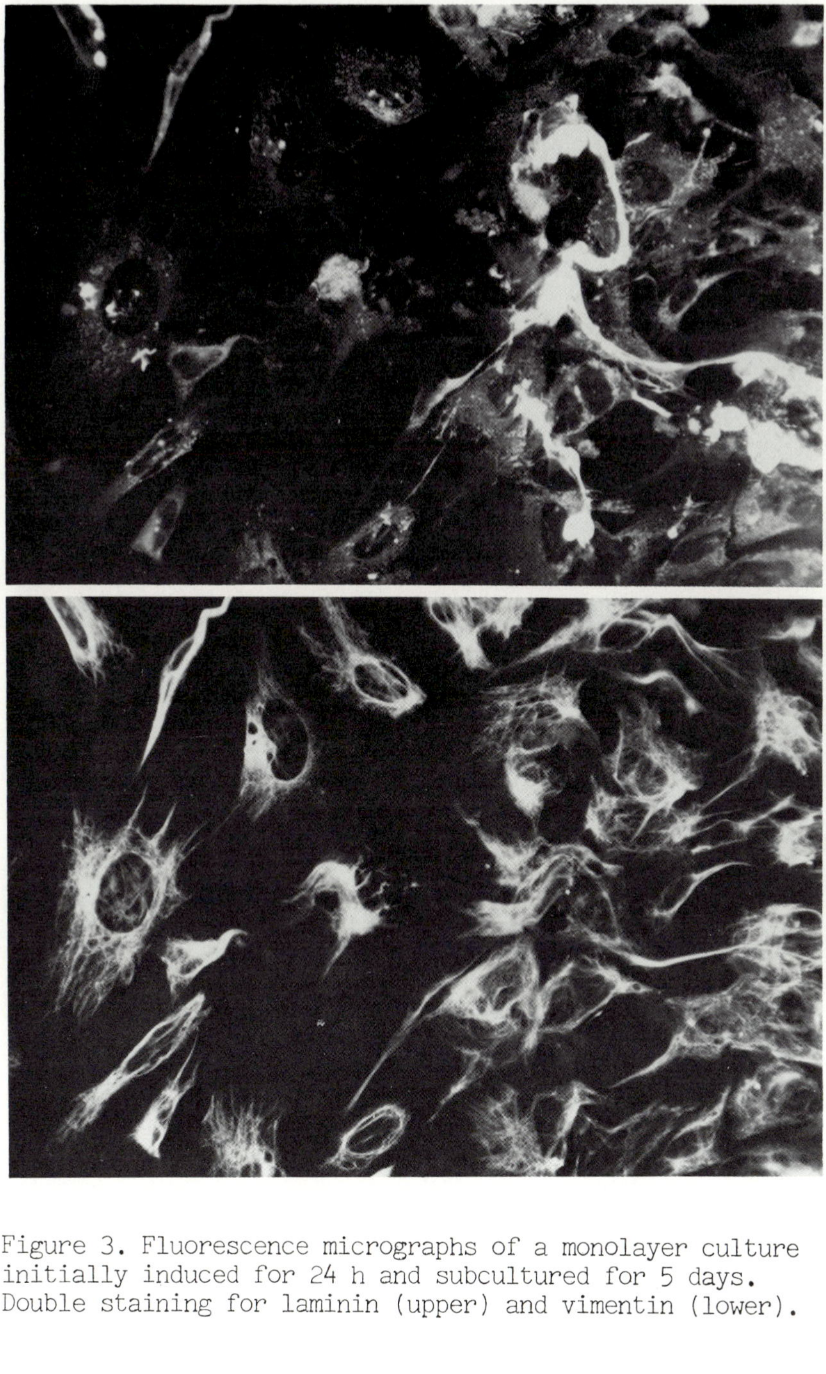

Figure 3. Fluorescence micrographs of a monolayer culture initially induced for 24 h and subcultured for 5 days. Double staining for laminin (upper) and vimentin (lower).

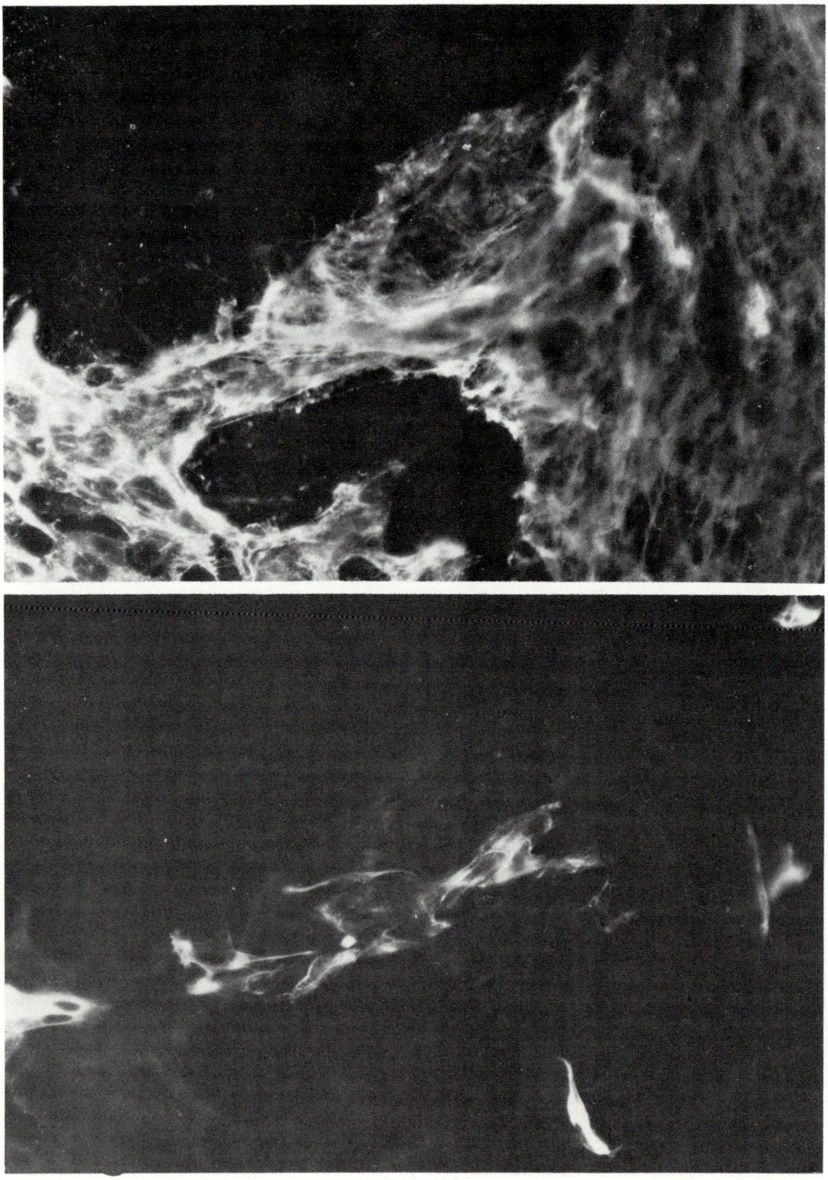

Figure 4. Fluorescence micrographs of monolayer culture initially induced for 48 h and subcultured for 7 days. Double staining for laminin (upper) and keratin (lower).

mesenchymal cells do not suppress their expression of fibronectin, and induction is not a prerequisite for the expression of laminin as shown by the vimentin-positive cells in the uninduced cultures. Concerning the changes in the two IF proteins, the findings in the monolayers are consistent with those in vivo with the exception of polarization which could not be detected in the monolayer cultures.

WORKING HYPOTHESES

The above findings in the monolayer cultures and their comparison with those in the three-dimensional conditions have led us to certain conclusions, many of which should still be considered merely working hypotheses:

1. Induction of the mesenchymal cells leads to an assembly-independent change in their gene expression (as shown by the two IF markers).
2. Synthesis of laminin is inherent of uninduced cells and might be merely enhanced by induction. Changes in its in vivo distribution seem to be associated with changing tissue architecture.
3. Fibronectin is lost in vivo apparently by local factors not present in the isolated monolayer cultures. An inductor-dependent activation of proteolytic enzymes would be a plausible, yet unproven mechanism for this phenomenon.
4. Polarization, not detectable in monolayer cultures, seems to require further three-dimensional interactions between cells and their micro-environment (cell-cell and cell-matrix interactions).

REFERENCES

Bernstein J, Cheng F, Roszka J (1981). Glomerular differentiation in metanephric culture. Lab Invest 45:183-190.

Ekblom P (1984). Basement membrane proteins and growth factors in kidney differentiation. In Trelstad RL (ed): The Role of Extracellular Matrix in Development, New York, Alan R. Liss, pp 173-206.

Ekblom P, Lehtonen E, Saxén L, Timpl R (1981a). Shift in collagen type as an early response to induction of the metanephric mesenchyme. J Cell Biol 89:276-283.

Ekblom P, Miettinen A, Virtanen I, Dawnay A, Wahlström T, Saxén L (1981b). In vitro segregation of the metanephric nephron. Dev Biol 84:88-95.

Grobstein C (1956). Trans-filter induction of tubules in mouse metanephric mesenchyme. Exp Cell Res 10:424-440.
Grobstein C (1967). Mechanisms of organogenetic tissue interaction. Natl Cancer Inst Monogr 26:279-299.
Holthöfer H, Miettinen A, Lehto V-P, Lehtonen E, Virtanen I (1984). Expression of vimentin and cytokeratin types of intermediate filament proteins in developing and adult human kidneys. Lab Invest 50:552-559.
Lehtonen E (1976). Transmission of signals in embryonic induction. Med Biol 54:108-128.
Lehtonen E, Saxén L (1985). Control of differentiation. In Falkner F, Tanner JM (eds): Human Growth, New York, Plenum Press (in press).
Lehtonen E, Jalanko H, Laitinen L, Miettinen A, Ekblom P, Saxén L (1983). Differentiation of metanephric tubules following a short transfilter induction pulse. Wilh Roux's Arch Dev Biol 192:145-151.
Lehtonen E, Virtanen I, Saxén L (1985). Reorganization of intermediate filament cytoskeleton in induced metanephric mesenchyme cells is independent of tubule morphogenesis. Dev Biol 108:481-490.
Saxén L, Lehtonen E (1978). Transfilter induction of kidney tubules as a function of the extent and duration of intercellular contacts. J Embryol Exp Morphol 47:97-109.
Saxén L, Wartiovaara J (1966). Cell contact and cell adhesion during tissue organization. Int J Cancer 1:271-290.
Saxén L, Koskimies O, Lahti A, Miettinen O, Rapola J, Wartiovaara J (1968). Differentiation of kidney mesenchyme in an experimental model-system. Adv Morphog 7:251-293.
Saxén L, Ekblom P, Lehtonen E (1981). The kidney as a model-system for determination and differentiation. In Ritzen M et al. (eds): The Biology of Normal Human Growth, New York, Raven Press, pp 117-127.
Wartiovaara J (1966). Cell contacts in relation to cytodifferentiation in metanephrogenic mesenchyme in vitro. Ann Med Exp Biol Fenn 44:469-503.

Progress in Developmental Biology, Part A, pages 419–422

TUBULIN AND ASSOCIATED PROTEINS FROM DEVELOPING BRINE SHRIMP, (ARTEMIA)

Rafiee, P. and MacRae, T.H.

Department of Biology, Dalhousie University

Halifax, N.S. B3H 4J1 Canada

INTRODUCTION

Tubulin, the major structural protein of microtubules, is often the product of multi-gene families and exists in most cells as electrophoretically distinct isoforms which are influenced in their activity by microtubule-associated proteins (MAP) (Borisy et al., 1984). It was our purpose to determine if the microtubule proteins in the brine shrimp could be induced to assemble with taxol (Vallee, 1982) and to determine if these proteins are subject to developmental regulation (MacRae and Ludueña, 1984).

RESULTS AND DISCUSSION

Incubation of *Artemia* cell-free extracts with 10 μM taxol (kindly provided by Dr. M. Suffness), followed by centrifugation through sucrose cushions (MacRae, 1984) yielded pellets which consisted of one major and numerous minor proteins (Fig. 1). The major protein comigrated with purified *Artemia* tubulin and interacted with a polyclonal anti-tubulin antibody (Fig. 2). This evidence, coupled with the electron microscopic observation of microtubules in the taxol-induced pellets (Fig. 3) showed that the major protein in the pellets was tubulin and that characterization of *Artemia* tubulin and MAP was possible by using taxol.

The nontubulin pellet proteins (Fig. 1) are classified as MAP since their sedimentation through sucrose cushions

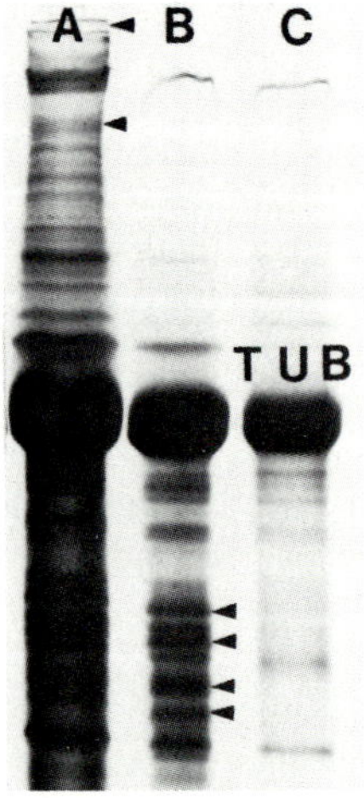

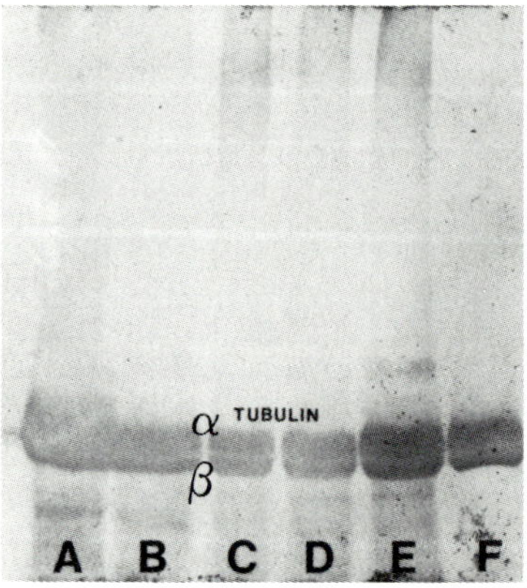

Fig. 1 SDS polyacrylamide gel (above left). Lanes A-C: taxol-induced microtubule pellets from organisms developed 10, 15 and 24 h respectively. Results with extracts from 0 and 5 h embryos were identical to those with 15 h extracts. TUB = tubulin. This and subsequent figures were adapted from Rafiee, P., MacKinlay, S.A. and MacRae, T.H., Taxol-induced asembly and characterization of microtubule proteins from developing brine shrimp (Artemia), (submitted for publication).

Fig. 2. Immunoperoxidase staining of western blot. Artemia microtubule proteins were electrophoresed on a 7.5% SDS polyacrylamide gel, transferred to nitrocellulose and stained by the immunoperoxidase method using a rabbit anti-bovine brain tubulin antibody (kindly provided by Dr. V. Kalnins) as primary antibody. Lanes A-E contained microtubule proteins from organisms developed 24, 15, 10, 5 and 0 h whereas lane F contained purified Artemia tubulin.

is dependent on microtubule formation, they can be separated from microtubules by increasing the NaCl concentration and they will reassociate with the microtubules when the NaCl is removed (unpublished data). During the first 15 h of development there is no change in the MAP composition but heavily loaded lanes (compare A and B of Fig. 1) exhibit high molecular weight bands (arrow-heads, lane A) not apparent in other lanes. After 24 h of development (lane C) several proteins indicated by arrow-heads (lane B) have disappeared and no new MAP appear.

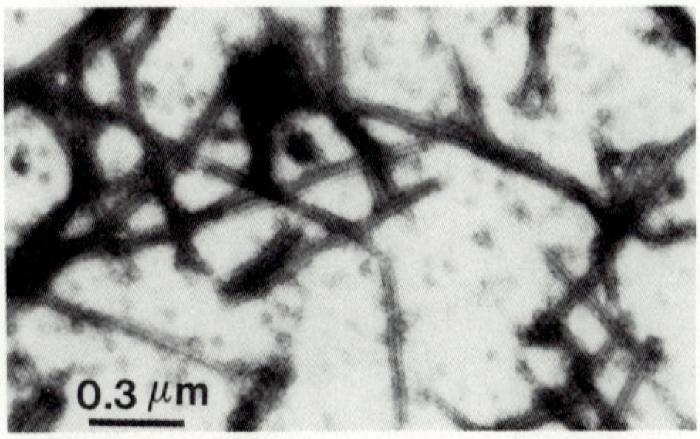

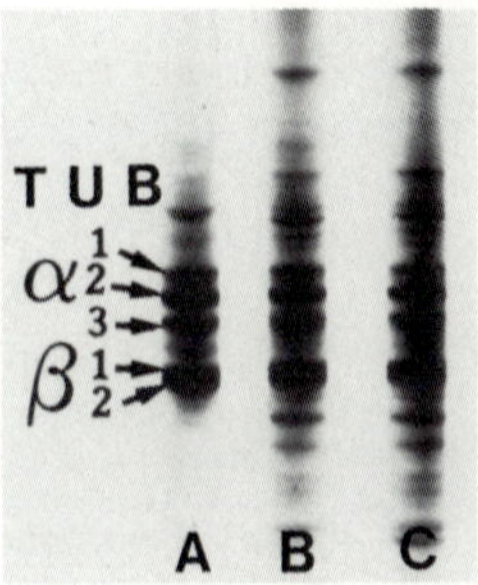

Fig. 3. Electron micrograph of taxol-induced Artemia microtubules negatively stained with 1% uranyl acetate.

Fig. 4. Isoelectric focusing gels. Gel A shows purified Artemia tubulin whereas B and C show proteins from pellets of microtubules induced to form with taxol in extracts from organisms after 0 and 15 h of development respectively.

Artemia tubulin, prepared either by taxol-induced assembly of microtubules or by biochemical methods, consists of 3α-tubulins and 2β-tubulins, along with several minor bands (Fig. 4, 5). Not only is this the first demonstration that taxol yields an isotubulin population identical to that obtained by other methods, it is also apparent that the isotubulin population does not change during the developmental period examined. It was necessary to add inhibitors of proteolytic enzymes to 36 h extracts in order to prevent degradation of the tubulin.

In summary, Artemia contains tubulin which can be induced to assemble with taxol and which can be electrophoretically resolved into a family of isotubulins. Within the limits of our techniques, the isotubulin population remains unchanged during development. The MAP population decreased during development and no new MAP were apparent despite taxol-dependent enrichment.

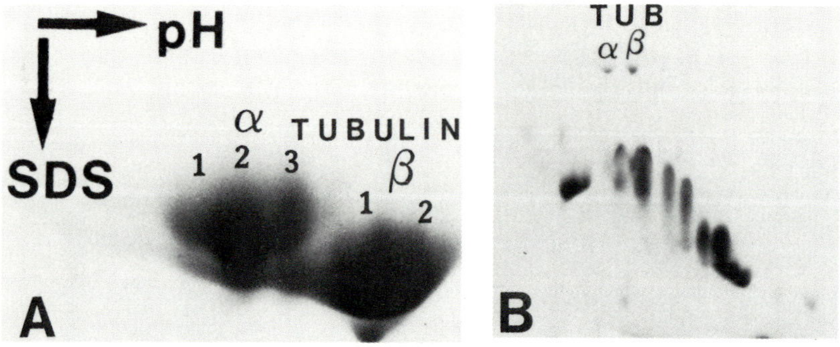

Fig. 5. Two-dimensional gels. Microtubule proteins, prepared by taxol-induced assembly of microtubules in extracts from organisms at 0 (A), and 36 (B) h of development (MacRae and Ludueña, 1984). Gels of microtubule proteins from 15 and 24 h organisms were identical to A, as were gels of biochemically purified tubulin. There was extensive degradation of tubulin in 36 h extracts in the absence of leupeptin, pepstatin A, phenylmethylsulfonyl fluoride and trypsin soy bean inhibitor (B)

REFERENCES

Borisy GG, Cleveland DW, Murphy DB (1984). "Molecular Biology of the Cytoskeleton". Cold Spring Harbor Laboratory.

MacRae TH (1984). Bundling of bovine and brine shrimp (Artemia) microtubules in vitro. Cell Biol Int Rep 8: 423-431.

MacRae TH, Ludueña RF (1984). Developmental and comparative aspects of brine shrimp tubulin. Biochem J 219: 137-148.

Vallee RB (1982). A taxol-dependent procedure for the isolation of microtubules and microtubule-associated proteins (MAPs). J Cell Biol 92: 435-442.

Progress in Developmental Biology, Part A, pages 423–426

EXPERIMENTAL ANALYSIS OF RHODNIUS PROLIXUS (INSECTA, HEMIPTERA) EMBRYOGENESIS

Gregory M. Kelly and Erwin Huebner

Department of Zoology, University of Manitoba, Winnipeg Manitoba, Canada R3T 2N2.

INTRODUCTION

Insect embryos are ideal models to study developmental mechanisms. Recent investigations into insect embryology have focussed mainly on the dipterans, thus neglecting other insect orders. Since most of the previous analysis of hemipteran embryogenesis failed to utilize contemporay techniques, the normal development of Rhodnius prolixus was extensively studied by SEM, TEM, whole mount fluorescent microscopy and light microscopy (Kelly and Huebner, 1984; and in prep.). Only with a thorough understanding of the normal developmental events can the underlying mechanisms of embryonic perturbation by juvenile hormone (JH) analogues and insect growth regulators (IGR) be explained.

Disruption of embryonic events by JH analogues has been reported in various insect orders including the Hemiptera (Enslee and Riddiford, 1977); however, little is known of the cellular mode of action of these compounds. Here we report preliminary evidence which suggests the perturbation of Rhodnius embryogenesis by fenoxycarb (Ro13.5223), a non-neurotoxic, JH-active IGR (Dorn et al, 1981) may in part be due to the disruption of microtubule elements in the amnion, one of the two extra-embryonic membranes.

NORMAL DEVELOPMENT

Following oviposition, the zygote nucleus located

deep within the yolk mass begins dividing. The energids which form migrate to the mitochondria-ribosomal rich cortex, divide and subsequently populate the entire surface of the yolk, thereby forming the blastoderm (Fig. 1a). Invagination of cells at the posterior pole of the yolk, occurring approximately 3½ days postovipostion gives rise to the germ band. By day 7, following the segmentation of the germ band the embryo begins blastokinesis.

Blastokinesis, characterized by a series of radical morphogenetic movements results in the 180° revolution of the embryo around the posterior pole of the yolk mass (Fig. 1b). Subsequent development involves the internalization of the yolk (dorsal closure, Fig. 1c), a continuation of tissue morphogenesis and eventual eclosion of the 1st instar nymph on day 15.

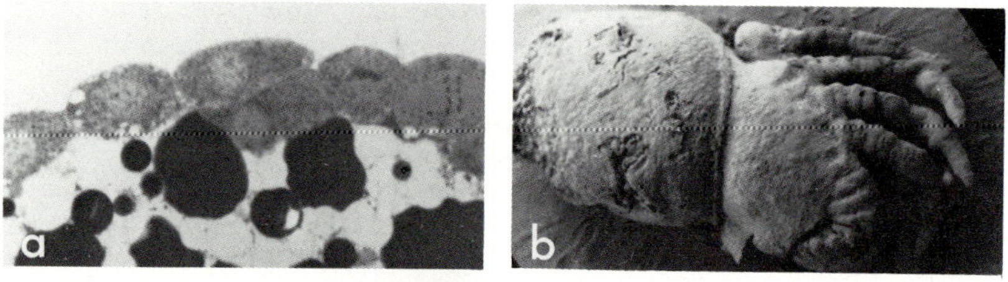

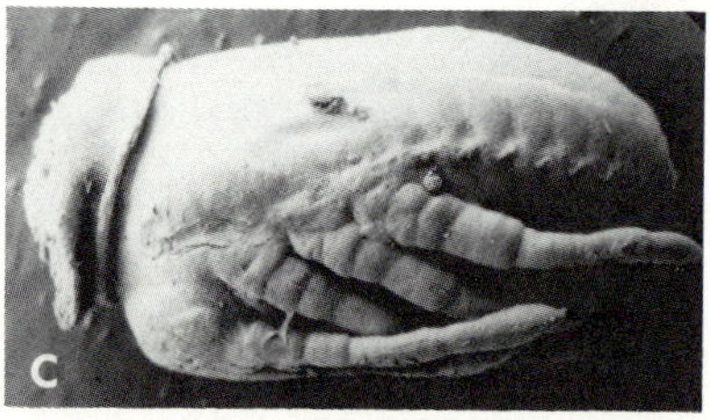

Figs. 1a-c. Three stages of *Rhodnius* embryogenesis. (a) Light micrograph of blastoderm cells; b,c) SEM of an embryo at mid-blastokinesis (b) and at late dorsal closure (c).

EXPERIMENTAL PERTURBATION

A detailed study on the effects of fenoxycarb, ethyl [2-(*p*-phenoxyphenoxy) ethyl] carbamate on *Rhodnius* development revealed that blastokinesis was the first morphologically recognizable stage affected (Kelly and Huebner submitted). This is similar to the findings of Enslee and Riddiford (1977) where exogenous JH disrupted blastokinesis in *Pyrrhocoris apterus*. From morphological analysis Enslee and Riddiford (1981) hypothesize that the extra-embryonic membranes (serosa and amnion) play a key role in the blastokinetic movements. Evidence suggests the serosal cells may be the target of the exogenous JH (Enslee and Riddiford, 1977).

The initial investigation of the extra-embryonic membranes of Rhodnius focussed on the amnion. Due to the limitation of space, details of this study will appear elsewhere. The normal amnion cells on day 5, two days prior to blastokinesis develop long cellular processes containing aggregates of microtubules (MT's) (Fig. 2a). However by day 6, although many MT exist juxtaposed to cell junctions, the number of microtubule aggregates in the cell processes diminish (Fig. 2b).

To evaluate the effects of fenoxycarb on the amnion, embryos were treated with a concentration (10μg/μl acetone) known to perturb development (Kelly and Huebner submitted). Preliminary results reveal that the cellular processes of the amnion appear as normal, on day 5 (Fig. 3a). However, by day 6 these processes appear to contain more MT aggregates than the normal, control condition (Fig. 3b).

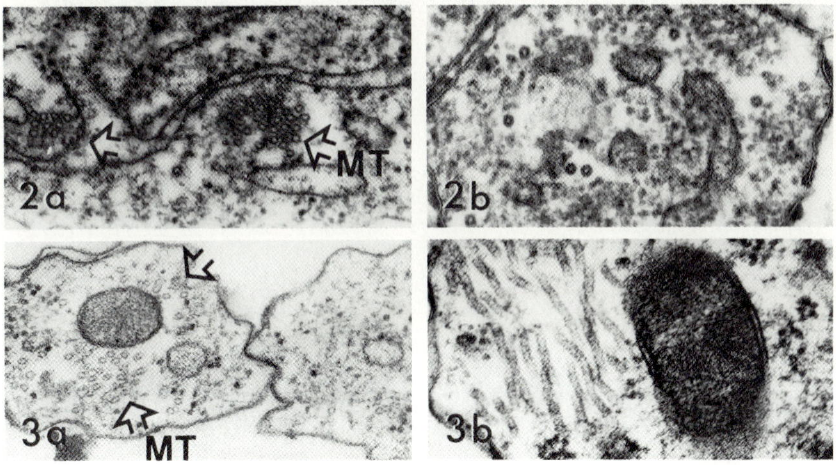

Figs. 2a-b. TEM's of normal day 5 (a) and day 6 (b) amnion cell processes.
Figs. 3a-b. These micrographs illustrate the fenoxycarb perturbed amnion cell processes on day 5 (a) and day 6 (b). Note the appearance of the microtubule aggregate in longitudinal section in 3b.

CONCLUSION

The amnion cells prior to blastokinesis, undergo a dramatic shape change. The underlying mechanisms of this are unclear although morphological analysis suggests the involvement of the cytoskeleton particularly the microtubules. Preliminary evidence indicates that an interference of the cell's cytoarchitecture by fenoxycarb results in a cessation of subsequent development. Elucidation of the precise nature of fenoxycarb perturbation requires further research into the involvement of microfilaments in the amnion cell in addition to the investigation of the serosal cell cytoskeleton.

REFERENCES

Dorn S, Frischknecht ML, Martinez V, Zurfluh R, Fischer U (1981). A novel non-neurotoxic insecticide with a broad activity spectrum. Z PflKrankh PflSchutz 88:269-275.

Enslee EC, Riddiford LM (1977). Morphological effects of juvenile hormone mimics on embryonic development in the bug, Pyrrhocoris apterus. Wilhelm Roux's Arch 181:163-181.

Enslee EC, Riddiford LM (1981). Blastokinesis in embryos of the bug, Pyrrhocoris apterus. A light and electron microscopic study 1. Normal blastokinesis. J Embryol exp Morph 61:35-49.

Kelly GM, Huebner E (1984) Cellular aspects of early-to mid-embryogenesis in the insect Rhodnius prolixus. J Cell Biol 99:270a.

Kelly GM, Huebner E (1985) The effects of a JH-active insect growth regulator on Rhodnius prolixus embryogenesis (Insecta, Hemiptera). Wilhelm Roux's Arch (submitted).

ACKNOWLEDGMENTS

This research has been supported by a Canadian Natural Sciences and Engineering Research Council Postgraduate Scholarship to GMK and an NSERC Strategic Grant to EH. We thank Drs. H. Frohlich and M. Frischknecht of Dr. R. Maag Ltd. for their generous gift of Ro13.5223.

Progress in Developmental Biology, Part A, pages 427–434

BLASTOCYST-UTERUS RELATIONSHIP IN THE INDIAN RHINOPOMATID, RHINOPOMA HARDWICKEI HARDWICKEI (GRAY)

K.B. Karim

Department of Zoology, Institute of Science, Nagpur 440 001, India

INTRODUCTION

The mammalian blastocyst bears a specific topographical relationship with the morphology of the uterus at the time of implantation, and this is usually constant within the family and sometimes within the order (Mossman, 1937, 1953, 1971). Recent reports on this aspect of the embryology of several bats have revealed that the orientation of the embryonic mass during implantation varies not only among different members of the family (Gopalakrishna and Karim, 1973; Jeevaji, 1979) but amongst different members of the same species (Gopalakrishna and Khaparde, 1972; Gopalakrishna and Karim, 1973; Ramakrishna, 1977; Sapkal, 1981).

While making a detailed study of the early embryology of the rat-tailed bat, Rhinopoma h hardwickei (Gray) the author examined several implanted blastocysts and observed several interesting features not reported so far for any rhinopomatid bat. This paper deals with the blastocyst-uterus relationship in the Indian rhinopomatid, Rhinopoma h hardwickei (Gray).

Females of Rhinopoma h hardwickei (Gray) were collected from several places such as Sikandra, mosque at Taj Mahal and Chini-ka-Rauza at and near Agra, India at frequent intervals with a view to obtaining closely graded developmental stages. For the present report altogether 33 females were examined out of which 17 females had free bilaminar blastocyst in the uterus, 8 had implanted bilaminar blastocysts, 5 trilaminar blastocysts and 3 had neural groove stage of development.

RESULTS

Nidation occurs in a preformed implantation chamber near the tubo-uterine junction. The free bilaminar blastocyst (Fig. 1) expands enormously and establishes contact with the uterine epithelium on its entire surface (Fig. 2). Hence, there is no remnant of the uterine lumen at the level of nidation. The implantation of the blastocyst is superficial and circumferential. The endometrial stromal tissue can be differentiated into two distinct zones: a darkly stained more or less compact zone immediately below the uterine epithelium occupying about three fourths the thickness of the endometrium and a lightly stained zone of endometrial stromal tissue below the myometrium. The dark zone shows numerous spaces below the uterine epithelium which are the remnants of the lumina of the uterine glands. Some of the uterine glands still persist in the form of thin streaks traversing across the thickness of the dark zone of the endometrium (Fig. 3). The trophoblast layer comes into contact with the uterine epithelial cells at several places and in these regions there is pilling up of the cells of the trophoblast layer in the form of small protruberances and the cytoplasm of the trophoblast cells and that of the uterine epithelial cells appears confluent with one another (Fig. 4). A total of eleven bilaminar and trilaminar blastocysts were examined. The orientation of the disc varied in different specimens. Out of the six implanted bilaminar blastocysts examined, the embryonic disc in three specimens was oriented in a direction between the lateral and mesometrial sides of the uterus (Fig. 5) while in three specimens the disc faced the lateral side of the uterus (Fig. 6). Five trilaminar blastocysts progressively at advanced stages of development were available for the present study. In three specimens the embryonic disc faced the lateral side of the uterus (Fig. 7), in one specimen the disc was oriented in a direction between the lateral and mesometrial sides of the uterus and in one specimen the disc was oriented slightly eccentric to the mesometrium. In all the 3 specimens at the neural groove stage of development the embryonic disc was oriented slightly eccentric to the mesometrium.

The orientation of the embryonic disc although variable during early stages of development, it tends to become oriented slightly eccentric to the mesometrium at the neural groove stage of development (Fig. 8) where the definitive chorio-allantoic placenta is established.

A significant feature of the uterine changes during these stages of implantation referes to the fact that the uterine glands undergo disintegration progressively from the distal towards the fundic regions (Figs. 2 and 3) and the cells of the glands become merged with the uterine stromal tissue. Hence, during the early implanting bilaminar stage, the uterine endometrium contains numerous large lacunae (Fig. 5) which are the remnants of the lumina of the glands. With the expansion of the blastocyst, the uterine wall becomes compressed thereby obliterating the lacunae left by the disintegrating glands - the lacunae being filled by the cells of the stroma and the cells of the disintegrated glands. Hence, at more advanced bilaminar blastocyst stage, the uterine wall appears compact (Fig. 6).

CONCLUSIONS

Nidation occurs in a preformed implantation chamber near the tubo-uterine junction.

Implantation of the blastocyst is superficial and circumferential.

The uterine epithelium is destroyed as an interaction with the trophoblast.

The orientation of the embryonic disc although variable during early stages of development, it tends to become oriented slightly eccentric to the mesometrium at the neural groove stage of development where the definitive chorio-allantoic placenta is established.

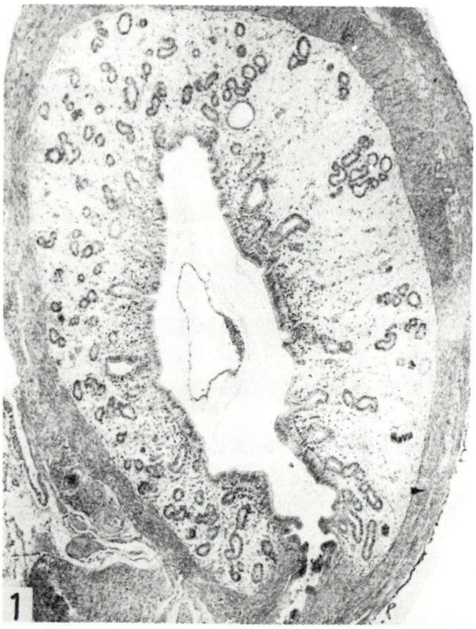
1

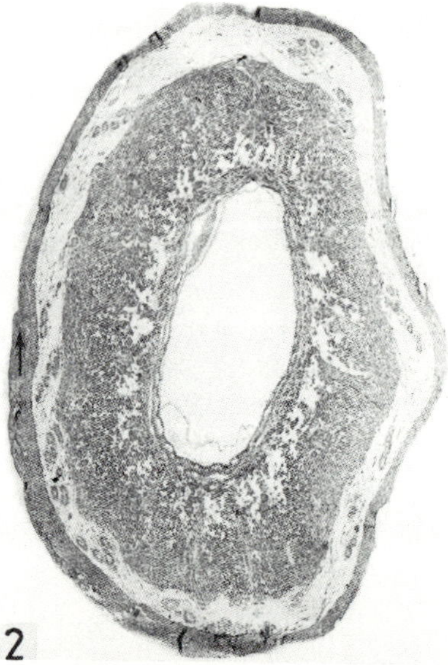
2

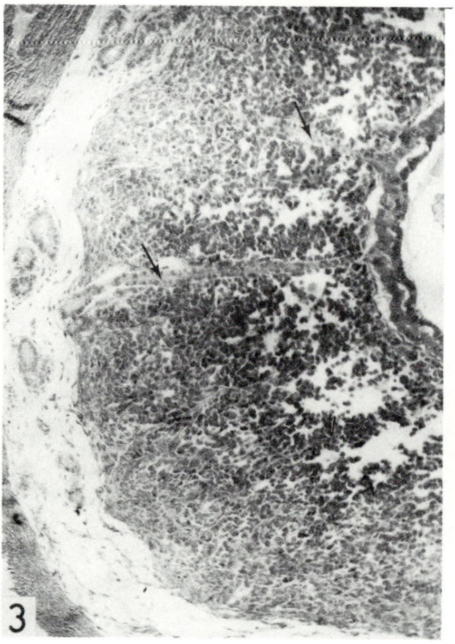
3

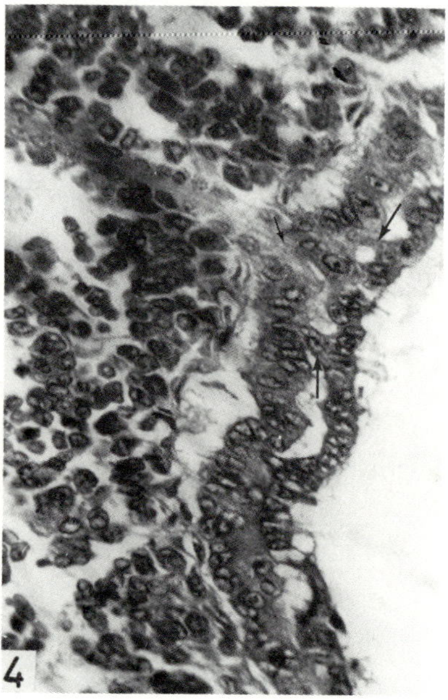
4

Fig. 1. Transverse section of the uterus containing a free bilaminar blastocyst. The blastocyst is present in an expanded spherical implantation chamber formed near the tubo-uterine junction. Note the marked progestational changes. x 70

Fig. 2. Transverse section of the uterus containing an early implanting bilaminar blastocyst. The embryonic disc is oriented in a direction between the lateral and the mesometrial sides of the uterus. Note the dark and light zones of the endometrium and the presence of numerous spaces below the uterine epithelium - remnants of the lumina of the uterine glands. Arrow points towards the mesometrium. x 28

Fig. 3. Part of the abembryonic segment of figure 2 magnified to show the dark and light zones of the endometrium. Some of the uterine glands still persist in the form of thin streaks (arrow) traversing across the thickness of the dark zone of the endometrium. x 180

Fig. 4. Part of figure 3 magnified to show that the streak-like remnants of the uterine glands appear to be still connected to the uterine epithelium (short arrow). The trophoblast layer has come in contact with the uterine epithelial cells at several places (long arrow) and in these regions there is pilling up of the cells of the trophoblast layer in the form of small protruberances and the cytoplasm of the trophoblast cells and that of the uterine epithelial cells appear confluent with one another. x 560

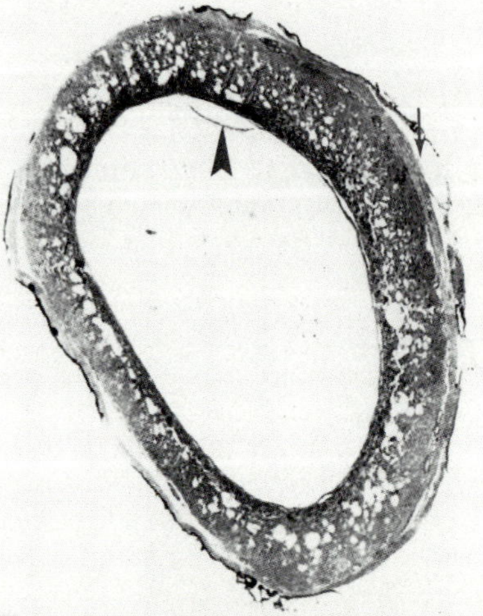

5

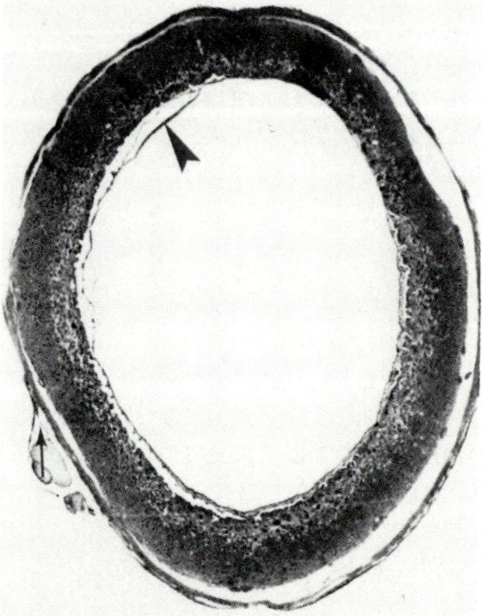

6

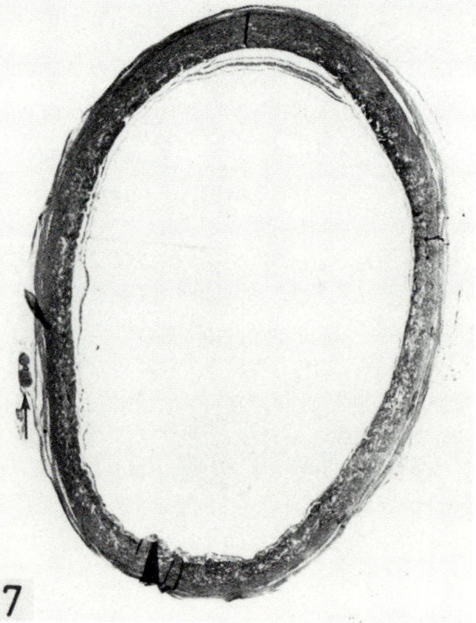

7

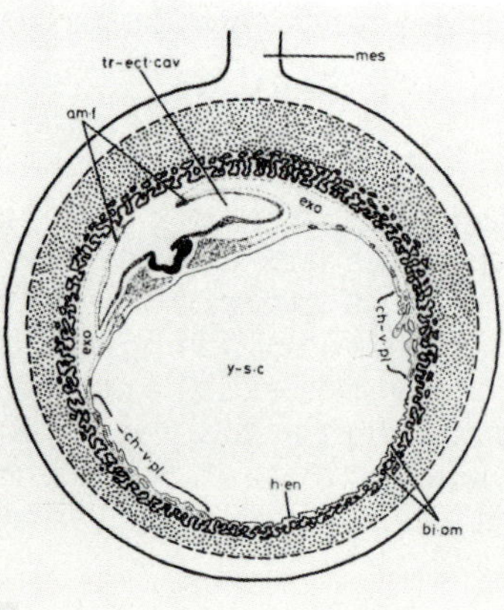

8

Fig. 5. Transverse section of the uterus containing an implanting bilaminar blastocyst with the embryonic disc (arrowhead) oriented in a direction between the lateral and mesometrial sides of the uterus. The entire dark zone of the endometrium shows the presence of numerous large lacunae - the remnants of the lumina of the distal parts of the uterine glands. Arrow points towards the mesometrium. x 22

Fig. 6. Transverse section of the uterus containing an implanted bilaminar blastocyst with the embryonic disc (arrowhead) oriented towards the lateral side of the uterus. Arrow points towards the mesometrium. Note the uterine wall appears compact because the lacunae left by the disintegrating uterine glands have become filled by the cells of the stroma and the cells of the disintegrated uterine glands. x 20

Fig. 7. Transverse section of the uterus containing an implanted trilaminar blastocyst. The embryonic disc is oriented towards the lateral side of the uterus. Arrow points towards the mesometrium. x 15

Fig. 8. Schematic representation of the arrangement of the foetal membranes at the early neural groove stage of development. am.f: amniotic folds; bi.om: bilaminar omphalopleure; ch-v.pl: chorio-vitelline placenta; exo: exocoelom; h.en: hypertrophied endoderm; mes: mesometrium; tr.ect.cav: tropho-ectodermal cavity; y-s.c: yolk-sac cavity.

The theory proposed by Kirby, Potts and Wilson (1967) for mouse that 'the embryonic mass may migrate actively inside the blastocyst covering in order to orient properly' is the only tangible explanation that can be offered for this species also.

REFERENCES

Gopalakrishna, A. and Karim, K.B. (1973) Blastocyst-uterus relationship in the Indian leaf-nosed bat, Hipposideros fulvus fulvus (Gray). Curr.Sci. 42: 860-862.

Gopalakrishna, A. and Khaparde, M.S. (1972) Variable orientation of the embryonic mass during the implantation of the blastocyst in the Indian false vampire bat, Megaderma lyra lyra (Geoffroy). Curr.Sci. 41: 738-739.

Jeevaji, I.H. (1979) Early development and implantation in the bat, Hipposideros speoris (Schneider)-Hipposideridae-Microchiroptera. Proc.Ind.Acad.Sci. B 88: 287-297.

Kirby, D.R.S., Potts, D.M. and Wilson, I.B. (1967) On the orientation of the implanting blastocyst. J.Embryol.Exp.Morph. 17: 527.

Mossman, H.W. (1937) Comparative morphogenesis of the foetal membranes and accessory uterine structures. Contrib.Embryol.Carn.Inst. Washington. 26: 127-246.

Mossman, H.W. (1953) The genital system and the foetal membranes as criteria for mammalian phylogeny and taxonomy. Jour.Mammalogy. 34: 289-298.

Mossman, H.W. (1971) Orientation and site of attachment of the blastocyst: A comparative study in The Biology of the Blastocyst. The University of Chicago Press p.p. 52-58.

Ramakrishna, P.A. (1977) Topographical relationship between the blastocyst and the uterus in the rhinolophid bat, Rhinolophus rouxi (Temminck). Curr.Sci. 46: 651-652.

Sapkal, V.M. (1981) Orientation of the embryonic mass in the blastocysts of the emballonurid bat, Taphozous melanopogon. Curr.Sci. 50: 100-101.

Index